International Symposium on Ultrasound in the Control of Industrial Processes

(UCIP 2012)

IOP Conference Series: Materials Science and Engineering
Volume 42

Madrid, Spain
18 - 20 April 2012

ISBN: 978-1-63439-853-4
ISSN: 1757-8981

Printed from e-media with permission by:

Curran Associates, Inc.
57 Morehouse Lane
Red Hook, NY 12571

CURRAN ASSOCIATES INC.
proceedings
.com

Some format issues inherent in the e-media version may also appear in this print version.

Printed by Curran Associates, Inc. (2015)

For permission requests, please contact the Institute of Physics at the address below.

Institute of Physics
Dirac House, Temple Back
Bristol BS1 6BE UK

Phone: 44 1 17 929 7481
Fax: 44 1 17 920 0979

techtracking@iop.org

Additional copies of this publication are available from:

Curran Associates, Inc.
57 Morehouse Lane
Red Hook, NY 12571 USA
Phone: 845-758-0400
Fax: 845-758-2634
Email: curran@proceedings.com
Web: www.proceedings.com

Table of contents

Volume 42

International Symposium on Ultrasound in the Control of Industrial Processes (UCIP 2012)
18–20 April 2012, Madrid, Spain

Accepted papers received: 7 November 2012
Published online: 10 December 2012

Ultrasonic Measuring Techniques and Applications

PREFACE

The following describes most of the presentations (both oral and poster) given at the International Symposium of Ultrasound in the Control of Industrial Processes (UCIP 2012) celebrated in Madrid between 18 and 20 April 2012.

This event was intended to be a meeting point for scientists, engineers and professionals from all over the world in the field of ultrasonics applied to the characterization and control of materials and processes in the industry. More precisely, the topics included were:

1. Novel applications of ultrasound in the industry (including high-power ultrasound)

- Food science
- Biotechnology and microbiology
- Pharmaceutics and cosmetics
- Petrochemistry and civil engineering

2. New insights in the ultrasonic characterization of media:

- Fluids and emulsions
- Nano- and micro-particle dispersions
- Soft materials
- Porous bodies and inhomogeneous materials

3. New developments in ultrasonic measuring techniques:

- Acoustic microscopy
- Piezoelectric sensors
- Ultrasonic imaging
- Signal processing
-

The symposium was organized by the Centro de Acústica Aplicada y Evaluación No Destructiva (CAEND, UPM-CSIC) in collaboration with the Universidad Politécnica de Cataluña, the Universidad Politécnica de Valencia and the University of Leeds. During the conference, 32 posters and 33 oral communications were presented. In addition, 4 invited lectures were imparted:

- 'Acoustic microscopy, spectroscopy and nanoparticle detection' by Dr Malcolm Povey;
- 'Acoustic and electroacoustic spectroscopy' by Dr Andrei Dukhin;
- 'High-Resolution Ultrasonic Spectroscopy and its application for material analysis by Dr Vitaly Buckin;
- 'Ultrasonic sensors for process applications – state of the art' by Dr Bern Henning;

and three tutorials were given:

- 'PZFlex – Finite Element Analysis for Virtual Prototyping' by Weidlinger Associates;
- 'SITAU – A flexible architecture controlled by MATLAB for the development of ultrasonic applications' by DASEL;
- 'Ultra-SCATTERER™ (Acoustics Suite) – The R&D Tool for Multiple Scattering of Generic Waves in Particulate Composites' by Félix Alba Consulting.

A total of 93 delegates from 21 countries participated actively in this event, and more than ten companies were involved through oral communications, tutorials, exhibition or sponsorship.

Conference photograph

Organizers

SCIENTIFIC COMMITTEE

- Dr Luis Elvira, Centro de Acústica Aplicada y Evaluación No Destructiva (CAEND), UPM-CSIC, Spain
- Dr Pablo Resa, Centro de Acústica Aplicada y Evaluación No Destructiva (CAEND), UPM-CSIC, Spain
- Dr Oscar F Martinez-Graullera Centro de Acústica Aplicada y Evaluación No Destructiva (CAEND), UPM-CSIC, Spain
- Dr Jordi Salazar Department of Electronic Engineering Universitat Politècnica de Catalunya, Spain
- Dr Jose Benedito Universitat Politècnica de València, Spain

ORGANIZING COMMITTEE

- Dr Luis Elvira, Centro de Acústica Aplicada y Evaluación No Destructiva (CAEND), UPM-CSIC, Spain
- Prof Malcolm J W Povey, Procter Department of Food Science University of Leeds, United Kingdom
- Dr Jordi Salazar, Department of Electronic Engineering Universitat Politècnica de Catalunya, Spain
- Dr Jose Benedito, Universitat Politècnica de València, Spain
- Dr Oscar F Martinez-Graullera, Centro de Acústica Aplicada y Evaluación No Destructiva (CAEND), UPM-CSIC, Spain
- Carmen Duran, Centro de Acústica Aplicada y Evaluación No Destructiva (CAEND), UPM-CSIC, Spain
- Dr Pablo Resa, Centro de Acústica Aplicada y Evaluación No Destructiva (CAEND), UPM-CSIC, Spain

- Dra Cristina Aparicio, Centro de Acústica Aplicada y Evaluación No Destructiva (CAEND), UPM-CSIC, Spain
- Prof Francisco Montero de Espinosa, Centro de Acústica Aplicada y Evaluación No Destructiva (CAEND), UPM-CSIC, Spain
- Jaime Rodríguez-López, Centro de Acústica Aplicada y Evaluación No Destructiva (CAEND), UPM-CSIC, Spain
- Dr Carlos Sierra, Centro de Acústica Aplicada y Evaluación No Destructiva (CAEND), UPM-CSIC, Spain
- Shiva Kant Shukla, Centro de Acústica Aplicada y Evaluación No Destructiva (CAEND), UPM-CSIC, Spain
- Pedro Castro Blazquez, Centro de Acústica Aplicada y Evaluación No Destructiva (CAEND), UPM-CSIC, Spain
- David Romero Laorden, Centro de Acústica Aplicada y Evaluación No Destructiva (CAEND), UPM-CSIC, Spain
- Javier Rodrigo Villazón Terrazas, Centro de Acústica Aplicada y Evaluación No Destructiva (CAEND), UPM-CSIC, Spain
- Patricia Nevado, Centro de Acústica Aplicada y Evaluación No Destructiva (CAEND), UPM-CSIC, Spain
- Sofia Aparicio, Centro de Acústica Aplicada y Evaluación No Destructiva (CAEND), UPM-CSIC, Spain
- Dr Montserrat Parrilla Romero, Centro de Acústica Aplicada y Evaluación No Destructiva (CAEND), UPM-CSIC, Spain
- Dr Luis Gómez-Ullate Alvear, Centro de Acústica Aplicada y Evaluación No Destructiva (CAEND), UPM-CSIC, Spain
- Dr Alberto Ibáñez Rodríguez, Centro de Acústica Aplicada y Evaluación No Destructiva (CAEND), UPM-CSIC, Spain

Sponsors

www.DASELSISTEMAS.com
Dasel sistemas
info@daselsistemas.com

FAC

Sonas Technologies

CORPORACIÓN ALIMENTARIA PEÑASANTA S.A.

TecniTest

Innerspec
High-performance NDT solutions

IOP Conf. Series: Materials Science and Engineering **42** (2012) 012001 doi:10.1088/1757-899X/42/1/012001

Application of High-Resolution Ultrasonic Spectroscopy for analysis of complex formulations. Compressibility of solutes and solute particles in liquid mixtures

V Buckin

School of Chemistry & Chemical Biology, University College Dublin, Belfield, Dublin 4, Ireland

E-mail: vitaly.buckin@ucd.ie

Abstract. The paper describes key aspects of interpretation of compressibility of solutes in liquid mixtures obtained through high-resolution measurements of ultrasonic parameters. It examines the fundamental relationships between the characteristics of solutes and the contributions of solutes to compressibility of liquid mixtures expressed through apparent adiabatic compressibility of solutes, and adiabatic compressibility of solute particles. In addition, it analyses relationships between the adiabatic compressibility of solutes and the measured ultrasonic characteristics of mixtures. Especial attention is given to the effects of solvents on the measured adiabatic compressibility of solutes and on concentration increment of ultrasonic velocity of solutes in mixtures.

1. Introduction

Compressibility of mixtures is extremely sensitive to the state of its components and, therefore, measurements of compressibility can be applied for analysis of variety of molecular processes in mixtures. In the past ultrasonic measurements of compressibility were successfully used for characterisation of state of solutes in liquid mixtures as well as for analysis of a broad range of molecular processes including conformational transitions in polymers, ligand binding, aggregation and crystallisation, chemical and biochemical reactions and others (see for example [1], [2], [3], [4], [5], [6], [7], [8], [9]). Modern high-resolution ultrasonic techniques for liquids allow very accurate measurements of compressibility (down to 10^{-5} %, see for example [10], [11]). High accuracy is especially important for dilute solutions and dispersions, where the contributions of solutes to the total compressibility of mixtures are very small. As compressions of materials are accompanied by transformations of energy, including release of heat, compressibility is affected by the dynamics of these processes. This results in a dependence of compressibility on the frequency of ultrasonic measurements, which determines the time scale of compression in ultrasonic wave, and on the size of the particles, which determines the time scale of the heat exchange between the particles and their environment. This also results in a dependence of the measured compressibility of solute particles on the thermophysical characteristics of solvent.

Our paper examines interpretation of high-resolution ultrasonic measurements of compressibility of dispersions. Specifically it analyses the effects of heat exchange within liquid mixtures on the fundamental relationships between the structural and physical characteristics of solutes in mixtures and the contributions of solutes to compressibility of mixtures expressed through apparent adiabatic compressibility of solutes and adiabatic compressibility of solute particles. In addition, it derives the relationships for calculations of apparent adiabatic compressibility of solutes and adiabatic compressibility of solute particles from the measured ultrasonic characteristics of mixtures. The equations discussed provide tools for analysis of effects of solvent on the measured compressibility of solutes and on ultrasonic characteristics of solutes in mixtures, which is especially important when ultrasonic characterisation of the same process but in different solvents is performed.

2. Compressibility of solutes and solute particles in liquid mixtures. Apparent adiabatic compressibility of solutes and compressibility of solute particles

Ultrasonic measurements provide adiabatic compressibility of liquids, $K_S (\equiv -\left(\frac{\partial V}{\partial P}\right)_S)$ which represents a change of volume of the liquid, V, per unit of pressure, P,, applied at adiabatic conditions , e.g. constant entropy, S, of volume V. Contribution of solute to the adiabatic compressibility of a mixture composed of the mass of solute m_{solute} and the mass of solvent m_0 can be characterised by the specific apparent adiabatic compressibility, φK_S, and the specific apparent volume, φV , of the solute:

$$\varphi K_S \equiv \frac{K_S - K_{S_0}}{m_{solute}}; \qquad \varphi V \equiv \frac{V - V_0}{m_{solute}} \tag{1}$$

Below in the text the lowercase symbols '*solute*' and '*0*' referrer to the properties of the solute and of pure solvent. The above equations provide the relationships linking the specific volume $v\,(\equiv \frac{V}{m} = \frac{1}{\rho}$, where m is the mass of volume V, ρ is its density) and the specific compressibility, $k_S\ (\equiv \frac{K_S}{m})$, of the mixture and of the pure solvent (v_0 and k_{S_0}) with the apparent characteristics of solute:

$$\begin{aligned} v &= (1-w)v_0 + w\varphi V; & \varphi V &= \frac{v - v_0}{w} + v_0 \\ k_S &= (1-w)k_{S_0} + w\varphi K_S; & \varphi K_S &= \frac{k_S - k_{S_0}}{w} + k_{S_0} \end{aligned} \tag{2}$$

where $w\,(\equiv \frac{m_{solute}}{m_0 + m_{solute}})$ is the concentration (weight fraction) of the solute in the mixture. The equations (2) are often used for calculations of φK_S and φV from the measured compressibility and volume (or density) of the mixture and of pure solvent.

The apparent properties represent the change of compressibility and volume of liquid caused by addition of solute to solvent. They can be measured experimentally and do not depend on the initial state of the solute (prior to its addition to solvent). However, in many cases they do not represent the state of the solute in the mixture due to solvation effects, which affect the state of the solvent. These effects include a transfer of components of solvent from the free (bulk) to the solvation state during solubilisation process, which contributes to the measured apparent characteristics. Below we will apply the term 'solute particle' for particulates of solute, which are dispersed in the bulk solvent, and which may include some components of solvent incorporated into the particles or/and adsorbed their surface. In case of solutions this term shall be applied to solvated solute molecules (molecule + its solvation shell where the physical properties of solvent are different of the properties of the bulk solvent). To characterise the compressibility and the volume of the solute particle in its physical state in the mixture the specific apparent volume and adiabatic compressibility of solute particle, v_{sp} and k_{sp}, can be introduced in the same way as the apparent characteristics of solute:

$$v_{sp} \equiv \frac{V - V_{0sp}}{m_{sp}}\;;\quad v = (1-w_{sp})v_{0sp} + w_{sp}v_{sp}\;;\qquad v_{sp} = \frac{v - v_{0sp}}{w_{sp}} + v_{0sp}$$
$$k_{S\,sp} \equiv \frac{K_S - K_{S_0\,sp}}{m_{sp}};\quad k_S = (1-w_{sp})k_{S_0\,sp} + w_{sp}k_{S\,sp};\quad k_{S\,sp} = \frac{k_S - k_{S_0\,sp}}{w_{sp}} + k_{S_0\,sp} \tag{3}$$

where V and K_S are the volume and the adiabatic compressibility of the mixture containing the solvent and solute particles, m_{sp} $(\equiv m_{solute} + m_{solvent\,sp})$ is the mass of solute particles in the mixture, which includes the mass of the solvent incorporated into the solute particle, $m_{solvent\,sp}$; V_{0sp} and $K_{S_0\,sp}$ are the volume and the adiabatic compressibility of the bulk solvent in the mixture.
The specific apparent volume and adiabatic compressibility of solute particles can be measured directly by adding solute particles as a whole or all the components of solute particle to the solvent and measuring the change of volume and compressibility caused by this (measuring of v, v_{0sp}, k_S and $k_{S_0\,sp}$, figure 1). The specific apparent volume and adiabatic compressibility of solute particles can also be calculated from the measured specific apparent volume and adiabatic compressibility of the solute as:

$$v_{sp} = \frac{\varphi V + w_s v_0 + \delta v_0 (1 + w_s - \frac{1}{w})}{(1+w_s)};\; k_{S\,sp} = \frac{\varphi K_S + w_s k_{S_0} + \delta k_{S_0}(1 + w_s - \frac{1}{w})}{1+w_s} \tag{4}$$

where $w_s (\equiv \frac{m_{solvent\,sp}}{m_{solute}})$ is the solvation level of the solute, $\delta v_0 (\equiv v_{0\,sp} - v_0, v_0$ is the specific volume of the bulk solvent in the mixture) is the difference between the specific volumes of the pure solvent and of the bulk solvent in the mixture and $\delta k_{S_0} (\equiv k_{S_0 sp} - k_{S_0}$, $k_{S_0\,sp}$ is the specific adiabatic compressibility of the bulk solvent in the mixture) is the difference in between the specific adiabatic compressibilities of the pure solvent and of the bulk solvent in the mixture. Equation (4) is a consequence of equations (2) and (3). The concentration of solute, w, in this equation reflects the dependence of δv_0 and δk_{S_0} on w at constant solvation level, w_s, when these values are not equal to 0. If solvation does not affect the composition of solvent, the values of δv_0 and δk_{S_0} are equal to zero. In this case:

$$v_{sp} = \frac{\varphi V + w_s v_0}{1+w_s};\quad k_{S\,sp} = \frac{\varphi K_S + w_s k_{S_0}}{1+w_s} \tag{5}$$

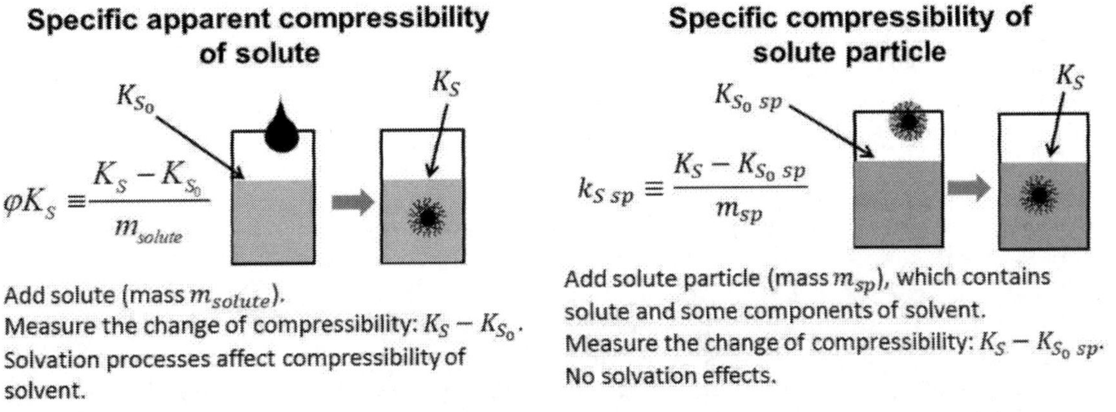

Figure 1. Specific apparent compressibility of solute and compressibility of solute particle.

Parameters φK_S, φV, v_{sp} and k_{sp} characterise the state of the whole solute or the whole pool of solute particles in the mixture. If the state of the solute depends on its concentration it is often useful to interpret the data in terms of specific partial characteristics of solutes, $\bar{v}$ and $\bar{k}_S$, defined as a change of volume, V, or compressibility, K_S, of the mixture caused by addition of infinitely small mass of solute and calculated per unit of mass of solute added:

$$\bar{v} \equiv \left(\frac{\partial V}{\partial m_{solute}}\right)_{m_0} = \varphi V + w(1-w)\left(\frac{\partial(\varphi V)}{\partial w}\right)_{m_0} ; \ \bar{k}_S \equiv \left(\frac{\partial K_S}{\partial m_{solute}}\right)_{m_0} = \varphi K_S + w(1-w)\left(\frac{\partial(\varphi K_S)}{\partial w}\right)_{m_0} \qquad (6)$$

were the symbol m_0 represents constant amount of other than solute components of the mixture. The partial characteristics can be calculated from the measured apparent characteristics, φK_S and φV, and the slope of their concentration dependence, $\left(\frac{\partial(\varphi V)}{\partial w}\right)_{m_0}$ and $\left(\frac{\partial(\varphi K_S)}{\partial w}\right)_{m_0}$. In the same way the specific partial volume and compressibility of solute particles, $\bar{v}_{sp}$ and $\bar{k}_{S\,sp}$, can be introduced:

$$\bar{v}_{sp} \equiv \left(\frac{\partial V}{\partial m_{sp}}\right)_{m_{0\,sp}} = v_{sp} + w_{sp}(1-w_{sp})\left(\frac{\partial v_{sp}}{\partial w_{sp}}\right)_{m_{0\,sp}} ; \ \bar{k}_{S\,sp} \equiv \left(\frac{\partial K_S}{\partial m_{sp}}\right)_{m_{0\,sp}} = k_{S\,sp} + w_{sp}(1-w_{sp})\left(\frac{\partial k_{S\,sp}}{\partial w_{sp}}\right)_{m_{0\,sp}} \qquad (7)$$

Similar to the above other apparent thermodynamic characteristics of solute and thermodynamic characteristics of solute particles can be introduced. This includes thermal expansibility, $E\,(\equiv\left(\frac{\partial V}{\partial T}\right)_P)$, heat capacity, $C_P\,(\equiv\left(\frac{\partial H}{\partial T}\right)_P$, H is the enthalpy of the mixture), and isothermal compressibility, $K_T\,(\equiv -\left(\frac{\partial V}{\partial P}\right)_T)$, thus producing the specific apparent thermal expansibility of solute, φE, specific apparent heat capacity of solute, φC_P, specific apparent isothermal compressibility of solute, φK_T, specific isothermal compressibility of solute particle, $k_{T\,sp}$, specific thermal expansibility of solute particle, e_{sp}, and specific heat capacity of solute particle, $c_{P\,sp}$, defined as:

$$\varphi E = \frac{E-E_0}{m_{solute}}; \quad \varphi C_P = \frac{C_P - C_{P_0}}{m_{solute}}; \ \varphi K_T = \frac{K_T - K_{T_0}}{m_{solute}}; \ e_{sp} = \frac{E - E_{0\,sp}}{m_{sp}}; \ c_{P\,sp} = \frac{C_P - C_{P_0\,sp}}{m_{sp}}; \ k_{T\,sp} = \frac{K_T - K_{T_0\,sp}}{m_{sp}}$$

where the abbreviations 0 and *sp* has the same meaning as above. Thus, the specific thermal expansibility, $e\,(\equiv \frac{E}{m})$, specific heat capacity, c_P and specific isothermal compressibility, $k_T\,(\equiv \frac{K_T}{m})$, of a mixture are presented as:

$$\begin{aligned} e &= (1-w)e_0 + w\varphi E; \ c_P = (1-w)c_{P_0} + w\varphi C_P; \ k_T = (1-w)k_{T_0} + w\varphi K_T \\ e &= (1-w_{sp})e_{0\,sp} + w_{sp}e_{sp}; \ c_P = (1-w_{sp})c_{P_0\,sp} + w_{sp}c_{P\,sp}; \ k_T = (1-w_{sp})k_{T_0\,sp} + w_{sp}k_{T\,sp} \end{aligned} \qquad (8)$$

3. Ultrasonic measurements of compressibility of solutes and solute particles in liquids

3.1. Ultrasonic wave and compressibility of liquids

Ultrasonic wave is a wave of longitudinal deformations where compressions occur in direction of wave propagation. In isotropic materials, which are homogeneous within the scale of ultrasonic wave, the differential equation for a plain ultrasonic wave propagating over the axis x is given by ([1], [12], [13], [14]):

$$\rho_l \frac{\partial^2 X}{\partial t^2} = M \frac{\partial^2 X}{\partial x^2} \tag{9}$$

where M is the longitudinal modulus of the medium, which represents the ratio of stress to strain in longitudinal deformation, X is the displacement of the medium at distance x, t is time, ρ_l is the inertial density of the medium, which represents the ratio between the force applied to a small part of the liquid and the acceleration caused by the force. For homogeneous within the length scale of the shear wave liquids ρ_l is equal to the gravimetric density, for non- homogeneous liquids ρ_l deviates from the gravimetric density and depends on the frequency. The equation (9) is valid for ultrasonic waves of small amplitude, e.g. relative longitudinal deformation in the wave, strain, is << 1. At this condition the speed of the wave is much higher than the speed of particles of the medium ([13]).

In general the strain (deformation) may occur with some time delay relative to the stress applied. Incorporation of these effects for the case of harmonically oscillating stress is often done through representation of strain, stress and other associated variables (e.g. change of pressure, δP, temperature, δT, and volume, δV) as complex values given by $\mathbf{Z}e^{i\omega t}$,where $\mathbf{Z}(\equiv|\mathbf{Z}|e^{i\varphi_Z} \equiv Z'+iZ"$, $|\mathbf{Z}|$ is the absolute amplitude and φ_Z the phase of $\mathbf{Z}$; Z' is the real and $Z"$ is the imaginary parts of $\mathbf{Z}$) is the complex amplitude of the corresponding parameter, $\omega(\equiv 2\pi f$, f is frequency of ultrasonic wave) is the angular frequency of oscillations, t is time, and $i\,(\equiv\sqrt{-1}\,)$ is the imaginary unit. In this case the longitudinal modulus can be represented by a complex constant, $\mathbf{M}\,(\equiv M'+iM")$. For most of liquids the elastic response dominates, $M' >> M"$ (often expressed as a small value of attenuation per wavelength, $\alpha\lambda << 1$, $\lambda = \frac{u}{f}$). In this case equation (9) provides the velocity, u, and attenuation, α of ultrasonic wave as ([1], [12], [13], [14]) :

$$u = \sqrt{\frac{M'}{\rho_l}}; \qquad \alpha = \frac{\omega M"}{2\rho_l u^3} \tag{10}$$

The modulus of longitudinal deformation is a combination of the modulus of volume deformation, $\mathbf{K}_\mathbf{V}\,(\equiv K_V'+iK_V")$, and the modulus of shear deformation, $\mathbf{G}(\equiv G'+iG")$: $\mathbf{M} = \mathbf{K}_\mathbf{V} + \frac{4}{3}\mathbf{G}$. For most liquids the shear storage modulus, G', is much smaller than its volume storage modulus, K_V', $(G' << K_V')$. In this case the above relationship for ultrasonic velocity can be further reduced to the well-known equation: $u = \frac{1}{\sqrt{\beta'\rho_l}}$, where β', which is the real part of compressibility of the liquid representing a relative change of the volume per unit of pressure applied in the ultrasonic wave: $\beta' \equiv -\frac{1}{V}\left(\frac{\delta V}{\delta P}\right)$.

3.2. Compressibility of liquids and effects of heat exchange

Compression of a volume V caused by an applied pressure results in production of heat, which changes the temperature of the compressed volume by δT. The volume change, δV, caused by the applied pressure, δP, is a sum of two terms, compression of volume V at constant temperature, $\delta V_T = -K_T\delta P$, and the change of volume caused by its heat expansion at constant pressure, $\delta V_P = E\delta T$,: $\delta V = \delta V_T + \delta V_P$, where K_T is the isothermal compressibility of volume V at constant temperature and E is its thermal expansibility at constant pressure. Therefore, the compressibility, $K\,(\equiv -\frac{\delta V}{\delta P}$), is given as $K = K_T - E\frac{\delta T}{\delta P}$. The change of temperature δT is determined by the balance of the heat production in the compression of the volume V and the heat exchange between this volume and its environment. Two (major) processes of the heat exchange exist in the compression (ultrasonic) waves propagating through liquids. The first one is the heat flow between different parts of the compression wave, which have at any given time different compression and thus different temperature. The length scale of the heat exchange process is determined by the characteristics of the thermal wave, its wavelength, $\lambda_T (= \sqrt{\frac{4\pi\kappa}{f\rho c_P}}$, κ is the heat conductance of the medium, ρ is the gravimetric density) and is attenuation with distance $\sim e^{-\frac{x}{\delta_T}}$, where $\delta_T = \frac{\lambda_T}{2\pi}$ is the exponential decay length of the thermal wave. Therefore, the extent of the heat exchange within the compression wave is defined by the ratio of the wavelength of compression (ultrasonic) wave, λ, which controls the gradient of temperature within the wave and λ_T, which controls the length scale of the heat exchange during the time interval of compression. The functional dependence of the real and imaginary parts of compressibility (and ultrasonic velocity and attenuation) in ultrasonic waves in liquids of the ratio of λ_T and λ was described previously ([13]). Its application for such liquids as water or oils at room temperature shows that for ultrasonic frequencies, up to hundreds of MHz, the deviations of ultrasonic velocity and the real part of compressibility coefficient, β', from their values at pure adiabatic conditions of compression do not exceed 10^{-7} of their absolute values. This is smaller than the existing limits of resolution in measurements of ultrasonic velocity. Thus, the compressibility of the media measured within normal ultrasonic frequency range, represents the adiabatic compressibility, K_S and the value of β' shall be taken as coefficient of adiabatic compressibility of the medium, $\beta_S (\equiv \frac{K_S}{V}$; $K_S \equiv -\left(\frac{\partial V}{\partial P}\right)_S$, S is entropy of the volume V). The imaginary part of compressibility at this limit provides the well-known equation for contribution to ultrasonic attenuation caused by the heat losses in homogeneous liquids, α_{TH}, ([15]): $\alpha_{TH} = \omega^2 T \frac{\kappa e^2 \rho}{2uc_P^{\,2}}$, where $e (\equiv \frac{E}{m}$, m is the mass of the volume V) is the specific thermal expansibility of the medium.

The second process of the heat exchange in the ultrasonic wave is realised in dispersions. It is associated with the heat flow between the continuous and dispersed media, if their thermophysical

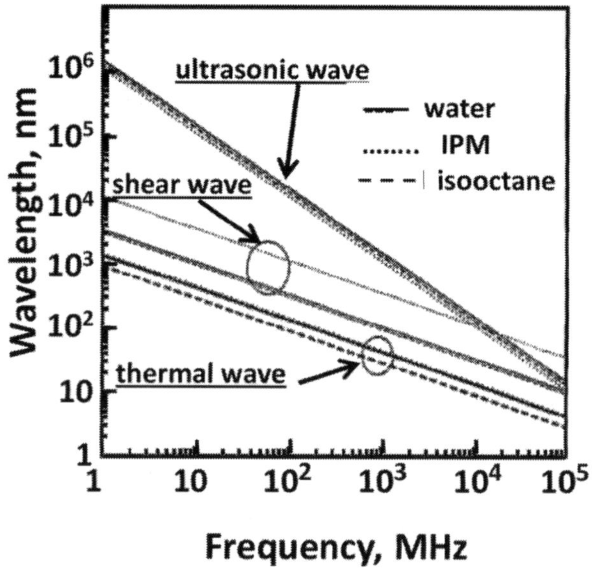

Figure 2. Wavelength as function of frequency for ultrasonic, thermal, and shear waves in pure water, isooctane and IPM solvent (isopropyl myristate: Epikuron 200 :n-propanol (6:2:2, w/w)) at 25°C calculated using physical parameters given in table 1. (High GHz frequency part of the plot shown for illustrative purposes only as the physical properties of the media at high frequencies are expected to deviate from those at low frequencies)

properties are different. This process and its effect on compressibility of liquids are discussed in the section 4.

3.3. Inertial density of liquid mixtures

The inertial density, ρ_I, of a liquid mixture may deviate from its gravimetric density (mass of unit of volume) , ρ, due to different accelerations of the dispersed compounds (particles) and of the continuous medium, in a gradient of pressure within the ultrasonic wave. For an oscillating gradient of pressure this difference is controlled by the ratio of the size of the particles to the wavelength of the shear wave, $\lambda_\eta (= \sqrt{\frac{4\pi\eta}{f\rho}}$, η is viscosity of the medium). The parameter $\frac{\lambda_\eta}{2\pi}$ (exponential decay length of the shear wave) determines the effective thickness of the transition layer around the border between the dispersed particle and the continuous medium, within which the speed of the medium changes from the speed at the border to the speed of the bulk medium. Dependence of λ_η on frequency (figure 2) results in a change of the value of ρ_I from the value equal to the gravimetric density at low frequencies, to a different value at high frequencies ([1]).

The effect of frequency on ρ_I can be illustrated qualitatively by considering a rigid container filled with a liquid dispersion, and a periodic force applied to the walls of the container. The effective inertial mass of the container is the ratio of the applied force to the acceleration of the container. At low frequencies, when the transition layer around the borders between the particles and the continuous medium exceeds the size of the container, the speed and acceleration of the particles and the continuous medium are the same. In this case the effective inertial mass of the container is the sum of the masses of its parts: the particles, the continuous medium and the walls. At high frequencies, when the transition layer is much smaller than the size of particles, the acceleration of the particles and the continuous medium are different. This results in a deviation of the effective inertial mass of the container from its gravimetric mass. In mixtures with the size of dispersed particles much smaller than the wavelength of the shear wave (approximately 10nm and below at ultrasonic frequencies, see figure 2) both densities are expected to be equal, $\rho_I = \rho = \frac{1}{v}$, which will be assumed below. In this case the equation for ultrasonic velocity in liquids is transformed to:

$$u = \frac{v}{\sqrt{k_S}} \tag{11}$$

3.4. Concentration increment of ultrasonic velocity of solutes and solute particles

Equations (2) and (11) allow calculations of apparent adiabatic compressibility of solutes from the measured values of ultrasonic velocity and density. However, for dilute mixtures, where the contribution of solutes to the specific compressibility and the volume of the mixture is much smaller than the contribution of solvent, this way of obtaining of φK_S (and k_{sp}) often possesses a problem as it requires very precise measurements of ultrasonic velocity and density in exactly the same mixture and at the same temperature. In addition, precision of measurements of the absolute values of ultrasonic velocity and density normally does not exceed 10^{-5} of the absolute value, however, the change in velocity, δu , and density, can be measured much better (down to 10^{-7} of absolute velocity (, [10], [16]) and 10^{-6} of density). Therefore, calculations of apparent adiabatic compressibility from the contributions of solutes to ultrasonic velocity and to specific volume of the mixture are often more accurate. These contributions are obtained by measurements of the change of ultrasonic velocity, $\delta u (\equiv u - u_0$, where u and u_0 are the velocity in the mixture and in the solvent), and of specific volume, δv, caused by addition of the solute to the solvent.

The contribution of a solute to the specific volume of the mixture is characterised by the specific apparent volume of the solute, φV. The contribution of a solute to ultrasonic velocity in the mixture is characterised by the concentration increments of ultrasonic velocity, a_u or a_{u^2}, defined as:

$$a_u \equiv \frac{\frac{u}{u_0}-1}{w} = \frac{\delta u}{w u_0}; \qquad a_{u^2} \equiv \frac{(\frac{u}{u_0})^2-1}{2w} = \frac{\delta u(2u_0+\delta u)}{2wu_0^{\ 2}} \tag{12}$$

The above concentration increments of ultrasonic velocity are linked with each other:

$$a_{u^2} = a_u(1+\frac{wa_u}{2}) \tag{13}$$

At infinite dilution, $w \to 0$, the increments become to be equal: $a_u = a_{u^2}$. If required, the ultrasonic velocity of liquid can be calculated from the known concentration increment of ultrasonic velocity as: $u = u_0(1+wa_u)$ or $u = u_0\sqrt{1+2wa_{u^2}}$.

When analysing the properties of solute particles the concentration increments of solute particles can be introduced, $a_{u\,sp}$ and $a_{u^2\,sp}$, which are defined in the same way as a_u and a_{u^2}:

$$a_{u\,sp} \equiv \frac{\frac{u}{u_{0\,sp}}-1}{w_{sp}} = \frac{\delta u_{sp}}{w_{sp}u_{0\,sp}}; \qquad a_{u^2\,sp} \equiv \frac{(\frac{u}{u_{0\,sp}})^2-1}{2w_{sp}} = \frac{\delta u_{sp}(2u_{0\,sp}+\delta u_{sp})}{2w_{sp}u_{0\,sp}^{\ 2}} \tag{14}$$

where $u_{0\,sp}$ is the ultrasonic velocity for the bulk solvent in the mixture and $\delta u_{sp} = u - u_{0\,sp}$. If the value of ultrasonic velocity in the bulk solvent is the same as in the pure solvent, $u_{0\,sp} = u_0$, the proportionality coefficient $\frac{w_{sp}}{w} = 1 + w_s$ allows recalculations of specific concentration increments of ultrasonic velocity:

$$a_{u\,sp} = \frac{a_u}{1+w_s}; \quad a_{u^2\,sp} = \frac{a_{u^2}}{1+w_s} \tag{15}$$

The concentration increments of ultrasonic velocity characterise the state of the whole solute in the mixture. If the state of the solute depends on its concentration it is often useful to obtain the partial concentration increments of ultrasonic velocity, $\bar{a}_u$ and $\bar{a}_{u^2}$, which represent the effect of addition of a small portion of solute to a solutions at a particular concentration:

$$\bar{a}_u \equiv \left(\frac{\partial}{\partial w}(\frac{u}{u_0})\right)_{m_0} = a_u + w\left(\frac{\partial a_u}{\partial w}\right)_{m_0}; \quad \bar{a}_{u^2} \equiv \frac{1}{2}\left(\frac{\partial}{\partial w}(\frac{u}{u_0})^2\right)_{m_0} = a_{u^2} + w\left(\frac{\partial a_{u^2}}{\partial w}\right)_{m_0} \tag{16}$$

The partial concentration increments of solute particles are introduced in the same way.

3.5. Calculations of apparent adiabatic compressibility of solutes and compressibility of solute particles from measured concentration increments of ultrasonic velocity

Combination of equations, (12), (14), (11), and (2) provide the following relationships for calculations of specific apparent adiabatic compressibility of solute and the specific compressibility of solute particle in a liquid from the concentration increments of ultrasonic velocity and the apparent specific volume:

$$\varphi K_S = k_{S_0}\left(2\frac{\varphi V}{v_0} - 2a_{u^2} - 1 + w\frac{\left(\frac{\varphi V}{v_0} - 2a_{u^2} - 1\right)^2}{1 + 2wa_{u^2}}\right)$$
$$k_{S\,sp} = k_{S_0\,sp}\left(2\frac{v_{sp}}{v_{0sp}} - 2a_{u^2\,sp} - 1 + w_{sp}\frac{\left(\frac{v_{sp}}{v_{0sp}} - 2a_{u^2\,sp} - 1\right)^2}{1 + 2w_{sp}a_{u^2\,sp}}\right) \tag{17}$$

At infinite dilution, $w \to 0$, the equations (17) are reduced to:

$$\varphi K_S = k_{S_0}\left(2\frac{\varphi V}{v_0} - 2a_u - 1\right); \qquad k_{S\,sp} = k_{S_0\,sp}\left(2\frac{v_{sp}}{v_{0sp}} - 2a_{u\,sp} - 1\right) \tag{18}$$

As discussed in the previous sections equation (11) is applicable at the visco-inertial low frequency limit only when the size of the solute particle is much smaller than wavelength of the shear wave. Calculations of the apparent adiabatic compressibility of solutes and compressibility of solute particles for the particles larger or comparable with the wavelength of the shear wave would require a substitution of the gravimetric density of the mixture in the equation (11) by an appropriate inertial density. However if the intrinsic density of the particles is close to the density of the solvent (low density contrast, emulsions for example) the inertial effects in the ultrasonic wave can often be neglected even for large particles ([1]) and equations (17) and (18) can be applied for the calculations of apparent adiabatic compressibility of solutes and compressibility of solute particles from measured concentration increments of ultrasonic velocity.

4. Effects of the heat exchange on the apparent compressibility of solutes and compressibility of solute particles

4.1. Thermal low and thermal high frequency limits for compressibility

Below we will consider mixtures of solute particles (molecules or molecular aggregates) surrounded by solvent. If the physical properties of the solute particles and the solvent are different, a change of pressure in the mixture results in a different change of temperature within the solute particles and within the surrounding solvent. This difference causes a heat flow between the particles and the solvent, which affects the volume change during compression. Thus, the effects of solute particle – solvent heat exchange shall be included in interpretations of measured compressibilities.

Two major structural characteristics control the change of the volume of the mixture when an excessive pressure is applied. The first is the length scale of homogeneity of pressure within the mixture, which is determined by the wavelength of the ultrasonic wave, λ. For the practical frequency range of precision measurements of compressibility, 1 to 100 MHz, λ is between 10^6nm and 10^4nm (figure 2). Therefore for the nano-scale particles of the size below 1000 nm the effects of non-homogeneity of pressure around them can be neglected. The second structural characteristic is the wavelength of the thermal wave, λ_T, (or the decay length, $\delta_T = \frac{\lambda_T}{2\pi}$), which determines the portions of the solute particle and the solvent involved in the heat exchange. In liquids λ_T varies between 100nm and 1000nm for the practical frequency range of precision measurements of compressibility (1 to 100 MHz, figure 2), which is within the range of the sizes of particles of dispersions. This results in a profound effect of the frequency of ultrasound on the measured compressibility of the micro particles in mixtures. Two limiting cases for particle – solvent heat exchange can be distinguished.

1) At thermal low frequency limit the period of compression is much longer than the time required for the heat exchange within the compressed volume. Under this condition the complete thermal

equilibrium within the compressed volume is achieved and all its parts have the same temperature at each time.

2) At thermal high frequency limit the period of compression is much shorter than the time required for the heat exchange between the solute particles and the bulk solvent. Under this condition the solute particles and the solvent are thermally insulated. In structural terms the thermal low frequency limit is realised when the size of the solute particles, l, is much smaller than the wavelength of the thermal wave, $l \ll \lambda_T$. For the thermal high frequency limit $l \gg \lambda_T$. In this case the volume (and the thermal mass) of the transition layer around the border between the solute particle and the solvent, where the solute-solvent heat exchange takes place, is negligibly small when compared with the volume of the solute particles and the solvent. Therefore, at ultrasonic frequencies compressibility of particles, polymers and molecular aggregates with the size below 10 nm shall be attributed to the thermal low-frequency limit, while compressibility of particles and aggregates with the size of microns and higher shall be described in terms of thermal high-frequency limit especially at the top of the frequency range (figure 2). Below in the text the uppercase symbol ^{0} will be used for variables at low frequency limit and $^\infty$ for high frequency limit. An absence of these symbols will presume general cases applicable for both limits, or their irrelevance.

4.2. Apparent compressibility of solutes and compressibility of solute particles at thermal low and thermal high frequency limits

At thermal low frequency limit all parts of the mixture are at the same temperature and the same value of $\frac{\delta T}{\delta P} = \frac{Te}{c_P}$. Thus, the specific adiabatic compressibility of the mixture at this limit, k_S^0, is given by:

$k_S^0 = k_T - \frac{Te^2}{c_P}$ where e and c_P are specific thermal expansibility and heat capacity of the mixture are given by equations (8). This and equations (2), (3) and (8) provide the relationship for specific apparent adiabatic compressibility of solute at thermal low-frequency limit, $\varphi K_S{}^0$, and specific adiabatic compressibility of solute particle at this limit, $k_{S\,sp}^0$:

$$\varphi K_S^0 = \varphi K_T - \frac{e_0^2 T}{c_{P_0}}\left(2\frac{\varphi E}{e_0} - \frac{\varphi C_P}{c_{P_0}} + w\frac{\left(\frac{\varphi E}{e_0} - \frac{\varphi C_P}{c_{P_0}}\right)^2}{1 + w\left(\frac{\varphi C_P}{c_{P_0}} - 1\right)}\right)$$

$$k_{S\,sp}^0 = k_{T\,sp} - \frac{e_{0\,sp}^2 T}{c_{P_0\,sp}}\left(2\frac{e_{sp}}{e_{0\,sp}} - \frac{c_{P\,sp}}{c_{P_0\,sp}} + w_{sp}\frac{\left(\frac{e_{sp}}{e_{0\,sp}} - \frac{c_{P\,sp}}{c_{P_0\,sp}}\right)^2}{1 + w_{sp}\left(\frac{c_{P\,sp}}{c_{P_0\,sp}} - 1\right)}\right) \tag{19}$$

The above equations demonstrate that at thermal low frequency limit the specific apparent adiabatic compressibility of solutes and the adiabatic compressibility of solute particles depend on the concentration even in ideal mixtures. The third terms of equations (19) represent a reduction of the solute–solvent heat exchange with concentrations, w and w_{sp}, due to the lowering of the amount of solvent in the mixture. At infinite dilution ($w, w_{sp} \to 0$) equations (19) are reduced to:

$$\varphi K_S^0 = \varphi K_T - \frac{e_0^2 T}{c_{P_0}}\left(2\frac{\varphi E}{e_0} - \frac{\varphi C_P}{c_{P_0}}\right); \quad k_{S\,sp}^0 = k_{T\,sp} - \frac{e_{0\,sp}^2 T}{c_{P_0\,sp}}\left(2\frac{e_{sp}}{e_{0\,sp}} - \frac{c_{P\,sp}}{c_{P_0\,sp}}\right) \tag{20}$$

At thermal high frequency limit no heat exchange between solute particles and the surrounding solvent is present. Therefore, the compressions of solute particles and the solvent occur independently on each other, and the adiabatic compressibility of the mixture at this limit, K_S^∞, is an additive thermodynamic characteristic.

At $w=1$ according to equations (19) $(\varphi K_S^0)_{w=1} = \varphi K_T - \dfrac{T\varphi E^2}{\varphi C_P}$, and at $w_{sp}=1$, $k_{S\,sp}^0 = k_{T\,sp} - \dfrac{Te_{sp}^2}{c_{P\,sp}}$. Also as at $w=1$ the no solvent is presented in the mixture the compressions occur at adiabatic condition for the solute at all frequencies. Therefore, $(\varphi K_S{}^0)_{w=1} = (\varphi K_S{}^\infty)_{w=1}$. Due to the additivity of K_S^∞ in ideal mixtures of solute and solvent the value of φK_S^∞ shall not depend on the concentration, the relationship $\varphi K_S^\infty = \varphi K_T - \dfrac{T(\varphi E)^2}{\varphi C_P}$ shall be valid for any concentration. At $w_{sp}=1$, $k_{S\,sp}^0 = k_{S\,sp}^\infty = k_{T\,sp} - \dfrac{e_{sp}^2 T}{c_{P\,sp}}$, where $k_{S\,sp}^\infty$ is the specific adiabatic compressibility of the solute particle at thermal high frequency limit. Following this the second part of equations (19) can be

$$k_{S\,sp}^0 = k_{S\,sp}^\infty + k_{HE\,sp}$$

rearranged as:

$$k_{HE\,sp} = \frac{e_{0sp}{}^2 T}{c_{P_0\,sp}}\left(\frac{c_{P\,sp}}{c_{P_0\,sp}} - \frac{e_{sp}}{e_{0\,sp}}\right)^2 \frac{c_{P_0\,sp}}{c_{P\,sp}} - w_{sp}\frac{\dfrac{e_{0sp}{}^2 T}{c_{P_0\,sp}}\left(\dfrac{c_{P\,sp}}{c_{P_0\,sp}} - \dfrac{e_{sp}}{e_{0\,sp}}\right)^2}{1 + w_{sp}\left(\dfrac{c_{P\,sp}}{c_{P_0\,sp}} - 1\right)} \tag{21}$$

where k_{HE} is the contribution of the heat exchange between the solute particle and its environment to compressibility. The above relationships demonstrate that the specific adiabatic compressibility of solute particle measured at thermal low frequency limit has a contribution from the heat exchange between the solute particle and its environment and, therefore, depends on the properties of solvent even in an absence of solvation effects. Figure 3 represents the compressibility $k_{S\,sp}^0$ of globular protein and of droplets of water in two solvents, pure isooctane and in IPM solvent used for microemulsion formulations (composition is given in figure 2), calculated according to equation (21). The physical properties of solute particles ($''_{sp}$) and bulk solvent ($''_{0sp}$) used in the calculations are given in table 1. As no solvation effects were included in the calculations ($w_s = 0$), the concentration w_{sp} represents the concentration of solute, $w_{sp} = w$.

Figure 3 shows that in non-aqueous systems the compressibility of solute particles at thermal low frequency limit depends significantly on the properties of the solvent. Aqueous mixtures, where water is the solvent, are different in this respect, due to the 'abnormal' physical characteristics of liquid water at room temperatures and below: very low value of thermal expansibility, e_{0sp}, and high value of heat capacitance, $c_{P_0\,sp}$. For aqueous mixtures $\left(\dfrac{c_{P\,sp}}{c_{P_0\,sp}} - \dfrac{e_{sp}}{e_{0\,sp}}\right)^2 \approx \left(\dfrac{e_{sp}}{e_{0\,sp}}\right)^2$ and therefore according to equation (21) at low concentrations the adiabatic compressibility of solute particle at thermal low frequency limit, $k_{S\,sp}^0$, is close to its isothermal compressibility, $k_{T\,sp}$: $k_{S\,sp}^0 \approx k_{S\,sp}^\infty + \dfrac{e_{sp}{}^2 T}{c_{P\,sp}} = k_{T\,sp}$.

4.3. Transition between the thermal low and thermal high frequency limits

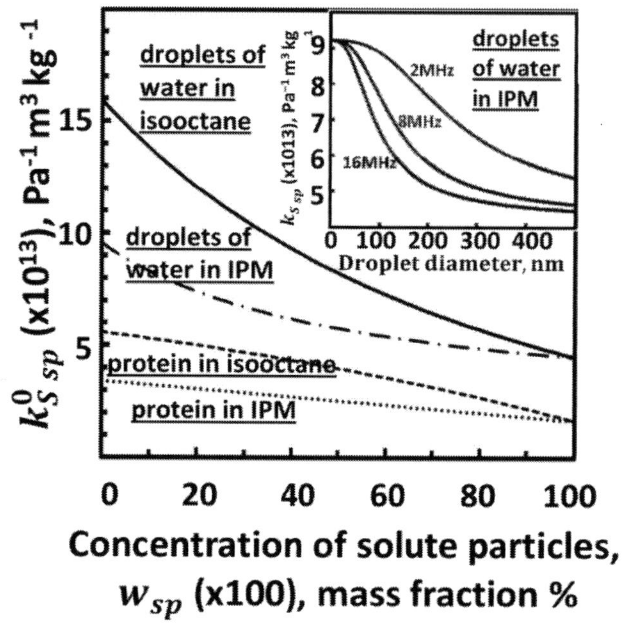

Figure 3. Specific adiabatic compressibility of protein particles and of droplets of water in IPM solvent (same composition as in Figure 2) and in isooctane at 25°C calculated for thermal low frequency limit using equation (21) and the physical properties given in table 1. **Insert.** Effect of droplet size on specific adiabatic compressibility of droplets of water in IPM solvent at different frequencies and at 25°C calculated as described in the text.

As frequency increases the compressibility of solute particle changes from its value at thermal low frequency limit to the value at thermal high frequency limit.The analytical dependence of compressibility on the frequency for a given size of the solute or on the size of the solute for a given frequency can be obtained through analysis of dynamics of the heat exchange between the solute particles and the solvent. The results depend on structural details of the mixture including particle geometry. Algorithms of analysis were described previously as a part of ultrasonic scattering theories ([1], [17], [18]) although specific calculations were normally limited to the case of spherical particles. Certain aspects of the results can be generalised. An increase of the frequency decreases the volumes of the solute particle and of the solvent participating in the heat exchange around the particle-solvent border. These volumes involve the whole particle and the solvent at low frequencies and become to be negligibly small at high frequencies. The dependence of compressibility on lg of frequency (or on lg of size at constant frequency) exhibits an 'S' shape curve. The main transition of compressibility between the thermal low and thermal high frequency limits occurs around frequency at which the size of solute particle is about the wavelength of the thermal wave. The major change of compressibility with frequency occupies approximately one decade on the frequency range. Figure 2 demonstrates the dependence of compressibility of aqueous droplets on their particle size in the IPM solvent at $w_{sp} = 0.01$ calculated using the PSize software module (v.2.28.01) provided with HR-US 102 spectrometers (Sonas Technologies Ltd.). This software utilises theoretical approaches developed previously for spherical particles in liquid continuous medium ([1], [19], [20], [21], [25]). The physical properties used in the calculations are given in table 1. Overall the calculated values of compressibility for zero and infinite diameter agree with values of adiabatic compressibility for thermal low and thermal high frequency limits, calculated according to equation (21).

4.4. Physical location of components of compressibility of solute particle

Compressibility of solute particle is defined as the difference between the compressibility of the mixture (solute particles in the bulk solvent) and the bulk solvent. At thermal high frequency limit the particle and the solvent are thermally insulated, and, in an absence of ideal and relaxation contributions, compressibility of solute particle represents the compression of the physical volume of solute particle per unit of pressure applied. At thermal low frequency limit, the solute – solvent heat exchange affects the temperature of the particle and the solvent and, therefore, their compressions. Thus, at this frequency limit a part of the measured compressibility of solute particle is physically located in the solvent. This part represents the difference between the compression of the solvent in the mixture and the compression of the pure solvent (per unit of pressure applied). A question can be raised on the portions of contributions of compressions of solute particles and of the solvent to the measured compressibility of solute particle. These portions can be obtained from the following considerations. The specific adiabatic compressibility of the bulk solvent is given as: $k_{S_0 sp} = k_{T_0 sp} - e_{0sp}\left(\frac{\partial T}{\partial P}\right)_S$. In pure bulk solvent (same composition as in the mixture) $\left(\frac{\partial T}{\partial P}\right)_S = \frac{Te_{0sp}}{c_{P_0 sp}}$.

In the mixture at thermal low frequency limit $\left(\frac{\partial T}{\partial P}\right)_S = \frac{Te}{c_P} = T\frac{(1-w_{sp})e_{0sp} + w_{sp}e_{sp}}{(1-w_{sp})c_{P_0 sp} + w_{sp}c_{P sp}}$. The difference between the compressibility of the bulk solvent in the mixture and the pure bulk solvent calculated per unit of mass of solute particle provides the 'solvent' part of compressibility of the solute particle located in the bulk solvent, $k_{S\,sp\,solvent}$: $k_{S\,sp\,solvent} = -e_{0sp}(\frac{Te}{c_P} - \frac{Te_{0sp}}{c_{P_0 sp}})$, which can be rewritten as:

$$k_{S\,sp\,solvent} = -(1-w_{sp})\frac{Te_{0sp}^{\ 2}}{c_{P_0 sp}}\frac{\frac{e_{sp}}{e_{0sp}} - \frac{c_{P sp}}{c_{P_0 sp}}}{1 + w_{sp}(\frac{c_{P sp}}{c_{P_0 sp}} - 1)} \tag{22}$$

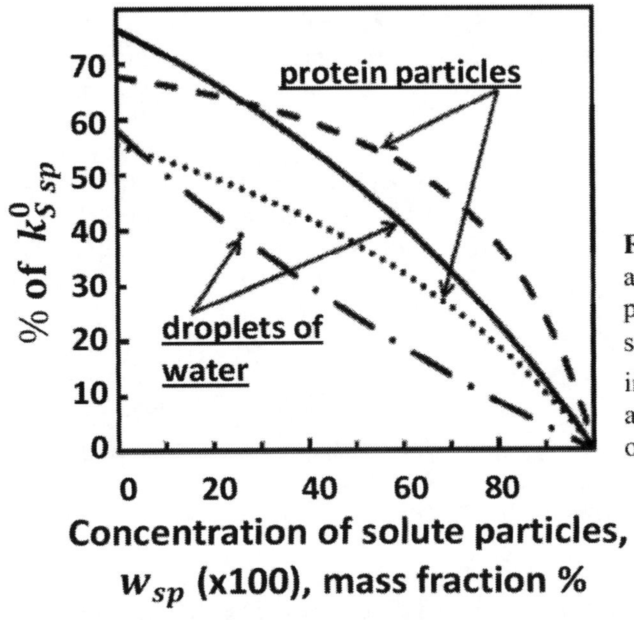

Figure 4. The 'solvent' portion of adiabatic compressibility of protein particles and of droplets of water in IPM solvent and in isooctane as % of $k^0_{S\,sp}$ shown in Figure 3; calculated using equations (22) and (21) at 25°C. The lines marking (IPM or isooctane) is the same as in figure 3.

It is interesting to note that the 'solvent' part of compressibility of solute particle can be positive or negative, depending on the ratios of thermal expansibility and heat capacity for the particles and the solvent. Figure 4 illustrates the portion ($100\frac{k_{S\,sp\,solvent}}{k^0_{S\,sp}}$) of the 'solvent' part of $k^0_{S\,sp}$ for protein and for droplets of water in isooctane and in IPM solvent calculated according to the equations (22) and (21).

The physical properties of solute particles ($''_{sp}$) and solvent ($''_{0sp}$) used in the calculations are given in table 1. As no solvation effects were included in calculations ($w_s = 0$), the concentration w_{sp} represented the concentration of solute, $w_{sp} = w$. Figure 4 demonstrates that the 'solvent' part of compressibility of protein particles and droplets of water in these solvents is significant and at low concentrations exceeds 50% of the total compressibility of solute particle.

4.5. Predictions of ultrasonic velocity and concentration increment of ultrasonic velocity of solutes from their apparent volume and apparent compressibility.

Utilisation of ultrasonic measurements in analysis of molecular processes often requires to relate the directly measured ultrasonic velocity or concentration increments of ultrasonic velocity with molecular characteristics of interest. For 'molecular' interpretation of concentration increments of ultrasonic velocity equations (19) can be rewritten in the following way, which allows calculations of concentration increment of ultrasonic velocity (as well as the ultrasonic velocity) from the specific apparent volume and compressibility of solute or the specific volume and compressibility of solute particles:

$$a_{u^2} = \frac{1}{2}\left(2\frac{\varphi V}{v_0} - \frac{\varphi K_S}{k_{S_0}} - 1 + w\frac{\left(\frac{\varphi V}{v_0} - \frac{\varphi K_S}{k_{S_0}}\right)^2}{1 + w\left(\frac{\varphi K_S}{k_{S_0}} - 1\right)}\right); \quad a_{u^2 sp} = \frac{1}{2}\left(2\frac{v_{sp}}{v_{0sp}} - \frac{k_{S\,sp}}{k_{S_0\,sp}} - 1 + w_{sp}\frac{\left(\frac{v_{sp}}{v_{0sp}} - \frac{k_{S\,sp}}{k_{S_0\,sp}}\right)^2}{1 + w_{sp}\left(\frac{k_{S\,sp}}{k_{S_0\,sp}} - 1\right)}\right) \quad (23)$$

At infinite dilution, $w \to 0$:

$$a_{u^2} = a_u = \frac{1}{2}\left(2\frac{\varphi V}{v_0} - \frac{\varphi K_S}{k_{S_0}} - 1\right); \quad a_{u^2 sp} = a_{u\,sp} = \frac{1}{2}\left(2\frac{v_{sp}}{v_{0sp}} - \frac{k_{S\,sp}}{k_{S_0\,sp}} - 1\right) \quad (24)$$

It is important to note that equations (23) and (24) are valid for the inertial low frequency limit only at which the inertial and gravimetric densities of the analysed mixture are the same. For larger particles these equations are applicable only in the case of low density contrast as discussed in the section 3.5.

5. Interpretation of compressibility of solute particles

Molecular interpretation of compressibility (and other thermodynamic characteristics) requires distinguishing contributions of different 'molecular' origin. For a mixture of solute particles and a solvent three major contributions are often considered: intrinsic, osmotic (ideal), and relaxation. These contributions are also presented in isothermal compressibility, heat capacity and thermal expansion of solute particles. As the adiabatic compressibility of solute particles at thermal low frequency limit is a non-linear function of these additive characteristics (equation (19)), in general, it is not a simple sum of the intrinsic, osmotic and relaxation contributions. However, at infinite dilution the linear equation (20) is correct and, therefore, the compressibility of solute particle, $k_{S\,sp}$, can be expressed as a sum of intrinsic compressibility of solute particle, $k_{S\,in}$, osmotic (ideal) contribution to compressibility, $k_{S\,id}$, and relaxation contribution to compressibility $k_{S\,rel}$, : $k_{S\,sp} = k_{S\,in} + k_{S\,id} + k_{S\,rel}$.

5.1. Contribution of osmotic effects to compressibility of solute particles

The osmotic (also called ideal) contribution to compressibility of solute particles is originated by the heat motion of the solute particles, which produces osmotic pressure in the mixture, thus leading to the expansion of solvent. The change of the volume of this expansion with pressure applied provides the ideal contribution to compressibility. Due to the origin (heat motion) of the ideal compressibility, its molar value does not depend on the molar mass of the solute particle, M. Therefore, its contribution to specific compressibility of solute particle is inversely proportional to molar mass, $k_{sid} \sim M^{-1}$, and shall vanish as the molar mass increases. Estimations of this contribution water and organic solvents, show that for polymers, and nano-size and larger particles this contribution provides only small portion of the measured compressibility of solute particle and normally can be neglected.

5.2. Relaxation contributions to compressibility of solute particles

The relaxation contribution to the apparent compressibility and the compressibility of solute particles arises if the particles in analyzed mixture have different states (conformational, protonated, etc.), and the distribution between the states oscillates as a result of oscillating pressure and temperature in ultrasonic wave. It represents the change of the volume of the mixture caused by a shift in the distribution of solute particles between different states, per unit of pressure applied. The relationships between the molecular characteristics of relaxation processes and their contributions to ultrasonic parameters were described previously (see for example [22]). The theories cover the thermal low frequency limit at which solute particles and solvent are at the same temperature at all times. They provide tools for estimation of relaxation contribution from known volume and enthalpy effects of the relaxation process and the equilibrium concentrations of its participants. The frequency dependence of the relaxation compressibility is determined by the factor $\frac{1}{1+(2\pi f\tau)^2}$ (τ is the relaxation time of the process). The relaxation compressibility vanishes at high frequencies.

At thermal high frequency limit each solute particle is thermally insulated from the rest of the mixture and oscillations of temperature within the particles of different states and within the solvent are different. Therefore, at this limit the relationships between the ultrasonic parameters and molecular characteristics of relaxation processes are complex and depend on the structural details of mixtures.

5.3. Intrinsic compressibility of solute particles

The intrinsic adiabatic compressibility of solute particles at infinite dilution represents the contribution of solute particles to the compressibility of mixture in an absence, or after subtraction, of the osmotic and the relaxation contributions. At thermal high frequency limit this compressibility represents the adiabatic compressibility of the material of particles (solutes). For low molecular weight molecules it is determined by compressibility of the solvation (hydration) shell and by the compressibility of the Van der Waals volume of solute molecules. The later represents the change of Van der Waals radii of atoms and the length of covalent bonds caused by applied pressure. The magnitude of this compressibility can be estimated from compressibilities of relevant crystals. For organic molecules it is often smaller than other contributions to compressibility and can be neglected. (see for example [23]). For large polymers and molecular aggregates the intrinsic adiabatic compressibility of solute particles is determined by their microstructural/molecular arrangement. At thermal low frequency limit the contribution of heat exchange is added to the intrinsic compressibility which is provided by equations (19) to (21).

6. Effects of solvent on compressibility and concentration increment of ultrasonic velocity of solute particles

Equations (19) , (22) and (21) demonstrate that at thermal low frequency limit solvent-solute hear exchange results in the dependence of compressibility of solute particles, $k^0_{S\,sp}$, (and the apparent compressibility of solute) on the values of the specific thermal expansibility and heat capacity of the solvent. Therefore, a substitution of solvent around the solute particles shall result in a change of adiabatic compressibility of solute particles and their concentration increment of ultrasonic velocity.

If the specific thermal expansibility and the heat capacity of the solvent ($e_{0\,sp}$ and $c_{P_0\,sp}$) and of the solute particles (e_{sp} and $c_{P\,sp}$) are known equation (19) or (22) can be applied for calculation of $k^0_{S\,sp}$ in solvent 2 from the value of $k^0_{S\,sp}$ measured in a solvent 1, provided that the isothermal properties of solute particles, $k_{T\,sp}$, e_{sp} and $c_{P\,sp}$ are not affected by the substitution of solvent and the osmotic contribution can be neglected. The same is true for apparent characteristics of solute (first part of equations (17) or (18)). For infinite dilution, $w, w_{sp} \to 0$, simplified equations for recalculations of φK_S^0, $k_{S\,sp}^0$, a_{u^2} and $a_{u^2\,sp}$ were obtained ([24]). Contrary to the above, for the thermal high frequency limit the solute-solvent heat exchange can be neglected. Therefore, a substitution of solvent shall not produce an effect on compressibility of solute particles, $k_{S\,sp}^{\infty}$, provided that the isothermal properties of solute particle, $k_{T\,sp}$, e_{sp} and $c_{P\,sp}$ (equation (19)) are not affected by the substitution and that the osmotic and the relaxation contributions can be neglected.

7. Conclusions

Liquid molecular solutions and nano scale particle dispersions at ultrasonic frequencies can be described in terms of thermal low and visco-inertial low frequency limits.

In non-aqueous liquid molecular solutions and nano scale particle dispersions the heat exchange between the components of mixtures may result in a substantial contribution to the measured adiabatic compressibility of solute particles and apparent adiabatic compressibility of solutes. Consideration of this contribution is important in the case of quantitative interpretation of the measured adiabatic compressibility of solute particles, or when the changes of compressibility (or ultrasonic velocity) caused by a particular molecular process are compared in different solvents.

For solutions and nano scale particle dispersions (thermal low frequency limit) contributions of heat exchange to compressibility of solute particles and to apparent compressibility of solute are provided by equations (19) and (21), which are valid for the whole concentration range and any particle shape.

The contribution of heat exchange to ultrasonic velocity and to the concentration increments of ultrasonic velocity in liquid solutions and nano scale particle dispersions is provided by equations (21) and (23).

Table 1. Physical properties of the media used in calculations.

Physical properties at 25°C	Water [a]	IPM [b]	Isooctane	Protein
Specific adiabatic compressibility ($\times 10^{13}$), $m^3\,kg^{-1}\,Pa^{-1}$	4.491	7.655	18.12 [c]	1.66 [d]
Specific volume ($\times 10^3$), $m^3 kg^{-1}$	1.003	1.158	1.457 [c]	0.662 [d]
Thermal conductivity, W $m^{-1}\,K^{-1}$	0.595	0.19	0.095 [a]	-
Specific thermal expansibility ($\times 10^6$) $m^3 kg^{-1}\,K^{-1}$	0.257	0.982	2.076	0.07 [d]
Specific heat capacity, J $kg^{-1}\,K^{-1}$	4180	1405	2037 [a]	1403 [e]
Viscosity, Pa s($\times 10^3$)	0.890	8.6	0.479 [c]	-

[a] [28]; [b]Calculated from [5] for composition given in Figure 2; [c][27]; [d] Taken as intrinsic characteristic of globular proteins [26] [e] Calculated from [29]as average value for three globular proteins: cytochrome C, myoglobin and lysozyme.

Equation (17) allows calculation of the apparent adiabatic compressibility of solutes (and adiabatic compressibility of solute particles) in solutions and dispersions of nano particles (visco-inertial low frequency limit) from the measured concentration increment of ultrasonic velocity at any concentration of solute.

Acknowledgments

This research was supported by the MASAF315 grant from SFI Ireland. Thanks to Shailesh Hallone and Margarida Altas for their assistance in preparation of manuscript.

References

[1] Povey MJ 1997 *Ultrasonic Techniques for Fluid Characterisation* (London:.Academic Press)
[2] Kharakoz DP and Sarvazyan, AP 1993 Hydrational and intrinsic compressibilities of globular proteins *Biopolymers* **33** 11-26
[3] Resa P, Elvira L, Sierra C and Montero de Espinosa F 2009 Ultrasonic velocity assay of extracellular invertase in living yeasts *Anal.Biochem*. **384** 68–73
[4] Resa P and Buckin V 2011 Ultrasonic analysis of kinetic mechanism of hydrolysis of cellobiose by beta-glucosidase *Analytical Biochemistry* **415** 1–11
[5] Hickey S, Lawrence MJ, Hagan SA and Buckin V. 2006 Analysis of the phase diagram and microstructural transitions in phospholipid microemulsion systems using High-Resolution Ultrasonic Spectroscopy *Langmuir* **22** 5575-83
[6] Hickey S, Lawrence M J, Hagan SA, Kudryashov E and Buckin V 2010 Analysis of phase diagram and microstructural transitions in an ethyl oleate/water/tween 80/Span 20 microemulsion system using High-Resolution Ultrasonic Spectroscopy *International Journal of Pharmaceutics* **388** 213–222
[7] Dwyer C, Donnelly L and Buckin V 2005 Ultrasonic analysis of rennet-induced pregelationand gelation processes in milk *J. Dairy Res*. **72** 303–310
[8] Smyth C, Kudryashov E and Buckin V 2001 High-frequency shear and volume viscoelastic moduli of casein particle gel *Colloids and Surfaces, A: Physicochemical and Engineering Aspects* **183–185** 517–526
[9] Van Durme K, Delellio L, Kudryashov E, Buckin V and Van Mele K 2005 Exploration of High-Resolution Ultrasonic Spectroscopy as an analytical tool to study demixing and remixing in poly(N-isopropyl acrylamide)/water solutions *J. of Polymer Science: Part B: Polymer Physics* **43** 1283–95
[10] Buckin V and O'Driscoll B 2002 Ultrasonic waves and material analysis: recent advances and future trends *Lab. Plus Int.* **16** (3) 17–21
[11] Kudryashov E, Smyth C, O'Driscoll B and Buckin V 2003 High-Resolution Ultrasonic Spectroscopy for analysis of chemical reactions in real time *Spectroscopy* **18** (10) 26–32
[12] Litovitz TA and Davis CM 1964 Structural and shear relaxation in liquids *Physical Acoustic* vol 2 part A ed W P Mason (NY: Academic Press) pp 282-349
[13] Thurston RN 1964 *Physical Acoustics: Principles and Methods* vol 1 part A ed W P Mason (New York: Academic Press) chapter 1
[14] Morse PM and Ingard KU 1986 *Theoretical Acoustic* (Princeton, NJ: Princeton University Press)
[15] Pierce A D 1991 *Acoustics* (NY: Acoustical Society of America)
[16] Buckin V and Smyth C 1999 High-resolution ultrasonic resonator measurements for analysis of liquids *Sem. Food Anal.* **4** 113–130
[17] Isakovich MA 1948 *Zh Eksp Teor.Fiz.* **18** 907-912
[18] Hemar Y, Hocquart R and Palierne JF 1998 Frequency-dependent compressibility in emulsions: Probing interfaces using Isakovich's sound absorption *Europhys. Lett.* **42** (3) 253-258
[19] Austin JC, Holmes AK, Tebbutt JS and Challis RE 1996 Ultrasonic wave propagation in colloid suspensions and emulsions: recent experimental results *Ultrasonics* **34** 369–374
[20] Epstein PS and Carhart RJ 1953 The absorption of sound in suspensions and emulsions.I. Water fog in air *J. Acoust. Soc. Am.* **25** 553–562
[21] Waterman PC and Truell R. 1961 Multiple scattering of waves *J. Math. Phys.* **2** 512–537
[22] Kaatze U, Hushcha T O and Eggers F 2000 Ultrasonic broadband spectrometry of liquids: a research tool in pure and applied chemistry and chemical physics *J. Solution Chemistry* **29** (4) 299-368
[23] Buckin V 1988 Hydration of nucleic bases in dilute aqueous solutions. Apparent molar, adiabatic and isothermal compressibilities, apparent molar volumes and their temperature slopes at 25 ◦C. *Biophys. Chem.* **29** 283–292

[24] Buckin V and Hallone SK 2012 Ultrasonic characterisation of W/O microemulsions – structure, phase diagrams, state of water in nano–droplets, encapsulated proteins, enzymes *Microemulsions - An Introduction to Properties and Applications* ed R Najjar (InTech) pp 33-66 ISBN: 978-953-51-0247-2

[25] Allegra JR and Hawley SA 1972 Attenuation of sound in suspensions and emulsions: theory and experiment *J. Acoust. Soc. Am.* **51** 1545–1564

[26] Chalikian T 2003 Volumetric properties of proteins *Ann. Rev. Biophys. Biomol. Struct.* **32** 207–35

[27] Gómez-Díaz D, Mejuto JC, and Navaza J M 2006 Density, viscosity, and speed of sound of solutions of AOT reverse micelles in 2,2,4-Trimethylpentane *J. Chem.Eng. Data* **51** 409-411

[28] 2004-2005 *Handbook of Chemistry and physics 85th edition* ed DR Lide (Boca Raton, Florida: CRC Press)

[29] Privalov P L and Dragan AI 2007 Microcalorimetry of biological macromolecules. *Biophysical Chemistry* **126** 16–24

Detection of tissue properties using a piezoelectric vibration-based syringe

C H Lin, Y C Huang, and M C Tsai[1]

Department of Mechanical Engineering, National Cheng Kung University, Tainan 701, Taiwan

E-mail: mctsai@mail.ncku.edu.tw

Abstract. For safety improvement purposes, detection of tissue properties at the tip of the needle in real-time is helpful in controlling the movement of the needle in the operation of invasive surgical tools. According to the energy-based force model, the mechanical impedance of tissue can be represented approximately by a spring-damping model. In this study, a vibration-based syringe which adopts piezoelectric elements to vibrate the needle is proposed to identify tissue properties. In principle, by measuring the input electrical impedance of the vibration-based syringe, the mechanical impedance of tissue can be detected indirectly while the syringe exerts a vibration on the tissue. The model of the piezoelectric driven syringe and the procedure for evaluating the electromechanical interaction are given, and the mechanical impedance of tissue is evaluated by simulation. The proposed detection method has potential for applications of identifying tissue properties in epidural punctures and estimation of the degree of ablation in thermal surgery systems.

1. Introduction

There are several medical applications which is required percutaneous ("through the skin") needle insertions such as biopsy, regional anesthesia, blood sampling, neurosurgery, brachytherapy, and thermal therapy [1] in which the needle is inserted into soft and inhomogeneous tissue to reach a certain target. In the operation of invasive surgical tools, tissue information at the needle tip is critical in order to control the movement of the needle [2]. Since tissue properties can be represented approximately by a spring-damping mechanical model [3], tissue can be identified by measurement of its equivalent mechanical impedance, which is defined as the ratio of the responded force to the displacement exerted on tissue. In this study, a piezoelectric vibration-based syringe which adopts piezoelectric elements to vibrate a needle is proposed to identify the mechanical impedance of the tissue due to the electromechanical interaction [4]. In principle, the mechanical impedance of the tissue can be indirectly detected by measuring input electrical impedance of the piezoelectric elements.

2. Modelling of piezoelectric vibration-based syringe

Consider a piezoelectric vibration-based syringe which utilizes a Langevin transducer to vibrate a hypodermic needle (27G x 1/2") as shown in figure 1 where the sandwich structure design is composed of two metallic pieces and stacks of piezoelectric elements (PZT-4). The electromechanical

[1] To whom any correspondence should be addressed.

interaction of the piezoelectric driven syringe can be demonstrated by a system block diagram as illustrated in figure 2 where capacitance C denotes the equivalent electrical property under constant stress, coefficient N represents its electromechanical interaction and M, B, and K are equivalent mass, damping coefficient, and stiffness of the syringe, respectively. Note that the electric current i and the external force F are the input variables of the system block diagram whereas the vibration displacement x and the voltage V are regarded as the output variables. The electromechanical interaction is demonstrated in figure 2 where the voltage V is converted into the force F_e and the vibration velocity $\dot{x}$ is converted back into the electrical current i by the coefficient N and the mechanical property is represented by an equivalent mass-spring-damper model.

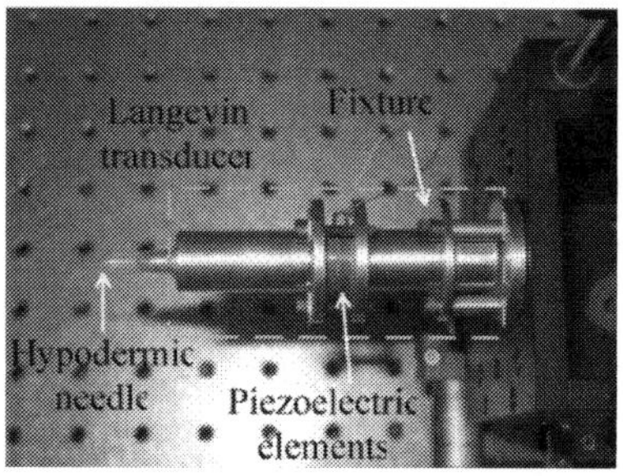

Figure 1. Entity of piezoelectric vibration-based syringe.

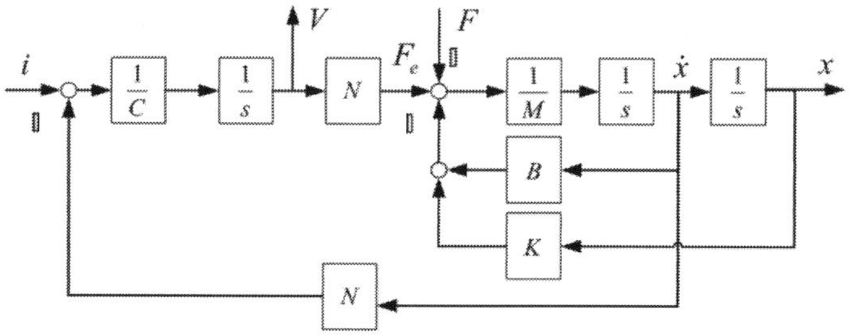

Figure 2. System block diagram of piezoelectric vibration-based syringe.

Let the input-output relationship of the system block diagram be represented by a two-port chain description matrix:

$$G(s)=\begin{bmatrix} G_{11} & G_{12} \\ G_{21} & G_{22} \end{bmatrix} \tag{1}$$

as depicted in figure 3 where mechanical variables and electrical variables are separated at each port. Then the relationships between the four variables are given by

$$\begin{bmatrix} V \\ i \end{bmatrix}=\begin{bmatrix} G_{11} & G_{12} \\ G_{21} & G_{22} \end{bmatrix}\begin{bmatrix} F \\ x \end{bmatrix}=\begin{bmatrix} \dfrac{1}{N} & \dfrac{Ms^2+Bs+K}{N} \\ \dfrac{Cs}{N} & \dfrac{Cs(Ms^2+Bs+K+\frac{N^2}{C})}{N} \end{bmatrix}\begin{bmatrix} F \\ x \end{bmatrix} \tag{2}$$

While the syringe exerts a vibration into the tissue, the equivalent mechanical impedance of the tissue, denoted by Z_t, is connected to $G(s)$ on the right-hand side. Then the electrical impedance of the syringe caused by Z_t of tissue, denoted by Z_e, is given by

$$Z_e=\frac{V}{i}=\frac{G_{11}F+G_{12}x}{G_{21}F+G_{22}x}=\frac{G_{12}+G_{11}Z_t}{G_{22}+G_{21}Z_t} \tag{3}$$

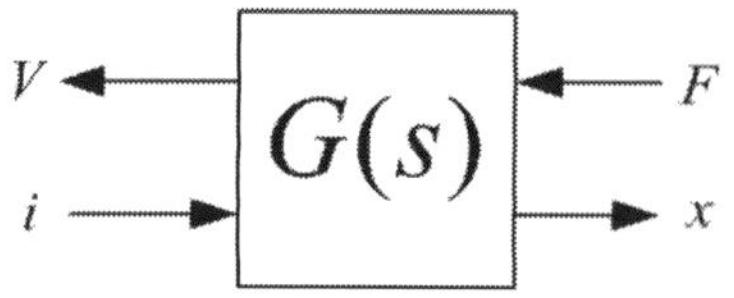

Figure 3. Two-port chain description matrix of piezoelectric vibration-based syringe.

3. Simulations of measuring mechanical impedance

To simulate the effects of tissue Z_t, let K_t and B_t denote the stiffness and damping coefficients, respectively. As can be found from equation (3), the mechanical impedance of tissue Z_t is embedded in the equivalent electrical impedance of the syringe Z_e. Then the mechanical impedance can be evaluated by a given Z_e as

$$Z_t = \frac{G_{12} - G_{22}Z_e}{G_{11} - G_{21}Z_e} \tag{4}$$

In the following, the system parameters of the syringe listed in table 1 will be used for simulation study. Also consider two different stiffness and damping coefficients of Z_t to simulate the variation of tissue, i.e., $K_{t1} = 5.0\times10^{9}$ (N s m^{-1}), $K_{t2} = 1.1\times10^{10}$ and $B_{t1} = 8.0\times10^{3}$ (N s m^{-1}), $B_{t2} = 3.0\times10^{4}$. Here the stiffness and damping coefficients of the mechanical impedance Z_t are considered respectively to investigate the effect on the electrical impedance.

As can be found in frequency responses of figure 4, the stiffness K_t increases both of the frequency of resonance and the frequency of anti-resonance. Moreover, the damping coefficient B_t reduces the peak value either at the resonance frequency or at the anti-resonance frequency as illustrated in figure 5. Reversely, for measured electrical impedance Z_e, the equivalent mechanical impedance of the tissue can be computed by equation (4) in the piezoelectric vibration-based syringe as depicted in figures 6 and 7 where the stiffness and the damping coefficient are considered respectively. In the case where only the stiffness is encountered, the mechanical impedance remains constant in the frequency response and it can be easily calculated by any frequency. Furthermore, for the case where only the damping coefficient is encountered, the mechanical impedance is proportional to the frequency and it can be calculated by its slope. However, the simplified model of tissue is often a combination of stiffness and damping coefficients, the result of this case is shown in figure 8 where the frequency response comprises both stiffness and damping coefficients and is given by

$$Z_t = \left(K_t^2 + B_t^2\omega^2\right)^{1/2} \tag{5}$$

Note that the two unknown parameters, K_t and B_t, can be calculated by two arbitrary frequencies.

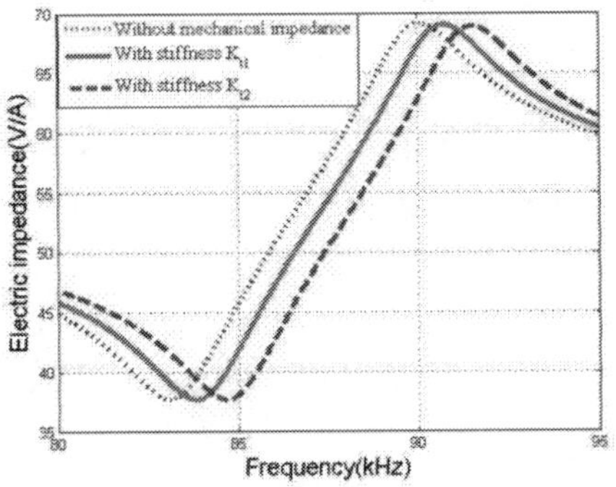

Figure 4. Frequency response of electrical impedance when the mechanical impedance is stiffness.

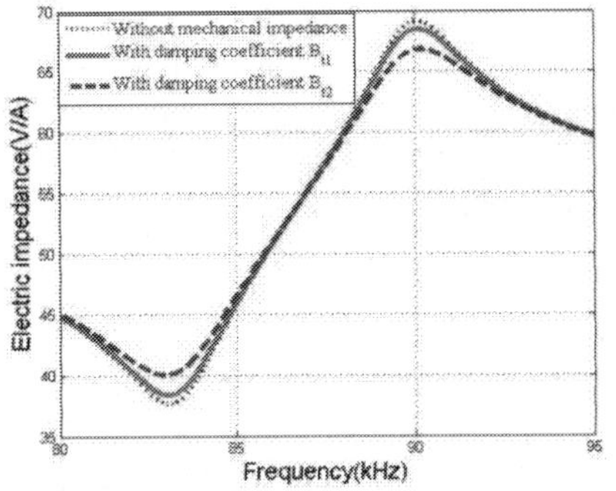

Figure 5. Frequency response of electrical impedance when the mechanical impedance is the damping coefficient.

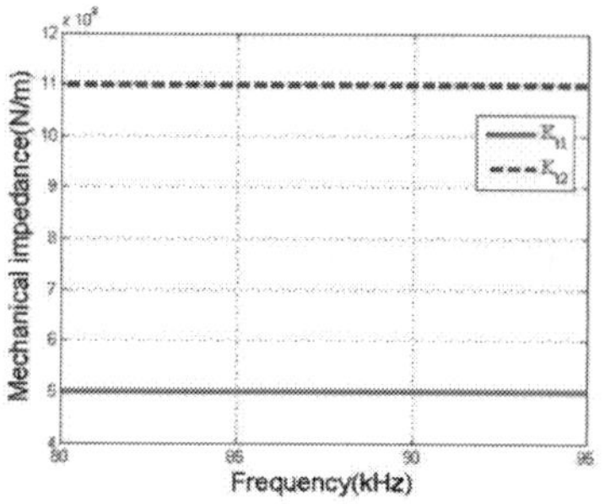

Figure 6. Stiffness transferred from the electrical impedance.

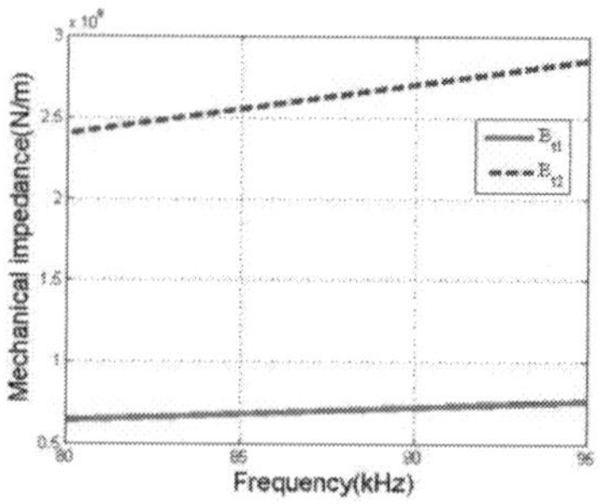

Figure 7. Damping coefficient transferred from the electrical impedance.

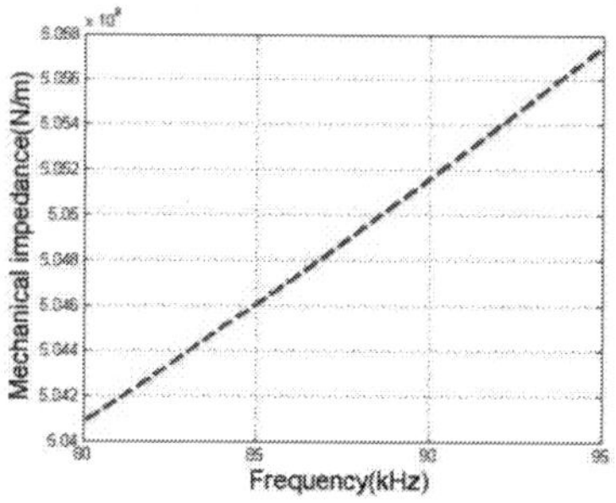

Figure 8. Mechanical impedance of the tissue transferred from the electrical impedance.

Table 1. Parameters of the piezoelectric vibration-based syringe.

Parameters	Values	Parameters	Values	Parameters	Values
M	41.68 (kg)	N	3.92 (N V^{-1})	K_{t1}	5.0×10^{9} (N m^{-1})
B	8.93×10^{4} (N s m^{-1})	C	23.753 (nF)	K_{t2}	1.1×10^{10} (N m^{-1})
K	2.89×10^{11} (N m^{-1})			B_{t1}	8.0×10^{3} (N s m^{-1})
				B_{t2}	3.0×10^{4} (N s m^{-1})

4. Conclusion

This paper presents a piezoelectric vibration-based syringe and a method to measure the mechanical impedance of tissue. The piezoelectric driven syringe not only can exert a vibration on the tissue but also identify variation of the tissue according to the electromechanical interaction. The syringe system was modeled by a two-port chain description matrix and the simulation results showed that the mechanical impedance of tissue can be calculated from the electrical impedance. This detection method has potential for identifying variation of tissue in epidural punctures and estimation of the degree of ablation in thermal surgery systems, especially for real-time detection during the insertion process.

Acknowledgement

This work was supported in part by the Department of Industrial Technology, Ministry of Economic Affairs,Taiwan under Contract 100-EC-17-A-05-S1-192. The authors would also like to thank Sunnytec for providing piezoelectric materials.

References

[1] Abolhassani N, Patel R and Moallem M 2007 *Med. Eng. Phy.* **29** 413-31
[2] Brett P N, Harrison A J and Thomas T A 2000 *IEEE T. Inf. Technol. B.* **4** 30-6.
[3] Tanaka N, Higashimori M, Kaneko M and Kao I 2011 *IEEE T. Bio-Med. Eng.* **58** 509-20
[4] Ling S F, Hou X and Xie Y 2005 *Sensor Actuat. A-Phys.* **120** 257-65

Internal cracks distribution in hot-pressed and spark plasma sintered nanostructured Bi_2Te_3-based thermoelectrics: acoustic microscopy versus optical microscopy

V M Prokhorov[1,3], **G I Pivivarov**[1], **V M Levin**[2] **and E S Morokov**[2]
[1]Technological Institute for Superhard and Novel Carbon Materials, Troitsk, Moscow Region, Centralnaya 7a, Russia,142190
[2] Emmanuel's Institute of Biochemical Physics, Russian Academy of Science, Moscow, Kosygina 4, Russia, 119334.

E-mail: pvm@tisnum.ru

Abstract. We have produced the samples of Bi_2Te_3-based nanostructured thermoelectric (TE) alloys by the method of high energy mechanotreatment from the mixture of the nano- and micro-sized powders (ball milling) followed by two methods of bulk samples preparing: (a) by the method of hot pressing sintering (HPS) and (b) by the method of spark plasma sintering (SPS). The microstructure of the obtained pellets was investigated by methods of acoustic and optical microscopy. With the crack/pore detection in acoustic microscopy images we marked that spot and cut the pellet through that spot perpendicular to the pellet's plane. The cut section, after polishing and etching, was examined on Olympus BX-51□ microscope. The obtained pictures for the HP sintering samples shows, that the planes of all macro- and of micro-cracks are oriented perpendicularly to the axis of the load of the pellet. By using acoustic microscopy we investigate cracks and pores distributions in bulk of nanostructured $Bi_{0.4}Sb_{1.6}Te_3$ alloy sintered by HPS and SPS methods.

1. Introduction

The appropriately doped Bi_2Te_3 alloys are one of the most important and competitive commercial thermoelectric (TE) materials that exist to date. The best heat-to-power conversion efficiency of conventional TE materials is just about 4% [1]. The efficiency of TE material is assessed with a figure of merit Z defined as $S^2\sigma/\kappa$, where S is the Seebeck coefficient (measured in $\mu V/K$), σ is the electric conductivity, and κ is the thermal conductivity. For practical applications, the nondimensional parameter, figure-of-merit, defined as ZT is used (where T is the absolute temperature).

Since Bi_2Te_3-based thermoelectric have a maximum performance at approximately 350 K, most of the previous works have been focused on the study of the TE properties at room temperature. Bi_2Te_3-based TE material takes a maximum value of ZT ~ 0.8 in bulk, while it could be greater than 2.5 in nanostructures.

Z can be improved for nanostructured materials by (i) reduction of lattice thermal conductivity by means of enhanced phonon scattering on the inclusions and on the interfaces between nanoparticles

[3] To whom any correspondence should be addressed.

and (ii) increase of S by electrons tunneling through interfaces between nanoparticles, with both goals helping to increase ZT [2].

Recently, a number of novel preparation procedures have been developed to fabricate Bi_2Te_3 nanocomposites with higher ZT values, such as the ball milling followed by HPS procedure [3] and SPS procedure [4]. In utilizing these various novel process procedures, the ZT values have been subsequently increased by 40-50%. In the previous studies, any possible degree of anisotropy of these nanocomposites has not attracted very much attention.

It was showed in [5] that only small changes in the sintering conditions of the SPS samples bear a large influence on the resulting sample compositions and on the anisotropy in the TE properties.

Recently, Xie et al. [6] investigated the correlation between the SPS pressure, the microstructure texture, and the anisotropy of the total thermal conductivity in these melt-spun spark-plasma-sintering $(Bi,Sb)_2Te_3$ compounds. The results show that the anisotropy of thermal conductivity is ~0, 2-3, 6-7, and 13-15% for the samples sintered at pressures of 20, 30, 45, and 60 MPa, respectively.

The influence of HPS conditions and the composition of the $(Bi,Sb)_2Te_3$ compounds on sizes of nanograins and TE properties was investigated by Bulat at al. [7].

In this work we study the effect of sintering methods, SPS and HPS, on the porosity and defects, generated in the sintering process of nanostructured TE material p-type $Bi_{0.4}Sb_{1.6}Te_3$.

2. Experimental details

Nanocrystalline specimens of the TE material $Bi_{0.4}Sb_{1.6}Te_3$ were prepared by the SPS and HPS methods at various temperatures and pressures and their defect properties were evaluated by acoustic and optic microscopy methods.

The installation SPS-511S manufactured by SPS SYNTEX INC (Japan) for researching and developing the technology of obtaining Bi_2Te_3-based TE materials of p-type with the nano-, micro- and sub-micron elements of the structure was used in this work.

For the HPS treatment under middle and high pressure we used a steel high-pressure "piston-cylinder" type chamber. The samples of Bi_2Te_3-based TE materials were sintered under 40-70 MPa pressure, at a temperature 380-480 °C with the duration of the isothermal holding of 5-20 minutes.

The samples in forms square plate with dimensions 10x10x(2-3) mm^3 and bars with dimensions 2.0x2.0x10 mm^3 were cut from pellets for pores distribution and elastic anisotropy studies.

The densities of samples were measured and these values were used to characterize the TE material porosity due to voids and cracks into the balk of sample.

2.1. Acoustic microscopy

Two type of impulse acoustic microscopes, fabricated in AM-laboratory of Emmanuel's Institute of Biochemical Physics, Russian Academy of Science, were used to measure local values of ultrasonic velocities (microacoustic technique) and to visualize bulk microstructure of a sample (scanning acoustic microscopy).

One of them is the wide-field pulse scanning acoustic microscope WFAM-1 (25 ÷ 100 MHz), designed in 1996. Another one is scanning impulse acoustic microscope SIAM (50 ÷ 200 MHz), designed in 2010. Ultrashort probing ultrasonic 30-40 ns pulses were used for measurements. 1D- or 2D- scanning of the probe beam over the specimen surface results in raster formation acoustic images (B-and C-scans, respectively). Experimental procedure was described in details previously [8].

3. Results and discussion

The obtained AM pictures for the HPS samples shows that the planes of all macro- and of micro-cracks are oriented perpendicularly to the axis of the load of the pellet. This testifies in favor to the assumption that the reason of the cracks is the presence of relatively large aggregates in the initial powder. They differ in their elastic properties from the bulk of nano-powder, and during the sintering their presence can lead to the known effect of zone detachment (due to the diffusion rate difference in aggregates and in nano-powders).

IOP Conf. Series: Materials Science and Engineering **42** (2012) 012003 doi:10.1088/1757-899X/42/1/012003

Figures 1(a) and 1(b) show the acoustic images of defects in pellet HP-sintered at 450 °C, 10 min, 50 MPa pressure; the cracks are marked by arrows. The observed lines below the cracks are "phantoms", which were formed as a result of the reflection of ultrasonic waves from flat surfaces of the models. The circle marks the defect which can be both a crack and some local disturbance of the material's density (the initial powder was not disaggregated). The "phantoms" are observed near those objects too. Figure 1(c) shows the sample obtained at 400 °C, 5 min, 60 MPa pressure.

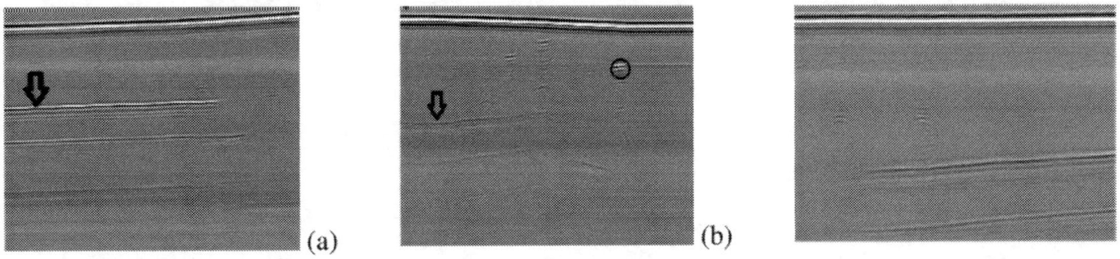

(a) (b) (c)

Figure 1. 25 MHz B-scans of acoustic microscopy. Images of the cracks (they are marked by arrows) in the HPS pellets of that nano-structured $Bi_{0.4}Sb_{1.6}Te_3$; the circle marks local structural imperfections of material (possibly pores).

The pictures show that the planes of all macro- and of micro-cracks are oriented perpendicularly to the axis of the load of the tablet. This testifies in favor to the assumption that the reason of the cracks is the presence of relatively large aggregates in the initial powder. They differ in their elastic properties from the bulk of nano-powder, and during the sintering their presence can lead to the known effect of zone detachment (due to the diffusion rate difference in aggregates and in nano-powders).

If you know in advance where and what are the defects in the sample, then by using optical microscopy they are found quite easily.

The Figure 2(a) shows the same specimen as the Figure 1(a), and Figure 2(b) – the same as 1(b). The plane of the cut is carried out on the defect marked by circle in figure 1(b), which is observed in the right lower corner in figure 2(b). As is evident, the defect is an aggregate with size approximately 100 microns covered with cracks. It cannot be asserted that these micro-cracks were contained in the aggregate - they may have appeared during the sawing of the specimen. But in any case, the presence of such defects is highly undesirable because they lead to spoilage in the manufacture of thermoelement branches.

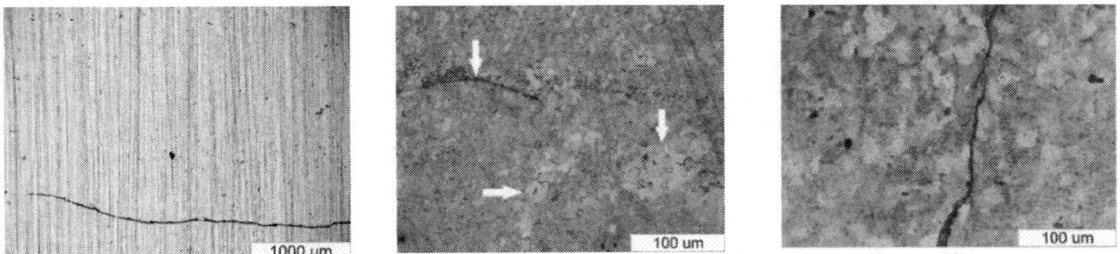

(a) (b) (c)

Figure 2. Optical microscopy images: (a) - the photograph of crack in the specimen observed in Figure 1(a); (b) - the etched section of the specimen shown in figure 1(b). Arrows mark the micro-crack in the upper right corner and the pores.

Cracks were observed in specimens made at 400°C (see the Figure of 1(b)) and at 420°C. It was possible to get rid of them with increasing sintering temperature to 440°C (pressure and time remain the same).

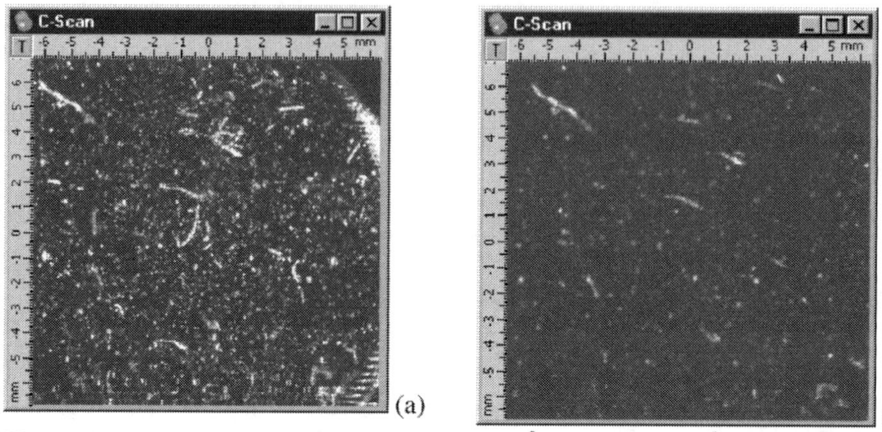

(a) (b)

Figure 3. The sample HPS disk Ø20x2.8 mm^3, ρ=6.774 g/cm^3. AM 100 MHz. C-scans: (a) integral bulk structure, (b) structure one of the internal layer.

Figure 3 shows AM images of defect structure of HPS prepared pellet (440 °C, 40 MPa, 10 min isothermal holding) obtained from disaggregated nanopowder $Bi_{0.4}Sb_{1.6}Te_3$. It is seen that in the structure of the specimen is completely defected in form of cracks and pores.

4. Conclusion

By means of acoustic microscopy measurements and structural investigations of specially processed TE alloys we revealed the following: Sintering by hot compression method as well as sintering of cold-pressed samples in all cases lead to Tellure precipitation second phase. This is accompanied with an increase of solid solution lattice parameter and increase of micro-deformations. The mean grain size and homogeneity of size-distribution depend not only on the temperature but on the compression time also. The increase of in cycle-work time makes these effects more pronounced.

The regular correlation was observed: the samples with reduced elastic properties have better TE figure of merit. Generally it is connected with thermal conductivity reduction of a nanostructured materials.

The acoustic microscopy data have been used to optimize synthesis parameters of nanostructured thermoelectrics.

Acknowledgments

The present work was supported through a research grant from Ministry of Education and Science of Russian Federation (Contract No.16.523.11.3002 and No.16.552.11.7014). The authors would like to thank eng. L. Solov'eva for the preparing the samples for optical studies.

References

[1] DiSalvo F J 1999 *Science* **285** 703
[2] Bell L E 2008 *Science* **321** 1457
[3] Poudel B *et al* 2008 *Science* **320** 634
[4] Tang X F, Xie W J, Li H, Zhao W Y, Zhang Q J and Niino M 2007 *Appl. Phys. Lett.* **90** 012102
[5] Xie W J *et al* 2010 *Nano Lett* **10** 3283
[6] Xie W J, He J, Zhu S, Holgate T, Wang Sh, Tang X, Zhang Q and Tritt T M 2011 *J. Mater. Res.* **26** 1791
[7] Bulat L P, Bublik V T, Drabkin I A, Karataev V V, Osvenskii V B, Parkhomenko Yu N,Pivivarov G I, Pshenai-Severin D A and Tabachkova N Yu 2010 *J. Electronic Mater.* **39** 1650
[8] Prokhorov V M and Pivovarov G I 2011 *Ultrasonics* **51** 715

IOP Conf. Series: Materials Science and Engineering **42** (2012) 012004 doi:10.1088/1757-899X/42/1/012004

Determination of the size of spherical inclusions smaller than the width of the sound beam

S Kümmritz, M Wolf, E Kühnicke and N Gust

Technische Universität Dresden, Institut für Festkörperelektronik, Helmholtzstraße 18, D-01187 Dresden (Germany)

E-mail: sebastian.kuemmritz@tu-dresden.de

Abstract. The current presentation introduces a new method for the determination of the size of balls with a diameter from nearly a wavelength up to the width of the sound beam (and higher). The basic approach is to analyse the directional pattern of the reflectors, which depends on size and shape. If the reflector is centered on the acoustic axis of the transducer, small objects cause approximately the same omni-directional echo, so that no differentiation of objects on the receiver is possible. For non-axial acoustic irradiation of the reflectors, differences of the directional pattern appear. While scanning spherical reflectors, these variations can be detected with a special structured array. Therefore a 3 MHz annular array is used, whose outer rings are divided in additional segments.

In this work first measurements of spheres with different sizes where done. The sphere diameters varied about one up to six wavelengths and the lateral size of the sound beam in the scan depth was 2.8 wavelengths. All measurements are combined with sound field simulations, to provide algorithms for the determination of the size of reflectors.

1. Introduction

Ultrasound imaging is used in many branches of industry. It will be found amongst others in the electronic industry for inspecting of conductor plates or semiconductor units, in the material science for characterization of materials and analysis of structural conditions, as well as in the medicine for analysis of bones and tissues. As in light microscopy the wavelength of the used signal limits the resolution. Beside the wavelength, the lateral size of the sound beam is also a limiting factor. For this reason, highly focusing transducers are used, which can achieve a lateral sound field size approximately the same as the wavelength by reducing the focal length. Often the region of interest is greater than the focal length. In this case several scans have to be performed, or it has to be worked behind the focus-area. However the lateral size of the sound beam also increases with increasing distance. Another way to improve the resolution could be the use of higher frequencies. Unfortunaly the penetration depht decreases with an increasing of the frequency. Therefore it is desirable to improve resolution also at lower frequencies. A first approach for determination of sphere sizes by analyzing time differences was allready presented by the authors in [1]. In this work another approach will be discussed.

2. Simulations and approach

The basic approach for determining size and shape of inclusions is to analyse the sound pressure distribution on the receiver. This can be calculated by using a half-analytical method, based on time-harmonic GREEN's functions [2]. First, the source area is discretized and uniformly covered with point sources. Every point source acts as a single origin for a sound field. The sound field, that is irradiated from a finitely extended source, is the superposition over the fields of all point sources. Next, the stress distribution on the sphere, caused by the irradiating sound field, has to be calculated. For big spheres, the GREEN's function for planar boundary surfaces can be used [3]. First theoretical research about advanced source terms for strongly curved boundary surfaces was done. This is necessary to calculate the sound field, reflected from the sphere and therefore the stress distribution on the receiver too.

A central alignment of the sphere with respect to the main sound beam induce no noticable effect in the simulations, but by a small displacement a distinctive radiation pattern appears. This causes an appreciable magnitude difference between the left and the right side of the sensor. For this reason a structured annular array (see figure 1 left) was used. The idea is, to arrange line scans over the spheres and to analyze this magnitude difference. In the right part of figure 1 the simulated magnitude difference of such a line scan between two oppsosite positioned elements (blue marked area in the structure, denoted by receiver 1 and 2) for different spheres is shown. The irradiating sound pressure field was generated by the first ring (red marked area, denoted with Ring 1). It can be seen, that the minimum of the magnitude differences decreases with an increasing of the sphere size.

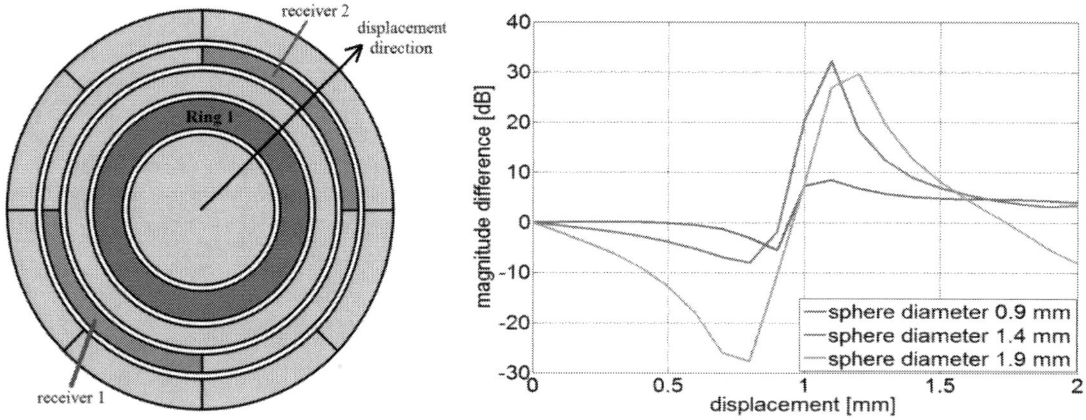

Figure 1. Left: transducer design; Right: simulated magnitude difference between two quarter elements (blue marked area in the transducer desgin) while ring 1 is used for sending (red marked) for different sphere sizes as a function of the displacement.

The directional pattern of small (about one wavelenght) and big (several wavelengths) spheres can explain this effect (see figure 2). The spheres were displaced about 1 mm with respect to the acoustic irradiation. The small sphere has an approximately omni-directional radiation pattern, whereas the big sphere has a directed one. Two receivers (denoted by receiver 1 and 2) were positioned on the left and on the right side of the acoustic axis. In the first case both receivers will enregister nearly the same magnitude. The second case will lead in a measurably magnitude difference between the receivers.

3. Measurements

Different spheres with diameters from 0.4 up to 3 mm were scaned in a grid like manner with a resolultion of 0.1 mm in immersion technique in water. The central and the ring elements were used for sending and the elements of the segmented outer rings for receiving. For all

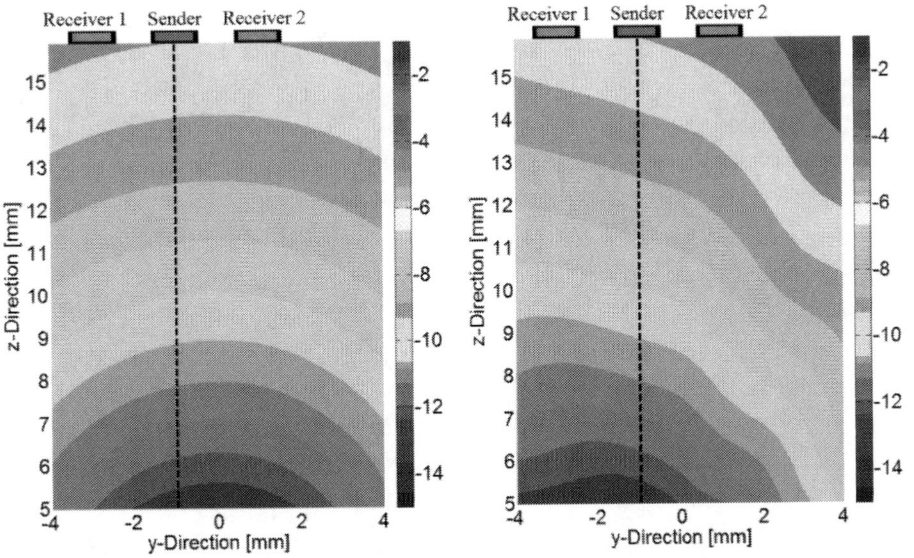

Figure 2. Directional pattern of different spheres with a displacement of 1 mm according to the acoustic irradiation; Left: small sphere; Right: big sphere.

measurements a delayline of 15 mm was used. In this region, the sound beam has a width of about 2.8 wavelengths. In order that the available, manageable sphere sizes have a size not much smaller the the wavelength up to the sound beam width, a operating frequency of about 3 MHz for the transducer was chosen. The spheres where positioned on the surface of a hydro gel. Because of its waterlike properties, it can be assumed as acoustical invisible. The magnitude of the reflected signal from the hydro gel was about 30 dB smaller then that of a sphere with 0.7 mm diameter. This was necessary to analyze the reflected signals caused by the spheres without reflections from the bearing.

From the measurement data line-scans over the spheres in the direction of the opposite positioned receivers were extracted (the displacement direction is denoted in the left hand of figure 1). For all measurements ring 1 was used as the sending element. In the left part of figure 3, the magnitude difference between the receivers for the line scans is shown as a function of sphere diameter and transducer displacement. According to the simulations, the magnitude difference decreases with the sphere diameter. The right graph of figure 3 shows the minimal magnitude for every line scan as a function of the sphere diameter (this conforms with the valley bottom of the left figure). An almost linear dependence can be found between magnitude difference and sphere diameter

4. Conclusions and perspectives

With the presented approach, sphere sizes down to 0.7 mm could be determined, what corresponds with the half of the sound beam width. The usage of different combinations of sending rings and receiving opposite positioned elements can be used to determine a plurality of calibration curves, to raise the accuracy. Furthermore this method has a great advantage: it is independent of the acoustic properties of the propagation medium. The sound path from the spheres to both receiving elements is nearly the same and therefore undesireable effects like attenuation are compensated.

Since calibration measurements require excessive time and effort (the registrated magnitude difference can depend on measurement depth, transducer design, object shape and size, ...),

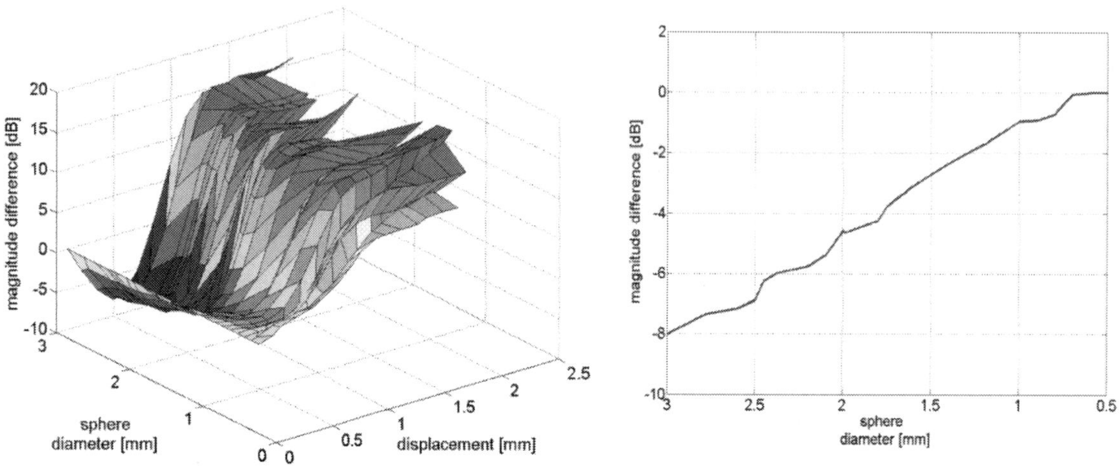

Figure 3. Overview for all measured magnitude differences between receiver 1 and receiver 2 (figure 1) extracted from the line scans (ring 1 was used for sending); Left: as a function of sphere diameter and transducer displacement; Right: minimal difference magnitude for every line scan (valley bottom of the left figure).

quantitative simulations are recommended. At the moment there is a difference in the magnitude of the analyzed effect between simulations and measurements. Therefore the modelling has to be improved in the following points:

(i) Better modelling of the sound pressure field from the transducer by characterisation of the sensor (excitating an element yields to a resonance of the neighbour elements and for this reason to a bigger oscillating area with non constant amplitude).

(ii) Improvement of the source functions for the modelling of curved boundary areas for the GREEN's functions.

Besides the investigation of the presented effect, further analysis criteria have to be found. The phase distribution at the receiver could be one possibillity. Since the simulations can calculate a C-Scan of the probes, a direct comparison with the measurement could be done too.

Acknowledgement

The authors would like to thank Deutsche Forschungsgmeinschaft (DFG) for financial support of the ongoing research project KU1075/12-1, and Sonaxis SA for transducer manufacturing.

References

[1] Gust N, Kühnicke E, Wolf M and Kümmritz S 2011 Determination of size and shape of small inclusions from sound-field information *IEEE International Ultrasonic Symposium* (Orlando, 18-21 October 2011) pp 764-766

[2] Kühnicke E 1996 Three-dimensional waves in layered media with nonparallel and curved interfaces: A theoretical approach *J. Acoust. Soc. Am.* **100**(2) 709-716

[3] Kühnicke E, Lenz M, Sorber J, Gerlach G and Trier H G 2008 Method for Curvature Measurements with Ultrasound *IEEE International Ultrasonics Symposium* (Beijing,China, 2-5 November 2008) pp 66-69

IOP Conf. Series: Materials Science and Engineering **42** (2012) 012005 doi:10.1088/1757-899X/42/1/012005

An automated data processing method dedicated to 3D ultrasonic non destructive testing of composite pieces

A Osman[1, 4], U Hassler[1], V Kaftandjian[2] and J Hornegger[3]

[1]Fraunhofer Development Center X-ray Technology EZRT, Dr.-Mack-Strasse 81, 90762 Fuerth, Germany
[2]National Institute of Applied Science INSA-Lyon, Laboratoire Vibration et Acoustique, 25 Avenue Capelle, 69621 Villeurbanne, France
[3]Head of the Pattern Recognition Lab of the Friedrich-Alexander University Erlangen-Nuremberg, Martensstrasse 3, 91058 Erlangen, Germany

E-mail: ahmad.osman@iis.fraunhofer.de

Abstract. State-of the art Non Destructive Testing using ultrasound is based on evaluation of C-scan images, which is done mainly visually. The development of the new Sampling Phased Array technique SPA by IZFP Fraunhofer provides a fast three-dimensional reconstruction of inner object structures. This new inspection technique is to be complemented with fully or semi-automated evaluation of ultrasonic data, providing maximum support to the operator. We present in this contribution a processing method for SPA ultrasonic data, where the main focus of this paper will be on speckle noise reduction. The evaluation method is applied on carbon fibre composite where it demonstrates robust and successful performance in recognition of defects.

1. Introduction

Ultrasound phased array inspection systems are replacing the conventional manual testing technique. Their imaging modes range from A-scans to 3D ultrasound reconstructions. This progress on level of data acquisition and reconstruction needs to be complemented with an automated evaluation of the received data in order to have a complete inspection chain without expert intervention. This task is the main focus of our research activity. To achieve this objective, we propose an evaluation strategy formed of a segmentation procedure and classification procedure. Indeed ultrasound images segmentation has been well studied, however they are still considered to be one of the hardest imaging modalities to segment. Their main drawback is the comparatively poor image quality due to the appearance of a sort of granular texture called speckle noise which has a higher contrast than background. Hence this speckle can in fact mask the presence of low contrasted defects. Thus it is mandatory to use special segmentation procedure in order to keep all potential defects present in the image and to reduce to the minimum regions corresponding to artefacts which can be caused by speckle noise, by reconstruction algorithm and the structure of the inspected specimen. In the next section we will present our evaluation method. The third section will show the experimental results obtained using the proposed method.

[4] To whom any correspondence should be addressed.

2. Evaluation method

The principle of the evaluation chain is presented in this section and resumed in figure 1. It can be divided into phases: segmentation and classification.

2.1 Segmentation method

The first step of the segmentation is to correct bad voxels. The correction is based on a modified median filter where an NxN window applied over each slice. If the central pixel in the window is null then it is replaced by the median value of all non null pixels inside the window. After bad voxels correction, image enhancement takes place where we compared three different filters: median, M3 [1] and NLM filter [2]. Each filter is applied slice by slice in the depth direction. The optimal filter was chosen according to quantitative measures. After image enhancement, the entrance (EE) and backwall (BWE) echoes are extracted based on the second derivative variation of grey level in depth direction. The extraction of the EE and BWE is critical. In fact defects may exist near the surface or the backwall of the inspected object, therefore if the corresponding slice(s) is (are) considered as entrance or backwall echoes slice (s), then those defects will not be detected. After localizing the EE and BWE, the internal volume is extracted. Next a threshold is applied on the voxel values, dividing between foreground and background. We applied Hysteresis thresholding approach to divide the internal volume between background and foreground. Voxels belonging to the same region are connected to each other into one label (region). After this step, features are measured to characterize each label. Features extraction is the first step required to distinguish the false alarms from the defects. Features extraction is centered principally around the measurement of geometric properties and on the intensity characteristics of regions. The measured set of features characterizing each suspicious region form the input of the classification method (see subsection 2.2) to decide about the nature of this region.

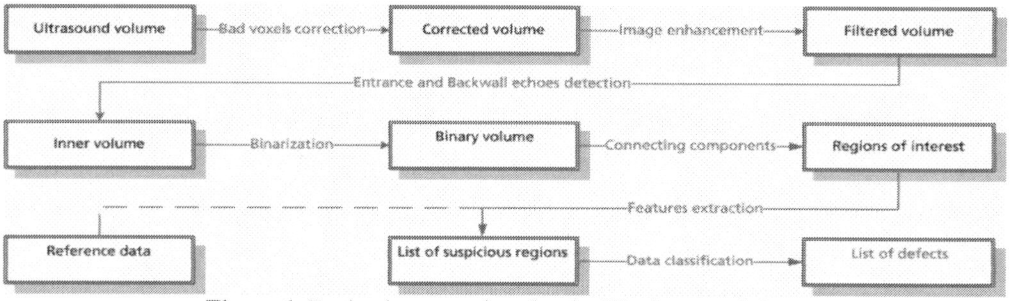

Figure 1. Evaluation procedure for the 3D ultrasound data.

2.2 Classification method

The input data for this step are the segmented regions of interest (objects) among which are true defects (TDs), while others are false defects (FDs). Each region is described by a list of geometrical features (volume, shadow etc.) and intensity characteristics (contrast etc.) Single features are considered as information sources in the proposed classification approach. This approach is based on data fusion theory where features are fused using Dempster-Shafer theory [3, 4] and statistical methods (mean and median values) in order to achieve better recognition rates of TDs and false defects FDs. The classification algorithm is divided into two phases:

- Learning phase: where features values are replaced by mass values in order to make the data fusion [5]. Then mass values are combined using the orthogonal rule of combination and statistical mean and median. Performance of each combination is measured via performance measures PTD, PFD. The output of this step are combination of sources which best classify the TDs and FDs.
- Validation phase: where a validation dataset is classified using the combination of sources obtained in the learning step.

The issue is to classify an object as a TD (i.e., positive) or FD (i.e. negative). Let: P be the total number of positives, N be the total number of negatives, TP be the total number of positives correctly classified, TN be the total number of negatives correctly classified. The performance measures are defined as follow:

- TDs classification rate: $PTD=\frac{TP}{P}$; FDs classification rate: $PFD=\frac{TN}{N}$

For more information about the classification method, readers can refer to [5].

3. Experimental results

The proposed approach is tested on a carbon fibre composite specimen of 14 mm thickness. This anisotropic material contains 24 defects at different depths: 1 mm, 7mm and 13 mm. Defects close to the surface and the back of the specimen are considered difficult to detect due to the influence of the entrance echo and backwall echo signals. Thus this specimen is considered as a good example to show the performance of SPA technology [6] and our evaluation method for detecting and classifying defects which are present in the reconstructed volume.

3.1 Data acquisition

The specimen is inspected via 5 MHz array of sensors with different gain, speed and resolution settings. We dispose of two volumes with different resolution: volume V_1 has 316x301x341 (XYZ) voxels and a voxel size of 1x1x0.05 mm^3 (see figure 2). In this volume appear 22 defects from 24 existing in the specimen. Volume V_2 has 310x265x31 voxels with a voxel size of 1x1x0.5 mm^3. SPA technique revealed the presence of 23 from 24 defects inside this volume.

(a) (b)

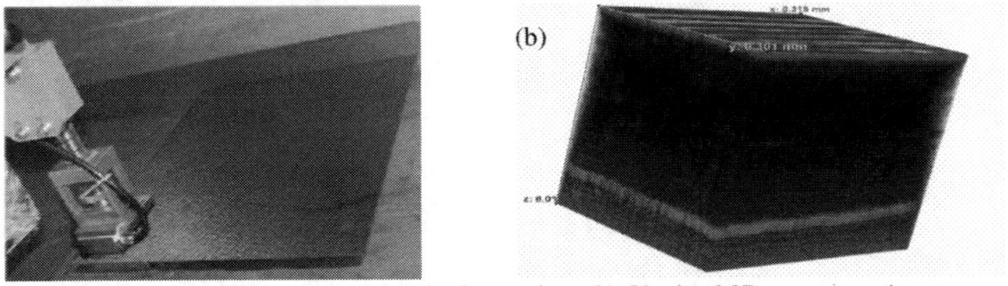

Figure 2. (a) Carbon fibre specimen under inspection. (b) Obtained 3D reconstruction.

3.2 Segmentation process

First bad voxels correction is carried out. Then we filtered the volume by considering the 3 above cited filters with different kernel sizes. Median and M3 filters are tested with 3x3, 5x5, 7x7, 9x9 and 11x11 kernel sizes. Due to time consuming and based on optimisation work presented in [7], NLM filter is tested with a fixed searching window size (search volume size M=5 [7]) of 11x11 and block size changing between 3x3 (NLMW5B1), 5x5 (NLMW5B2) and 7x7 (NLMW5B3). Quantitative comparison of filtering quality established that NLMW5B1 performed better than the other filters. In figure 3, an illustration is shown using a slice before and after application of this filter. We notice the reduction of speckle noise while the boundaries of defects are enhanced. After filtering, we automatically detect the entrance and backwall echoes slices and extract the inner volume between the end of the entrance echo and the start of the backwall echo. The inner volume is then binarized via hysteresis thresholding. In the next step, connected voxels are clustered together to form labelled regions in the volume. Figure 4.a shows results for the suspicious regions inside the volume V_1. Each region is characterized by a list of features (see feature Contrast in figure 4 (b).

3.3 Classification process

The number of detected suspicious regions in V_1 is 74 from which 22 are TDs and the rest are FDs. In the volume V_2, the number of detected regions is 59 including 23 TDs. We merged the two datasets into one dataset and then divided it (Odd/Even) into: Learning dataset formed of 25 TDs and 42 FDs and Testing dataset: formed of 20 TDs and 46 FDs. The source which gave the highest PTD (PTD=1, PFD=0.97) on the learning dataset is the fusion between two features (Length in {y,z} plan and MeanIntensity). On the testing dataset, this source gave PTD=1, PFD=0.93. As for the highest PFD on the learning dataset, the fusion between two features (Width in {x,y} plan and Width in {y,z} plan) gave PFD=1 and PTD=0.96. On the testing dataset this PFD of this source decreased to PFD=0.87 and PTD= 0.95.

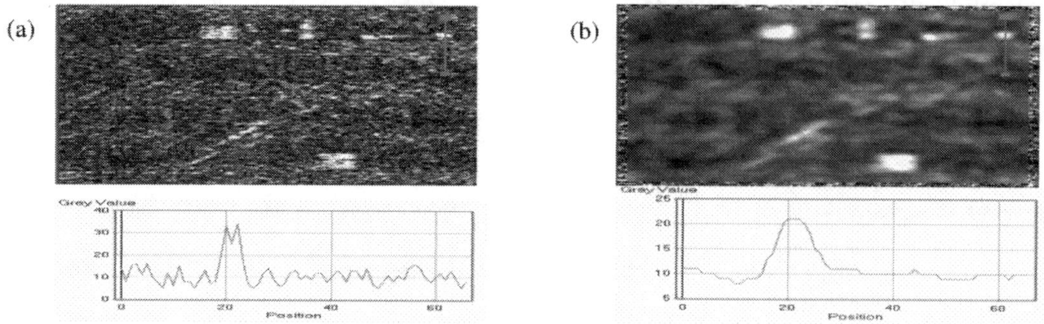

Figure 3. (a) Initial C-scan view from V_1 of a carbon fibre specimen with corresponding grey level profile along the red line tracker. (b) NLM filtered C-scan view with corresponding grey level profile.

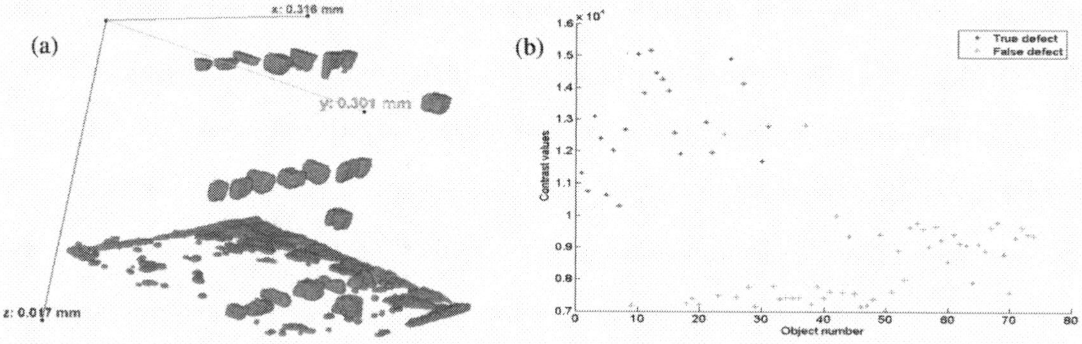

Figure 4. (a) Binary volume where defects and artefacts can be seen. (b) Contrast values repartition for true and false defects.

4. Conclusion

We proposed in this paper a robust segmentation approach. We presented our latest classification results obtained by means of data fusion approach using features as sources of information. The performance of our classifier slightly decreases for FDs on the testing dataset. Thus we are aiming to improve the classifier performance based on our experience gained during working with the same classifier on 3D Computed Tomography data.

References

[1] Thangavel K, Manavalan R and Aroquiaraj I 2009 Removal of speckle noise from ultrasound medical image based on special filters: comparative study *ICGST-GVIP Journal* ISSN 1687-398X **9**

[2] Coupé P, Hellier P, Kervrann C and Barillot C 2009 Nonlocal Means-Based Speckle Filtering for Ultrasound Images *IEEE Trans. On image processing* **18** (10) 2221-29

[3] Dempster A 1967 Upper and lower probabilities induced by multivalued mapping *Ann. Math. Stat.* **38** 325-339

[4] Shafer G 1976 *A Mathematical Theory of Evidence* (Princeton: Princeton University Press)

[5] Osman A, Kaftandjian V and Hassler U 2011 Improvement of X-ray castings inspection reliability by using Dempster-Shafer data fusion theory *Pattern Recog. Lett.* **32** 168-180

[6] Bulavinov A, Kroening M, Reddy K and Gabriel H Ribeiro J 2007 Real-Time Quantitative Ultrasonic Inspection *Proc. IV Conf. Pan American for NDT (Buenos Aires, October 2007)*

[7] Coupé P, Yger P, Prima S, Hellier P, Kervrann C and Barillot C 2008 An Optimized Blockwise Nonlocal Means Denoising Filter for 3-D Magnetic Resonance Images *IEEE Trans. on Medical Imaging* **27** (4) 425-441

Acoustic microscopy in the food industry

N Watson[1], M Povey[1], E Corona[2], J Benedito[2] and N Parker[3]

[1] Department of Food Science and Nutrition, University of Leeds, Leeds LS2 9JT, UK.
[2] Grupo de Análisis y Simulación de Procesos Agroalimentarios, Departamento Tecnología de Alimentos, Universidad Politécnica de Valencia, Camí de Vera s/n, E46022, Valencia, Spain.
[3] School of Mathematics and Statistics, Newcastle University, Newcastle NE17RU, UK.

E-mail: n.j.watson@leeds.ac.uk

Abstract. Acoustic microscopy has been used for many years to image and measure the elastic properties of materials across a wide range of scientific disciplines. However the application of this technique in the food industry is scarce. In this paper we outline the operation of a reflection-mode acoustic microscope and discuss some of the issues relevant to its operation in the food sector. We then present two relevant case studies in which we employ acoustic microscopy to analyse potato cells and the fat structure in Iberian ham and chorizo.

1. Introduction

Acoustic microscopy is the study of materials on a microscopic scale using focused sound waves, as first demonstrated by Lemons and Quate [1] in 1974 to study biological tissue. Benefits of acoustic microscopy include its potential to perform sub-surface imaging in optically-opaque materials, it does not require dyeing of samples and the image contrast is controlled by mechanical properties. Often the main challenge is in interpreting the complex data to give meaningful results. A scanning acoustic microscope (SAM) works by using a positioning system to move a focused acoustic transducer relative to a sample. Reflection-mode SAMs are the most common acoustic microscope and use a single transducer to both transmit and receive the sound waves. Although acoustic microscopes exist with different transducer arrangements, reflection-mode SAMs are the most popular due their relative simplicity and versatility, and will be considered in this paper. More detailed information on the theory, design and wider applications of acoustic microscopes can be found in several books and review articles [2,3,4].

Acoustic microscopes have been used extensively to study electrical components, composite material and biological systems, but their use in the food sector has been limited [5]. This paper will discuss the main features and components of a reflection-mode acoustic microscope and present two ongoing case studies which highlight the benefits and challenges of using an acoustic microscope in the food sector.

2. Operational considerations

2.1. Acoustic transducer

The most important component in the SAM is the focused acoustic transducer. The transducer generates the sound waves which interacts with the sample and receives the acoustic information which is used to provide images and measurements of the sample. The possibility of the generation of surface waves when the sound interacts with the surface will be neglected here. The important selection criterion for the acoustic transducer is the operating frequency. This frequency governs the imaging resolution of the transducer as well as the propagation depth, as shown in figure 1 for a range of frequencies generally used in acoustic microscopy. A trade off exists between resolution and propagation depth so the operator must select the most suitable frequency for the sample they are using and the data they require. Shown is a representation of the propagation distance through a homogeneous fluid; the presence of a sample will further modify the propagation distance depending on, for example, the sound transmission into the sample and the sample's acoustic attenuation. A more detailed description of the operation of our SAM can be found in [6].

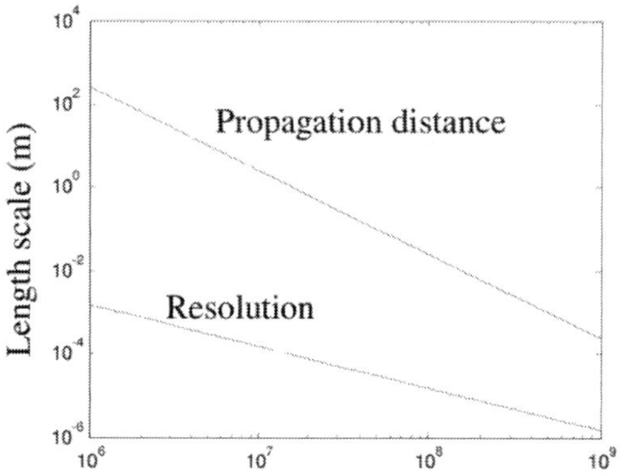

Figure 1. Propagation distance and resolution as a function of frequency, adapted from [6]. This assumes sound propagation in pure water at 30 °C.

2.2. Scan type

At each scan point, the returning sound waves received by the transducer generate a voltage which varies with time-of-flight. This voltage "trace" is the basis for constructing images of the sample. Depending on the imaging requirements by a user, the SAM can be operated in A-scan, B-scan and C-scan mode. An A-scan is a time-of-flight measurement performed at a single lateral position, a B-scan consists of multiple time-of-flight measurements performed laterally across the sample (X or Y) and a C-scan consists of multiple measurements across an area (X and Y) taken at constant time-of-flight. For surface imaging, C-scans are generally used. However, C-scans are the most time and data intensive and for certain requirements, e.g. monitoring properties which change rapidly with time, faster A-scans and B-scans may be preferable.

2.3. Sample preparation

An important part of acoustic microscopy which is little discussed is sample preparation. The top surface of the sample must be as flat and as horizontal as possible as any deviations will cause large angle reflections and reduced received signals. Furthermore, for high resolution imaging, for which the depth of focus may be of the order of tens microns, an uneven surface may lead to only certain sections being in focus. To obtain flat samples equipment such as a microtome can be used. Often biological samples are fixed and mounted in wax before slicing and imaging. These techniques can be used for acoustic microscopy work although care must be taken when acquiring the elastic properties of fixed cells as these processes can change the cell's properties. Furthermore, the fixing and mounting method may directly affect the sound propagation in the sample.

Acoustic microscopes require an immersion fluid to couple the acoustic wave between the transducer and sample. Air is unsuitable due to high attenuation and sensitivity to fluctuations. Water is commonly used due to its excellent transmission properties and well-calibrated acoustic behaviour. Food samples which are water soluble can be accommodated by either using another coupling fluid such as oil, or to cover the sample with a thin layer of microfilm which can remain essentially transparent to the sound waves.

3. Potatoes

Potatoes are one of the most important foods today yet 20% of crops can be compromised due to problems such as bruising [7]. Currently there is a limited understanding of why certain potatoes bruise more easily than others. Acoustics offer the potential to study the cellular properties of potatoes and we will present some preliminary work into this area.

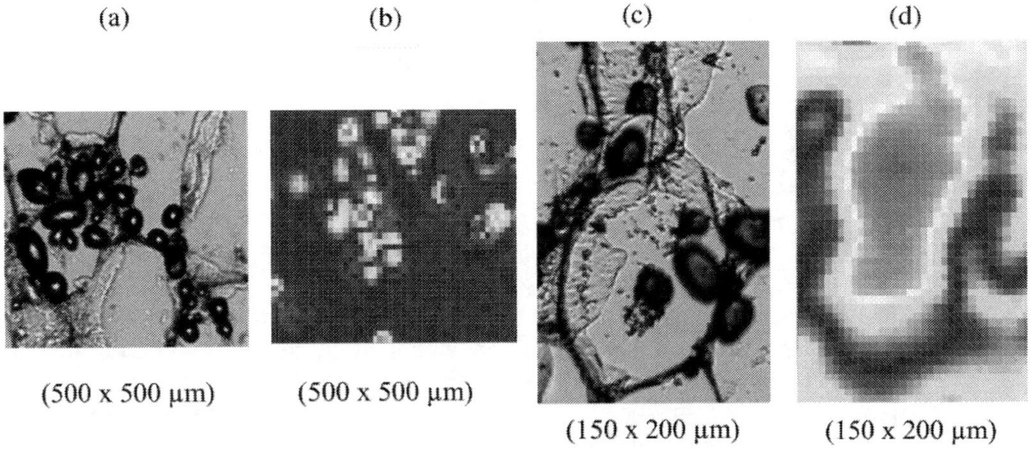

Figure 2. Optical (a), (c) and acoustic (b), (d) images of potato cells.

A SAM developed in our research group was used to image potato cells. A 100 MHz focused transducer with a 6mm focal distance generates a lateral imaging resolution of approximately 20 um [6]. Optical images were obtained with a Bresser LCD microscope. The samples were prepared by fixing, wax mounting and slicing using a Bright M3500 microtome.

From figure 2 it can be seen that the key components of potatoes are the cell walls and starch granules. The optical images (figures 2(a) and 2(c)) show both the starch and cell walls. Using the SAM (figures 2(b) and 2(d)) it was possible to image these components separately, depending on the C-scan slice chosen. The acoustic images are less clear than the optical counterparts due to the large difference in lateral resolution between the two methods (for the SAM the lateral resolution is around 20 microns, whereas for the optical microscope it is of the order of a micron). In the acoustic images, maximal signal (red regions) arises from the centre of the cells and the top of the starch granules, where the surface is locally horizontal and leads to maximal signal returning to the transducer.

4. Fat in Iberian products

Pigs can be reared using the intensive method (Cebo) or more natural method (Montanera). The method of rearing has an impact on the sensory quality of the fat in the pigs and methods are required to measure it, for example through their mechanical signatures. In this work we used our SAM to study samples of ham and chorizo. Penetration of sound into the samples was highly restricted at 100 MHz (as used in the previous section) due to viscous and scattering effects. Instead, here, we employ a 10 MHz transducer with a lateral resolution of approximately 300 microns.

The two meat samples are shown in figure 3 (a) and (b), with the corresponding acoustic B-scans shown in figure 3 (c) and (d). In the B-scans, the x-axis displays scan position and the y-axis displays the time-of-flight. The color scale reflects the magnitude of the reflected signal, with white/black correspond to zero/maximum signal, respectively. The first reflections received (~7.2 μs) corresponds to the top surface of the samples. The dashed lines represent the position of the glass slide and signals above this line are reflections and echoes emanating from the slide itself. Thus the region of sample is the intervening region in the plot between the first received signal and the dashed lines.

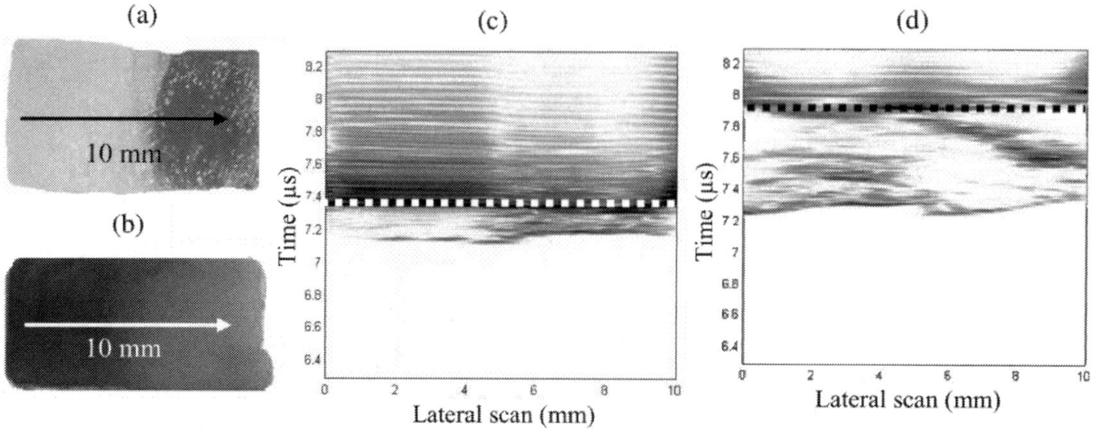

Figure 3. (a)Ham sample, (b) chorizo sample, (c) B-scan of ham, (d) B-scan of chorizo.

The ham sample (figure 3 (a)) has two distinct regions of fat and muscle. Consistent with this we see a crossover in the B-scan image (figure 3(c)), from a generally light region (left, fat) to a generally dark region (right, muscle). There is a significant difference in acoustic contrast between these two tissue types. Across the whole fluid-sample interface, a large amplitude signal arises due to the large acoustic impedance difference across the interface. Reflections from the subsurface region are caused by back-scattering from tissue inhomogeneities, e.g. cell walls, nuclei, fat and proteins. The different structure of fat (adipocytes) and muscle tissue (muscle fibres and connective tissue) leads to very different levels of back-scatter between the tissue types, with the muscle given the larger reflected signals. Within each tissue type, the back-scatter is relatively uniform.

For the chorizo sample (figure 3(d)) the back-scattering from within the sample indicates the very different structure to the ham sample. Here, the sub-surface reflections are inhomogeneous, consistent with the characteristic mixture of lean and fatty tissues within the chorizo. At least three fat globules are visible within the sample (light/white regions). These early results suggest that acoustic microscopy has great potential in the study and characterisation of fat/muscle food types.

5. Conclusion

In this paper we have briefly described reflection-mode acoustic microscopy and the issues faced by the operator. Our two case studies demonstrate the potential for applying acoustic microscopy within food science and the food industry.

Acknowledgements

The Authors would like to thank Regiane Scharf for the preparation of the potato samples and the BBSRC for funding (BB/FOF/302).

References

[1] Lemons R A and Quate C F 1974 *Appl. Phys. Lett.* **24** 163–165
[2] Yu Z and Boseck S 1995 *Rev. Mod. Phys.* **67** 863
[3] Gr Maev R 2008 *Acoustic microscopy fundamentals and applications* (NJ, USA: Wiley-VCH)
[4] Briggs G A D and Kolosov O 2010 *Acoustic microscopy* (New York: Oxford Science)
[5] Kaleb M, Allan-Wojm P, and Shea Miller S 1995 *Trends in Food Sci.Techol..* **6** 177-186
[6] Parker N G, Nelson P V and Povey M JW 2010 *Meas. Sci. Technol.* **21** (4) 045901
[7] Whitaker JR and Lee CY 1995 *Enzymatic browning and its prevention* vol 600 ed CY Lee and JR Whitaker (Washington: American Chemical Society) pp 2-7.

IOP Conf. Series: Materials Science and Engineering **42** (2012) 012007 doi:10.1088/1757-899X/42/1/012007

Comparison of spread spectrum and pulse signal excitation for split spectrum techniques composite imaging

L Svilainis[1], S Kitov[1], A Rodríguez[2], L Vergara[3], V Dumbrava[1], and A Chaziachmetovas[1]

[1]Kaunas University of Technology, Department of Signal Processing Studentu str. 50-340, LT-51368, Kaunas, Lithuania
[2]Universidad Miguel Hernández de Elche, Department of Communications Engineering, Avda. Universidad S/N, 03202 Elche, Spain
[3]Universidad Politécnica de Valencia, Department of Communications, Camino de Vera S/N, 46022 Valencia, Spain

E-mail: linas.svilainis@ktu.lt

Abstract. Ultrasonic imaging of composites was investigated. Glass and carbon fiber reinforced plastic produced by resin transfer molding and prepreg forming were analyzed. In some of the samples air bubbles were trapped during RTM (resin transfer molding) process and interlayer gaps were present in prepreg technology samples. One of the most expected techniques to apply in such case is the Split Spectrum processing. On the other hand such signals require specific processing to reliably reconstruct the temporal position of the defect reflection. Correlation processing can be used for signal compression or Wiener filtering can be applied for spectral content equalisation. Pulse signals are simple to generate, but lack the possibility to alter the signal's spectrum shape. Spread spectrum signals offer a powerful tool for signal energy over frequency band increase and resolution enhancement. CW (continuous wave) burst has high energy but lacks the bandwidth needed for SSP (spread spectrum processing). The aim of the investigation was to compare the performance of the above signals in case of composite imaging, when various Split Spectrum Processing techniques are used with preceding Wiener processing for spectral content compensation. Resulting composite signals and images obtained are presented. Structural noise removal performance was evaluated as Receiver Operating Characteristics (ROC).

1. Introduction

Ultrasonic imaging of composite materials can not use conventional techniques since signals scattered by composite structure are masking the useful signal [1]. Due to the presence of the structural noise, resolving the defect or even the backwall of the material is complicated.

Techniques used for composite imaging can be divided into two categories: i) using the frequency range least affected by the structure (usually low frequencies); ii) techniques exploring the diversity of the structure signals so averaging can be applied. Using low frequencies is not always acceptable since resolution of imaging depends on envelope bandwidth which usually is a direct function of operation frequency. Techniques exploring the diversity can be further divided into spatial (synthetic aperture focusing technique, SAFT [1]) and frequency domain (Split Spectrum Processing, SSP [2]).

We have concentrated on SSP technique. Assumption was made that SSP processing calls for smooth and as wide as possible signal bandwidth. Then, we aimed our investigation on Spread Spectrum (SS) signals performance comparison with conventional excitation signals when applied in SSP technique.

2. Split Spectrum Processing

SSP algorithm has been around for decades [2] and is considered as sub-optimal to the problem of detection of the ultrasonic pulse in a structural noise [3-8]. Technique is based on the assumption that the echoes coming from the scattering centers of the material poses diversity in frequency domain, i.e., the structure noise response is expected to vary significantly at different frequency channels. The echo from a large discontinuity, on the contrary, will act the same way in different channels. Hence, most of the SSP techniques implement a filter bank followed by a combination of the filter outputs (figure 1).

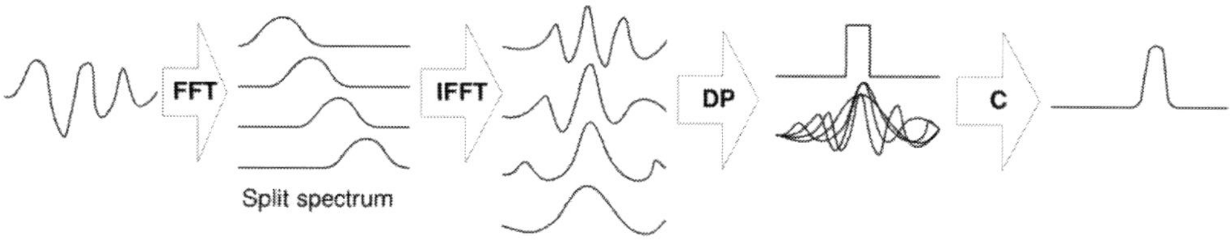

Figure 1. Standard data flow diagram for Split Spectrum Processing.

Main combination techniques used in our investigation were: [4] polarity thresholding (PT), [6] scaled polarity thresholding (SPT), [8] frequency multiplication (FM) and [3] minimization (MIN). Minimisation was chosen as a classical reference, its output is a minimum value of the filter banks. In PT, if polarity of the bank signals is the same, useful signal is assumed and average of the banks output is used, if at least one bank polarity does not match, zero mask is used. SPT is essentially the same as PT, only it uses linear mask, produced by ratio of same (N_+) and opposite polarity (N_-) banks to total banks number N. FM is a kind of geometric mean, where product of banks output is used.

2.1. Performance measures

Signals performance measurement was done using the Receiver Operating Characteristics (ROC). We have assumed that the goal of the SSP is the lowest probability of the false alarm (PFA) while maintaining maximum of the probability of detection (PD). PFA and PD were obtained as false alarm counts N_{FA} and detection counts N_D at variable threshold level V_{th} normalized to possible events N:

$$PFA(V_{th}) = \frac{N_{FA}(V_{th})}{N} \qquad PD(V_{th}) = \frac{N_D(V_{th})}{N} \tag{1}$$

Obtained PFA and PD were plotted against each other to produce ROC [9].

3. System setup

System (figure 2) is composed from control part, excitation pulser and the signals acquisition units. Industrial standard PC104 modular construction was used as form factors and bus. The excitation pulser driver has 100 MHz sampling rate and is driving the totem pole topology pulser [10]. Code upload and control is accomplished by host PC via PC104 bus. The receiving part is preceded by high impedance preamplifier [11] with high voltage protection input limiter. Conditioned signal is supplied to the dual high speed ADC, consisting of two high speed 10-bit monolithic sampling ADC operating at programmable sampling frequency up to 100 MS/s. All the control tasks and PC104 bridge functions are implemented by Complex Programmable Logic Device (CPLD) chip.

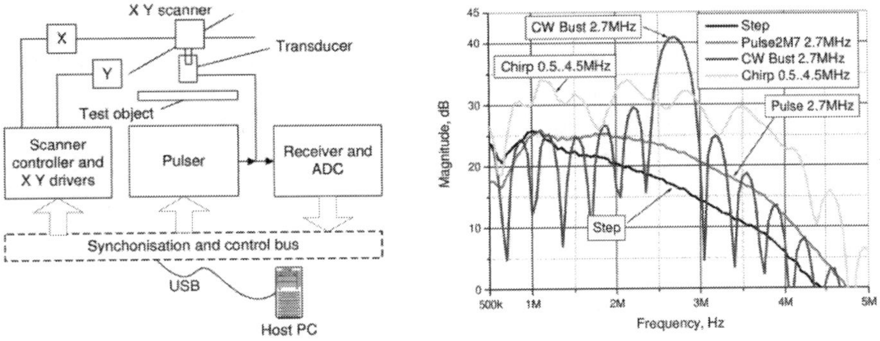

Figure 2. System setup (left) and spectral content of excitation signals used (right).

Five signals were uploaded into pulser memory for subsequent excitation: CW burst of 1 MHz frequency, CW burst of 2.7 MHz frequency (transducer center frequency), 0.5 MHz to 4.5 MHz spread spectrum signal implemented as linear chirp, rectangular pulse with duration 270 ns (corresponds to 2.7 MHz transducer center frequency), 3 ns steep step corresponding to frequency range up to 106 MHz. We deliberately avoided techniques exploring the phase correlation [7] since these techniques should be less sensitive to the amplitude variation among the banks. Since spread spectrum signals were used, SSP algorithm was modified, adding the Wiener processing [4] before SSP to compensate the spectral content and concentrate all components into one temporal position.

4. Experimental results

Three materials were used in investigation: thin (2.2 mm) sample of kayak wall, sputtered GFRP (glass fiber reinforced plastic) composite; thick (4.9 mm) multilayer sample of kayak wall: GFRP plus random carbon filament plus glass fiber net; and high porosity GFRP, 9 mm thick: randomly oriented chipped glass fibers. Spherically focused ultrasonic transducer, TS 12PB2-7P30, diameter 12 mm, 1.5 mm focal spot was chosen for its narrow beam. Transducer was operated with focal spot positioned on material under investigation. In such case spatial averaging is minimal and effect of the structural noise is maximised. Resulting B-scan images for multilayer kayak are presented in figure 3.

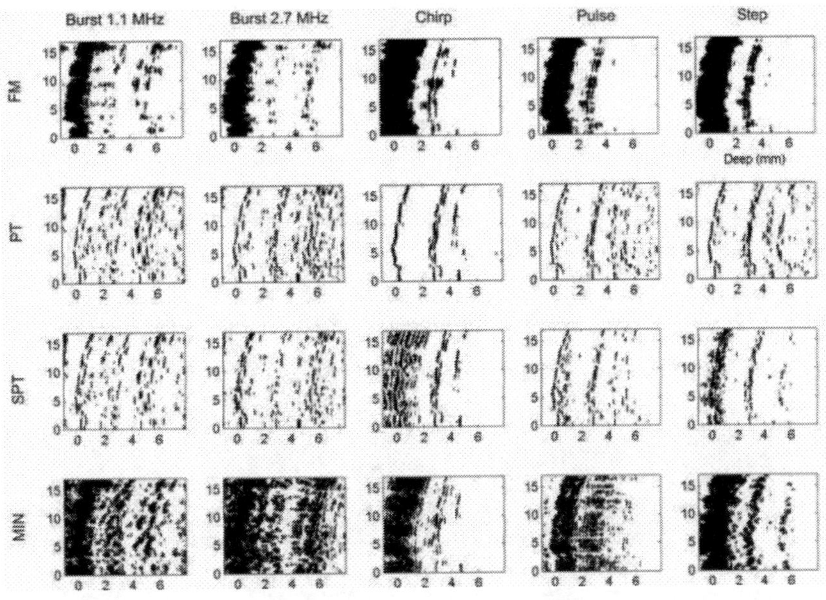

Figure 3. B-scan image obtained at various signal and combination techniques: multilayer kayak.

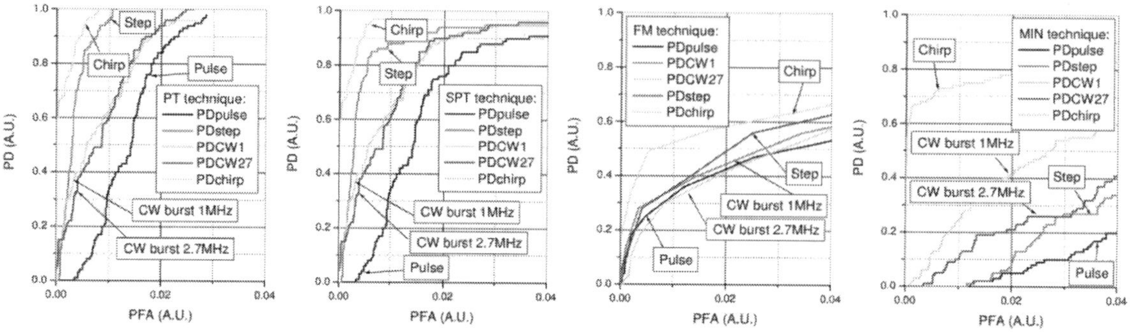

Figure 4. Signals' performance comparison: ROC for high porosity composite.

ROCs for the excitation signals used for high porosity composite imaging are presented in figure 4. Investigation indicates that best performance is obtained for spread spectrum (chirp) signal. Signal performs best in polarity thresholding SSP processing.

5. Conclusions

Use of spread spectrum excitation gives best performance for majority of materials and techniques. Best performance was achieved in conjunction with rectangular mask techniques in combination with minimization. Pulse and step excitation signals performance was comparable, though at narrower choice of techniques. CW burst signals sometimes are falsely assumed as good performers if duration after correlation compression is not taken into account. In some cases, when reflection contained not all frequency components, signals with larger low frequency content were in favor.

Acknowledgments

This research (acquisition system and spread spectrum signals) was funded by a grant (No. MIP-058/2012) from the Research Council of Lithuania. SSP part was supported by PROMETEO 2010/40.

References

[1] Li Y, Newhouse V L, Shankar P M, Karpur P 1992 Speckle reduction in ultrasonic synthetic aperture images *Ultrasonics* **30** 223–237

[2] Bilgutay N M, Saniie J, Furgason E S, and Newhouse V L 1979 Flaw-to-grain echo enhansment. *Proc.Ultrasonics International* (Austria, May 1979) p 157

[3] Karpur P, Shankar P, Rose J and Newhouse V L 1987 Split spectrum processing: optimising the processing parameters using minimisation. *Ultrasonics* **25** 204

[4] Saffari N, Svilainis L 1997 Signal processing for ultrasonic imaging of lossy composites *Ultragarsas* **28** 7-11

[5] Ericsson L, Stepinski T 2002 Algorithms for suppressing ultrasonic backscattering from material structure *Ultrasonics* **40** 733–734.

[6] Rubbers P, Pritchard C J 2003 An overview of split spectrum processing *NDT.net* **8** 8.

[7] Rubbers P and Pritchard C 2003 Complex plane Split Spectrum Processing: an introduction *NDT.net* **8** 11

[8] Rodríguez A, Miralles R, Bosch I, Vergara L 2012 New analysis and extensions of split-spectrum processing algorithms *NDT and E International* **45** 141–147

[9] Rodríguez A, Salazar A, Vergara L, 2012 Analysis of split-spectrum algorithms in an automatic detectionframework *Signal Processing* **92** 2293–2307

[10] Svilainis L, Chaziachmetovas A, Kybartas D, Jurkonis R 2011 Advanced Hardware for Fine-Tuning and Optimization of Sonoporation Efficiency In vitro *Proc. IDAACS* (Prague,15-17 September 2011) pp 65-70

[11] Svilainis L, Dumbrava V, Chaziachmetovas A 2010 Ultrasonic Preamplifier Performance Evaluation *Proc. IEEE ITI* (Croatia, 21-24 June 2010) vol 32 pp 663 - 668

IOP Conf. Series: Materials Science and Engineering **42** (2012) 012008 doi:10.1088/1757-899X/42/1/012008

Data fusion of ultrasound and GPR signals for analysis of historic walls

A Salazar, J Gosalbez, G Safont and L Vergara
Universitat Politècnica de Valencia, Institute of Telecommunications and Multimedia Applications, Camino de Vera s/n, 46022, Valencia, Spain

E-mail: asalazar@dcom.upv.es

Abstract. This paper presents an application of ultrasounds and ground-penetrating radar (GPR) for analysis of historic walls. The objectives are to characterize the deformation of a historic wall under different levels of load weights and to obtain an enhanced image of the wall. A new method that fuses data from ultrasound and GPR traces is proposed which is based on order statistics digital filters. Application results are presented for non destructive testing (NDT) of two replicates of historic ashlars' masonry walls: the first one homogeneous and the second one containing controlled defects such as cracks and nooks. The walls are measured separately using ultrasounds and GPR at different load steps. Time and frequency parameters extracted from the signals and different B-Scans for each of the NDT techniques are obtained. After this, a new fused representation is obtained, which results demonstrate the improvement of characterization and defect detection in historic walls using data fusion.

1. Introduction

The principal objective of this paper is to provide an approach for non-destructive testing (NDT) based on ultrasound and ground-penetrating radar (GPR) signal processing for diagnoses of historical walls. Common materials employed in heritage buildings are stone, mortar and brick. This kind of constructions resists well the pass of the time; nonetheless, degradation processes are inevitable and difficulties arise for measuring its degree of degradation. The degradation processes affect both structural (cracks, fissures, detachments, displacements…) and aesthetic (dirt, crusts, efflorescence…) characteristics of the historical buildings. Among the main pathologies that cause building breakdown are humidity damages caused by capillarity ascent, breeze or high humidity environments, successive freeze-thaw cycles that result in crystallization, and broken mortar joints. At present, to deal with structural evaluation, it is usual to use destructive testing by means of the extraction and characterization of tubes from the material that are characterized using classical morphological and physiochemical analyses. Thus, considering heritage value preservation, NDT is being increasingly applied in pathological and structural diagnoses of historical buildings.

In this paper, we present results of testing two scale models of historic ashlar masonry walls using through-transmission ultrasounds (two transducers, an emitter and a receiver each one located at opposite sides of the wall) and pulse-echo GPR (an antenna used as emitter and receiver located at one side of the wall). The walls were built with travertine ashlars from Godella's quarry in Spain and the dimension of the ashlars was 40x30x20cm (see figure 1). The final dimension of the two walls was 287cm length, 220cm height and 20cm thickness. Impoverished mortar with a low compression resistance (<4MPa) was used as binding material. One of the walls was homogeneous and the second one was mechanized including artificial defects (two drill holes, a vertical flaw and a crack or nook filled with mortar, see figure 2). The walls were divided in 7 rows and 7 columns as shown in figure 2. A total of 42 ultrasound measurements were taken in the positions indicated by blue crosses and the radar signals were collected using a survey cart with encoder through the trajectories depicted by red dash-dot lines in figure 2. In order to evaluate the sensitivity of the NDT methods to tensional states of the walls, they were introduced in a hydraulic press and loaded with 10Tn, 50Tn and 80Tn.

Figure 1. Ashlar inhomogeneous masonry wall.

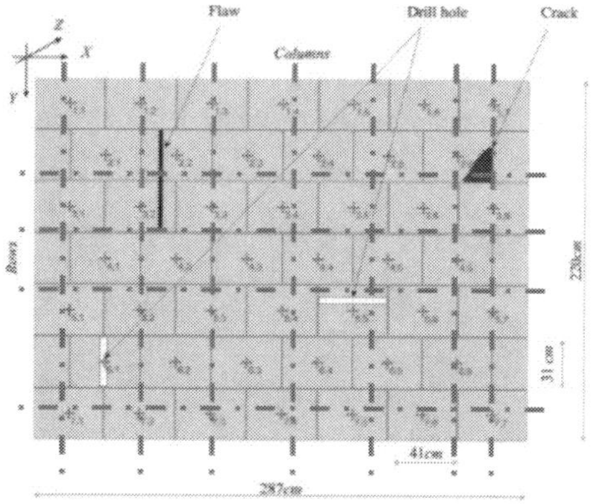

Figure 2. Defect and measurement outline.

Ultrasonic and GPR inspection are based on mechanical and electromagnetic waves, respectively. These waves propagate inside the material structure and its response is measured by signals that contain the reflections produced by the material grain microstructure plus the echoes caused by the in-homogeneities inside the material [1][2]. Signal processing allows extraction of information for characterization of the propagation medium and for detecting material in-homogeneities [3]. In this paper, we present results in detection of detects in the wall scale models by ultrasounds and GPR separately (Section 2 and 3) and enhanced results (Section 4) obtained by fusing both results. Data fusion is a new method that is progressively finding more applications in NDT, such as fusion of X-ray and ultrasound images in welds; and infrared thermography imagery in nuclear applications. The fusion methods employed have been based on classification systems and recently data fusion was done using possibility theory and fuzzy sets [4]. The fusion method presented here combines ultrasound and radar images estimated from the propagation velocity and signal power, respectively. From the original measurements, 2D images are formed using interpolation and morphological transformations, and after fusion, the fused image is filtered using an order statistical digital filter.

2. Ultrasonic tomography

An ultrasonic signal (A-Scan) was measured for each spatial position described in figure 2. The following parameters were extracted from ultrasonic signals: propagation velocity (V), signal power (P) and maximum frequency (fmax), see table 1. The temporal window used for their estimation begins at 30μs and finish at 100μs. Once the ultrasonic parameters were extracted for each, a 2D image usually called B-Scan can be obtained. In this ultrasonic image or tomography X axis is associated to the length and Y axis is associated to height of the wall. The color of the image represents the value of the ultrasonic parameter. The variation of the color represents the variation of the ultrasonic parameters that can be related to the shape and spatial position of the defects. In order to obtain realistic visualization (smoothed eliminating sharp edges) of the wall tomography, a classical linear interpolation algorithm called ordinary Kriging was applied [5].

Several tomographies were estimated for each of the ultrasonic parameters and load steps. The variations of the tomography values that were expecting to associate with defects, were masked by variations of the own material and measurement variance. Thus, differential tomographies were obtained between subtracting tomographies corresponding to different load steps (e.g., subtracting

10Tn tomography from 50Tn tomography). By this way, the variations between the ultrasonic parameter and the variations of the load were compared (ultrasonic parameters are sensitive to tensional state; velocity and signal power suffer an increment with load whereas central frequency is kept constant). Figure 3 shows a differential 50Tn-10Tn tomography, wall outline over imposed, where some defects are detected. The rationale is that ultrasonic dependence with load is different between sound and unsound zones and ultrasonic parameters suffer different percentage of variation, as reflected in figure 3. The vertical flaw and horizontal drill hole were not detected because the load was not so high to suppose an important variation of the ultrasonic parameter, in this case, velocity. The crack is detected and it is possible to note how this defect affects adjacent ashlars.

Table 1. Parameters extracted from the ultrasonic signals.

$V = \frac{Distance}{\text{Time of flight}} = \frac{d}{\Delta t}$	$P = \frac{\sum_{N_{Samples}} \lvert x[n] \rvert^2}{50 N_{Samples}}$	$f_{max}[Hz] = \underbrace{\max}_{f}\{X_{N_{samples}}(f)\}$ $X_{N_{samples}}(f) = TF\{x_{N_{samples}}(n)\}$ $x_{N_{samples}}(n) \rightarrow N_{samples}$ truncated from $x(n)$

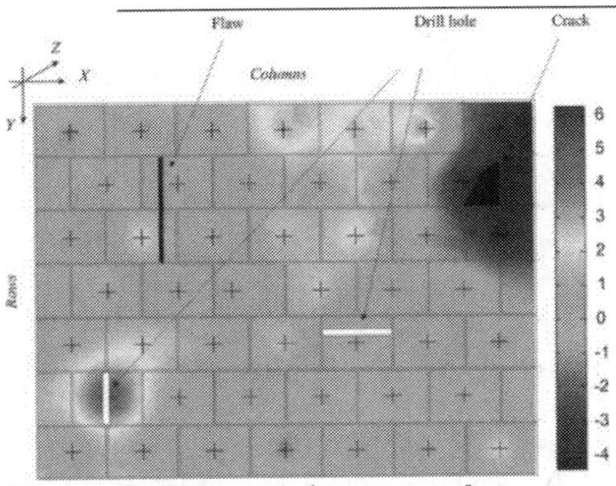

Figure 3. Ultrasound tomography.

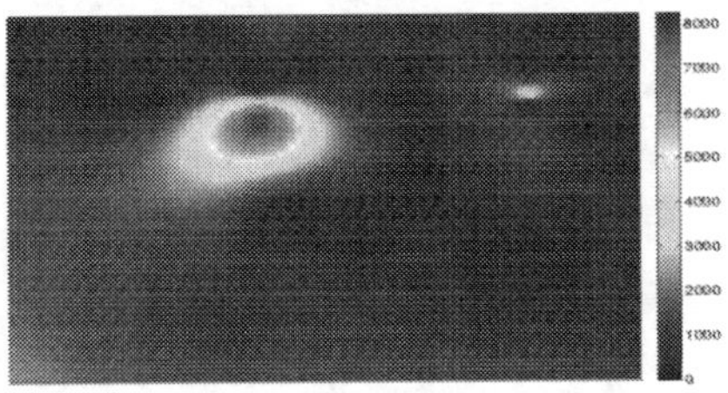

Figure 4. GPR tomography.

3. GPR tomography

Processing of GPR signals included three common processing steps: background signal removal, depth resolution enhancing and Kirchoff migration. The result of this processing is a radargram, which is an image (also called B-Scan) that represents values of the measured signal (in this case we used signal power) at different depths of the material through the points of a trajectory (defined in figure 2). The objective of the several processing steps was to highlight the zones in the radargram where the inhomogeneities were located. Thus, envelope estimation and automatic gain control were also implemented for data interpretation and to enhance the image contrast. A tomography was obtained by merging the 11 radargrams obtained for each of the GPR trajectories. This image was cleaned applying morphological operations using disk-shaped structuring element with a radius of 5 pixels and thresholding power values to .7 of the maximum. This last guarantees only high reflections to be represented and filtering spurious reflections. After this, as in the ultrasound tomography processing, an interpolation algorithm was used to generate a smoothed tomography. Figure 4 shows the GPR tomography obtained for the inhomogeneous wall at 50Tn load step. The high reflection produced by the flaw and the power decrease caused by the crack are depicted. These defects affect wave propagation travelling depending on its shape in two different ways, in the case of the crack; the propagated waves have to surround the defects and therefore their energy decreases. In the case of the flaw, multiple reflections and diffraction with the defect borders arise.

4. Results and Conclusion

Figures 5 and 6 represent fused tomographies obtained from the ultrasound and GPR images explained in Section 2 and 3. The operators for combining the images were product and sample mean for figures 5 and 6, respectively. After fusion, the fused image was filtered using a median filter. Figure 6 shows an enhanced tomography were three of the four defects (flaw, crack and vertical drill) are shown and spurious defective areas seem more attenuated.

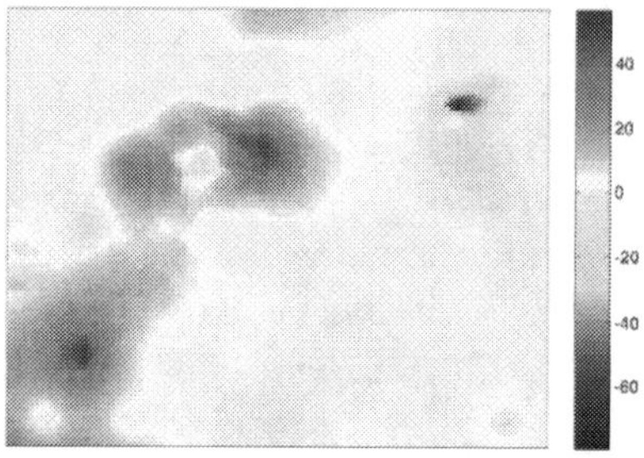

Figure 5. Fused tomography by product.

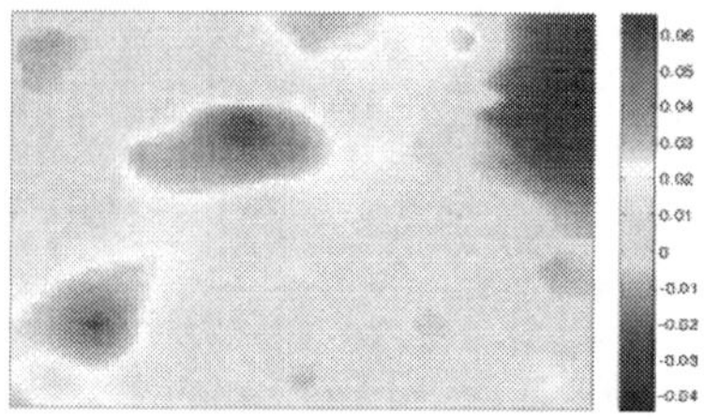

Figure 6. Fused tomography by sample mean.

We observe that the different fusion methods produce different results. An explanation for that may be done in the framework of optimal estimation theory. Thus we may consider that the final goal of the problem is to estimate a parameter, associated to every point in the image, represented by the amplitude value of every pixel. This parameter is supposed to change (normally increasing) due to the presence of a defect. The two values respectively measured by GPR and ultrasonic methods may be considered "observations" from which the parameter is to be estimated. In Bayes estimation theory [6], the parameter and the observations are considered random variables having some marginal and joint distributions, and the optimum estimator corresponds to the one minimizing some cost function. It is, for example, well-known, that the minimum mean-square error (MMSE) estimator is given by the expected value of the parameter conditioned to the observations. In general, this conditioned expected value is a nonlinear function of the observations, but if the marginal and joint distributions of the parameters and the observations are Gaussian the optimum estimator of the parameter is a linear combination of the observations. Actually, if the observations are uncorrelated and have equal cross-correlation to the parameter, the optimum estimator is just the sample mean of the observation, which is one the fusion method that we have implemented.

Following a similar interpretation, the fusion implementing the product is optimum when the marginal and joint distributions of the parameter and the observation are log-normal (equivalent to a linear combination after taking logarithms). In a general case the appropriate non-linear function will depend on the underlying distributions. Among them those ones based in order statistics [7] are very popular. Thus, we have experimented also with the fusion based in selecting the maximum value of the two observations, although other methods could be tried. Considering the obtained results it seems that the Gaussianity and uncorrelation assumptions are reasonable in our case, as the best performance has been obtained with the fusion implementing the sample mean of the two images.

Acknowledgments

This work has been supported by *Generalitat Valenciana* under grant PROMETEO/2010/040, and Spanish Administration and European Union FEDER Prog. under grant TEC2011-23403 01/01/2012.

References

[1] Cheeke J D 2002 *Fundamentals and Applications of Ultrasonic Waves* (Florida: CRC Press)
[2] Salazar A and Vergara L 2010 *Eurasip Journal on Advances in Signal Processing* 2010 125201
[3] Salazar A Vergara L and Llinares R 2010 *Mechan. Systems and Signal Proc.* **24** (6) 1870-1886
[4] Ploix M-A, Garnier V, Breysse D, and Moysan J 2011 *NDT&E International* **44** 442-448
[5] Journel A G and Huijbregts Ch J 2004 *Mining Geostatistics* (New Jersey :The Blackburn Press)
[6] Kay S M 1993 *Fund. of Statistical Signal Processing, Estimation Theory* vol 1 (New Jersey: Prentice Hall)
[7] Pitas I and Venetsanopoulos AN 1990 *Nonlinear Digital Filters* (Boston:Kluwer Academic)

Beamforming method based on phase dispersion Fourier analysis for SAFT images

D. Romero-Laorden, O. Martínez-Graullera, A. Ibañez and L. G. Ullate

Centro de Acústica Aplicada y Evaluación no Destructiva (CSIC-UPM)

E-mail: `david.romero@csic.es`

Abstract. A phase coherence descriptor based on the Fourier analysis of phase dispersion in ultrasound data is proposed. It can be easily computed and introduced as a ponderation factor during a traditional beamforming process. Theoretical and experimental results show that, for a linear array of 64 elements operating in Synthetic Aperture Techniques, the dynamic range is enhanced reaching up to 30dB.

1. Introduction

Synthetic Aperture Focusing Techniques (SAFT) have been a topic deeply studied in several areas such as radar, sonar or ultrasonic imaging. They are a very good solution to reduce the hardware requisites and consequently the consumption and cost of the systems, making possible their miniaturization and portability.

However, beamforming in SAFT is a highly data parallel process that can be easily pipelined and executed on a single GPU or several of them. Nowadays GPU beamformers are a reality [1], but their use in real time is limited by some factors. A Full Matrix Array capture (FMA) is used to obtain the complete set of signals and after the acquisition a Total Focusing Method (TFM) is applied to generate an image [2]. Supposing an array of N elements it means to process $N \times N$ signals. Therefore, TFM requires a high number of signals what increase the cost involved in the acquisition system, the data transfer and the computational cost. In order to develop real time imaging systems, the solution relative to this problem is to employ acquisition methods based on solutions of minimum redundancy (coarray-based). The simplest is the technique known as 2R-SAFT which significantly reduces the number of electronic channels and provides images with good quality and grating lobes free [3]. Furthermore, coarray-based implementations with GPGPU techniques achieves an speed up of 150x [1] over CPU beamforming.

The problem is that dynamic range is compromised because coarray-based methods involve only $(2N-1)$ signals compared to the TFM which involves $(N \times N)$. This paper proposes the use with minimum redundancy solutions, of a new coherence factor map based on a descriptor of Fourier Series to enhance the dynamic range, reaching the image quality of the TFM.

2. Minimum redundancy approach

We have chosen a method of minimum redundancy suggested in [4] (we will refer it as FnR SAFT as abbreviation) where each element of the "effective aperture" or coarray is composed by only a signal, working with a reduced set of parallel channels. Supposing $N_e = \frac{2N}{Nc}$ the number of elements in emission, the i-emitter is defined as

$$\hat{e}_i = \frac{N_c}{2} i \qquad i = [1 : N_e] \quad / \quad 1 \leq \hat{e}_i \leq N \tag{1}$$

where N_c is the number of parallel channels operating in reception and N is the number of array elements. Similarly, we describe the k-receiver as:

$$\hat{r}_k = \hat{e}_i + k - \frac{N_c}{2} \qquad k = [1 : N_c] \quad / \quad 1 \leq \hat{r}_k \leq N \tag{2}$$

For convenience, suppose $X_{\hat{e},\hat{r}}(t)$ the number of signals acquired expressed in-phase I and quadrature components Q:

$$X_{\hat{e},\hat{r}}(t) = I_{\hat{e},\hat{r}}(t) + jQ_{\hat{e},\hat{r}}(t) = \mathbf{X}_{\hat{e},\hat{r}}(t) e^{j\phi_{\hat{e},\hat{r}}(t)} \tag{3}$$

where $X_{\hat{e},\hat{r}}(t)$ is the signal, obtained by the emitter $\hat{e}$ and receptor $\hat{r}$. Then $\mathbf{X}_{\hat{e},\hat{r}}(t)$ is the modulus and $\phi_{\hat{e},\hat{r}}(t)$ its corresponding signal phase. According to this expression and relative to the method mentioned in this work, the FnR image $A(\vec{x})$ can be calculated as:

$$A(\vec{x}) = \sqrt{\left(\sum_{i=1}^{N_e} \sum_{k=1}^{N_c} I_{\hat{e}_i,\hat{r}_k}(T(\vec{x})) \right)^2 + \left(\sum_{i=1}^{N_e} \sum_{k=1}^{N_c} Q_{\hat{e}_i,\hat{r}_k}(T(\vec{x})) \right)^2} \tag{4}$$

where $T(\vec{x})$ is the focussing delay for the $\vec{x} = (x, z)$ point which compensates emission and reception distances:

$$T(\vec{x}) = \frac{\sqrt{(x - x_{\hat{e}})^2 + z^2} + \sqrt{(x - x_{\hat{r}})^2 + z^2}}{c} \tag{5}$$

being $x_{\hat{e}}$ and $x_{\hat{r}}$ the coordinates of the elements $\hat{e}$ and $\hat{r}$, and (x, z) the space point.

3. Coherence Factor Map Image

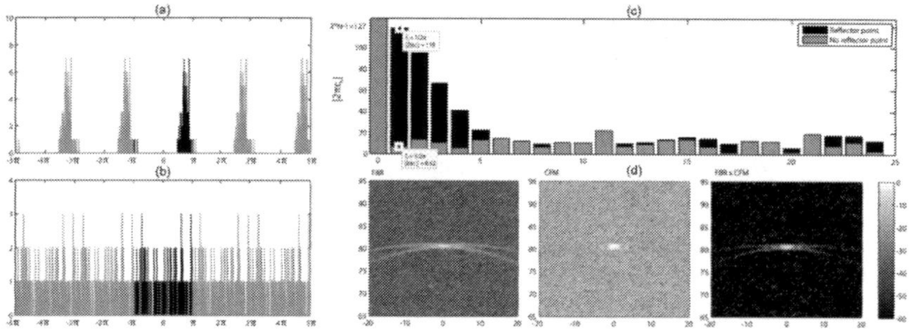

Figure 1. (a) $P(\phi)$ and $\Phi(\phi)$ in a point near to the reflector; (b) $P(\phi)$ and $\Phi(\phi)$ in a no reflector point; (c) 24 coefficients of Fourier series for both points, the component $f_\phi = \frac{1}{2\pi}$ has been remarked; (d) The images of the reflector point for F8R, CFM and F8R×CFM.

In general, coherence-based methods weight each image point according to the coherence estimated from the receive aperture domain data [5, 6]. As Delay and Sum beamforming, the coherence based beamformer is a non-adaptive tool based on the fact that element-signal-data from the mainlobe region are coherent, whereas those from the sidelobe region are incoherent.

IOP Conf. Series: Materials Science and Engineering **42** (2012) 012009 doi:10.1088/1757-899X/42/1/012009

Once the delays corresponding to each signal have been applied, at each image point a set of signal phases is obtained to compose a phase distribution function by a sum of Kronecker deltas:

$$P(\phi) = \sum_{i=1}^{N_e} \left(\sum_{k=1}^{N_c} \delta[\phi - \phi_{\hat{e}_i,\hat{r}_k} T(\vec{x}))] \right) \tag{6}$$

where $\phi_{\hat{e}_i,\hat{r}_k}(T(\vec{x}))$ is the phase of the signal emitted by $\hat{e}_i$-element and received by $\hat{r}_k$-element compensated at (x, z). These values are distributed in the $[-\pi, \pi]$ interval and can be extended to form the periodic signal using a comb-function based on Dirac deltas:

$$\Phi(\phi) = \sum_{m=-\infty}^{\infty} \delta(\phi - m2\pi) * P(\phi) \tag{7}$$

Figure 1 shows two image points in a simulated medium. A 64-element linear array has been simulated with 127 signals (F8R with 30% of noise).In a point near to the reflector position phases are concentrated (figure 1(a)), and a location where there is no reflector (figure 1(b)) showing an uniform distribution in the $[-\pi, \pi]$ range. The periodic extension of $P(\phi)$ can be decomposed by Fourier Series into a sum of simple oscillating functions:

$$c_n = \frac{1}{2\pi} \int_{-\pi}^{\pi} \Phi(\phi) e^{-j2\pi n f_o \phi} d\phi = \frac{1}{2\pi} \sum_{i=1}^{N_e} \left(\sum_{k=1}^{N_c} e^{-j2\pi n f_o \phi_{\hat{e}_i,\hat{r}_k}(T(\vec{x}))} \right) \tag{8}$$

where, according to the signal period, $f_o = \frac{1}{2\pi}$. Figure 1(c) shows the values of c_n for the first 24 sinusoidal components of the Fourier series. If the signal $P(\phi)$ is composed by a random distribution of phases its energy is spread over all the Fourier Coefficients which are low valued. However, when there is a concentration of phases it increases the presence of the sinusoidal components of period $\frac{1}{2\pi}$ than can be evaluated easily if $n = 1$ (equation (8)). Then, our coherence factor at point (x, z) is defined as:

$$C_1(\vec{x}) = (2\pi)c_1 = \sum_{i=1}^{N_e}\sum_{k=1}^{N_c} e^{-j\phi_{\hat{e}_i,\hat{r}_k}(T(\vec{x}))} = \sum_{i=1}^{N_e}\sum_{k=1}^{N_c} \frac{I_{\hat{e}_i,\hat{r}_k}(T(\vec{x}))}{\mathbf{X}_{\hat{e}_i,\hat{r}_k}(T(\vec{x}))} - j\sum_{i=1}^{N_e}\sum_{k=1}^{N_c} \frac{Q_{\hat{e}_i,\hat{r}_k}(T(\vec{x}))}{\mathbf{X}_{\hat{e}_i,\hat{r}_k}(T(\vec{x}))} \tag{9}$$

The maximum value of this function is achieved when all phase components are concentrated and is ($2N - 1$, 127 in this case), which is the total number of signals involved in the process. In figure 1 the coherence has been measured at the reflector point that is contributed by 118 signals (figure 1(a)), and in the point where there is no reflector the coherence is reduced to be composed by only 6 signals (figure 1(b)).

The complete image Coherence Factor Map (CFM) can be expressed then by:

$$H(\vec{x}) = \sqrt{\left(\sum_{i=1}^{N_e}\sum_{k=1}^{N_c} \frac{I_{\hat{e}_i,\hat{r}_k}(T(\vec{x}))}{\mathbf{X}_{\hat{e}_i,\hat{r}_k}(T(\vec{x}))} \right)^2 + \left(\sum_{i=1}^{N_e}\sum_{k=1}^{N_c} \frac{Q_{\hat{e}_i,\hat{r}_k}(T(\vec{x}))}{\mathbf{X}_{\hat{e}_i,\hat{r}_k}(T(\vec{x}))} \right)^2} \tag{10}$$

Using it with $A(\vec{x})$ (equation (4)) the new image equation is then $I(\vec{x}) = A(\vec{x})H(\vec{x})$.

4. Experimental Results

As an experimental procedure a tissue phantom (Model 040GSE - CIRS Inc.) with a 0.5dB/MHzcm attenuation, and 0.1mm diameter wires was examined using a 2.6MHz phased array transducer of $N = 64$ elements, 0.28mm of pitch. The figure 2 shows a line of 11 wires (covering an area starting at 17 to 122mm depth). The TFM image was composed for the DAS algorithm using the complete set of signals ($N \times N = 4096$). Next, F8R and Coherence Factor Map were calculated for the case of $N_c = 8$ using 127 signals ($2N - 1$). According to our

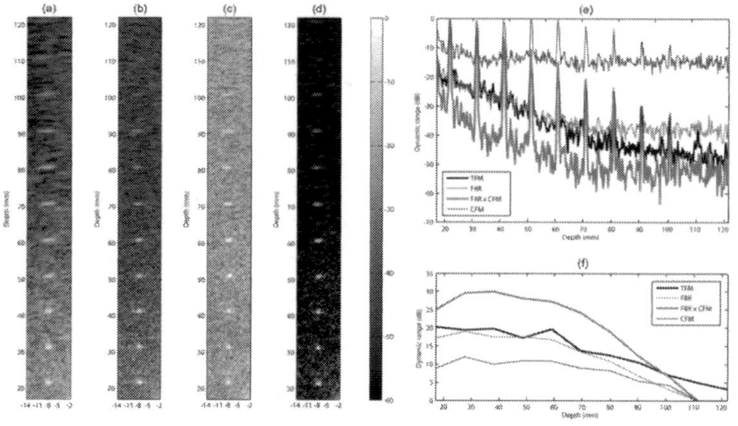

Figure 2. Experimental images of a line of wires in a tissue phantom. (a) TFM, (b) F8R, (c) CFM, (d) F8R×CFM; (e) Lateral profile in deep; (f) Variation of dynamic range in deep.

calculations and compared with 2R-SAFT, working with a strategy of minimum redundancy as F8R reduce 20% the acquisition time.

Corresponding images relative to TFM, F8R, CFM and F8R×CFM are represented in figures 2(a)(b)(c)(d) respectively. Original A-Scan has a SNR of 12dB for the highest echo. As we can observe, TFM image shows 11 wires, the SNR at each deep remains constant to 20dB until 60mm, then it decreases to 5dB at 120mm (figure 2(f)). The F8R image shows only 9 wires, with a constant local SNR of 17dB until 60mm, disappearing the echo response at positions upper to 100mm. The CFM image shows 9 wires however the coherence is reduced for positions upper to 80mm and it is lost for positions upper to 100mm. The combination of CFM and F8R produces a SNR that raises values of 30dB and is maintained upper than TFM until 100mm.

5. Conclusions

A new descriptor based on phase dispersion has been presented with the goal of increasing the dynamic range in SAFT images. Experimental results have validated our method and it has been shown that its implementation is extremely simple. The method has shown that it is robust against noise and attenuation. The CFM maintains a constant SNR in a large region of the image that produces, for the F8R×CFM, better results than TFM in the 80% of the image, being finally limited by the SNR in the F8R image. This problem can be coped with pulse code compression techniques, and different options to combine it with CFM are now under study.

Acknowledgements

This work has been supported by the Spanish Ministry of Economy in project DPI2010-19376.

References

[1] Romero-Laorden D, Martínez-Graullera O, Martín-Arguedas C J, Tokio Higuti R and Octavio A 2009 *IEEE Ultrason. Symp.* (Rome, Italy, 20-23 September 2009) pp 1334–1337
[2] Holmes C, Bruce W Drinkwater and Wilcox P D 2005 *NDT & E International* **38** 701–711
[3] Martín-Arguedas C J, Martínez-Graullera O and Ullate L G 2008 *IEEE Ultrason. Symp.* (Beijing, China, 2-5 November 2008) pp 721–724
[4] Lockwood G R, Li P C, O'Donnell M and Foster F S 1996 *IEEE Trans. Ultrason. Ferro. Freq.Contr* **43** 7–14
[5] Hollman K W, Rigby K W and O'Donnell M 1999 *IEEE Ultrason.Symp.* (Caesars Tahoe, NV, 17-20 October 1999) pp 1257–1260
[6] Camacho J, Parrilla M and Fritsch C 2009 *IEEE Trans. Ultrason. Ferro. Freq.Contr* **56** 958–74

Prospective of the application of ultrasounds in archaeology

A Salazar[1], A Rodriguez[2], G Safont[1] and L Vergara[1]
[1]Universitat Politècnica de Valencia, Institute of Telecommunications and Multimedia Applications, Camino de Vera s/n, 46022, Valencia, Spain
[2]Universidad Miguel Hernández de Elche, Avda. de la Universidad s/n, 03202, Elche, Spain

E-mail: asalazar@dcom.upv.es

Abstract. This paper presents a prospective analysis of non destructive testing (NDT) based on ultrasounds in the field of archaeology applications. Classical applications of ultrasounds techniques are reviewed, including ocean exploration to detect wrecks, imaging of archaeological sites, and cleaning archaeological objects. The potential of prospective applications is discussed from the perspective of signal processing, with emphasis on the area of linear time variant models. Thus, the use of ultrasound NDT is proposed for new ceramic cataloguing and restoration methods.

1. Introduction

The principal objective of this paper is to provide a prospective analysis of the use of non-destructive testing (NDT) by ultrasounds in archaeological applications. NDT by ultrasounds is a powerful technique of material diagnosis and characterization that has been applied in a wide variety of applications including distance gauging, corrosion monitoring, and flaw detection [1]. Briefly, this technique operates in two modes: pulse-echo (an ultrasound sensor or transducer is used both as emitter and receiver) and trough-transmission (two transducers, the first one is used as emitter and the second one is used as receiver). Figure 1 shows a chart summarizing the principal techniques of characterization and dating used currently in archaeology. Contrary to NDT methods, most of these techniques require some destruction of the analyzed materials, expensive equipment, and tube preparation, and thus, they are costly and time-consuming.There are relatively few references of the applications of ultrasounds in archaeology, and in general, they can be roughly organized into four classes: imaging, cleaning, flaw detection, and material characterization.

Imaging applications consists of obtaining tomographic representations of underground and underwater archaeological sites and objects. Some examples of this application are the following: imaging of the stern section of a shipwreck obtained by an ultrasound triangulation system composed of three receivers and two emitters [2]; an experimental setup for underwater artefact identification [3]; a cave internal structure based on a 3D model from ultrasound reflections, which was used for several simulations including the sediment flow inside the cave [4]; underground interface estimation to be used in exploration of comparatively large archaeological sites [5]; and a precise localizer and sketcher of archaeological findings after they are extracted by archaeologists [6]. The use of ultrasounds for cleaning in archaeology have covered applications such as ultrasound brushing of archaeological metal-made objects [7], isolating a suspension of binder carbonates from bulk mortars in mortar radiocarbon dating [8], and removal of contaminants in archaeological bone analysis [9].

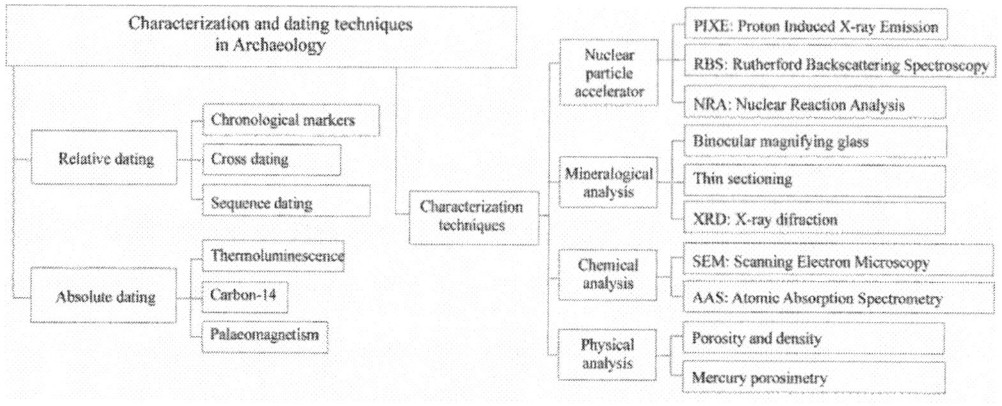

Figure 1. Classical archaeological characterization and dating techniques.

Some examples of flaw detection applications are the following: visualization of the internal structure of historic relics in wooden pillars of shrines from estimation of the time-of-flight in several measurements through the pillar surface using fan-beam geometry [10]; diagnosis of ancient wooden panel paintings (detection of delaminations between superimposed priming layers and wood support and cracks in the priming layers itself) by using air-coupled ultrasonic technique [11]; 3D microtomography for visualization and reconstruction of voids and cracks within, internal structures, restorative interventions, and the cuts through the original cast works in sculptures [12]; and integrating ground-penetrating radar (GPR) and ultrasounds to evaluate the condition of a Roman marble slab after a fall down during transportation [13]. Finally, ultrasounds also has been employed for archaeological material characterization, for instance, examining weathering processes and intensity acting of humidity and salt in the rocks of historical buildings at Alhambra, Spain [14] and determination of mechanical properties (Young's modulus and Poisson's ratio) for prehistory lithics from Armorican massif in western France [15].

Recently, we proposed an application of ultrasounds for cataloguing archaeological ceramics according to the chronological period of the analyzed pieces (for instance, Bronze age, Roman, Middle ages, etc.) [16]. The cataloguing method employed a versatile classifier based on mixtures of independent component analyzers [17] which has been successfully applied to NDT problems [18]. In the next section, we discuss prospectively the extension of this method to new applications based on modeling the ultrasound signal by means of a linear time variant model.

2. Ultrasound signal modelling as a linear time variant system

In general, exact knowledge of the behavior of and ultrasonic pulse travelling inside a material is a complex matter. This is especially true in ceramic materials, where the ultrasonic waves suffer strong effects of attenuation, dispersion and backscattering. Nonlinear wave acoustic propagation theory is hence claimed for a detailed prediction of the sensor recorded wave for a given ultrasonic excitation of the material [19]. Things become even more complicated in restored ceramic due to the interactions between the original material and the one used for restoration. However, from the perspective of classification or cataloguing of ceramics pieces into a finite (not too large) number of classes, simpler linear or nonlinear models may be enough to justify the use of some signal extracted features (see table 1).

Thus, a simple model may assume that the material is formed by many small reflectors (reflectivity of the material). When the ultrasonic pulse impinges in a reflector, a new pulse is generated. The recorded signal would be the superposition of the many pulses thus generated by the material microstructure. This essentially relates to consider a linear system model where the input is the reflectivity of the material and the impulse response is the ultrasonic pulse excitation. Hence the output of the linear system (recorded sensor signal) would be the convolution of the reflectivity and the excitation pulse. Figure 2 shows a scheme of the ultrasound through-transmission LTV model.

Table 1. Examples of features extracted from ultrasound signal.

Total signal attenuation	$\hat{x}(t) = Ae^{-\beta t} \Rightarrow Aten\_Total = \beta$	Attenuation curve initial value (dB)	$\hat{x}(t) = Ae^{-\beta t} \quad Po = 10\log(A)$
Propagation velocity	$v = \frac{piece\ thickness}{ultrasound\ time\ of\ flight}$	Centroid frequency	$fc = \frac{\int_{f_1}^{f_2} f \cdot \lvert X(f) \rvert df}{\int_{f_1}^{f_2} \lvert X(f) \rvert df}$
Principal frequency	$f_{max} / \lvert X(f_{max}) \rvert \geq \lvert X(f) \rvert \forall f$	Time-reversibility	$\frac{1}{\sigma_x^3} \left\langle \left(\frac{dx(t)}{dt} \right)^3 \right\rangle$
Principal frequency Amplitude	$\lvert X(f_{max}) \rvert$	Third order autocovariance	$\langle x(t) \cdot x(t-1) \cdot x(t-2) \rangle$
Signal power	$P\_Total = \frac{\int_0^T \lvert x(t) \rvert^2 dt}{T}$	Instantaneous centroid frequency	$fc(t = t_0)$

In the simplest case, the pulse (impulse response) may be considered to be constant, which means that the pulse arriving at every reflector of the microstructure is always the same and that the only effect of the reflector is to generate a new replica with given amplitude, depending on the reflector size. This corresponds to a strictly linear time invariant (LTI) system in the context of system theory. More complex models may be gradually considered depending on the complexity of the material. Thus, a frequency independent attenuation of the pulse can be easily introduced into the model by assuming a linear time variant (LTV) system. This can be generalized to frequency dependent attenuation by allowing a general distortion of the pulse as it propagates deeply into the material. Also, some nonlinear effects may be considered by, for instance, allowing some zero-memory nonlinear transformation before or after the LTV system.

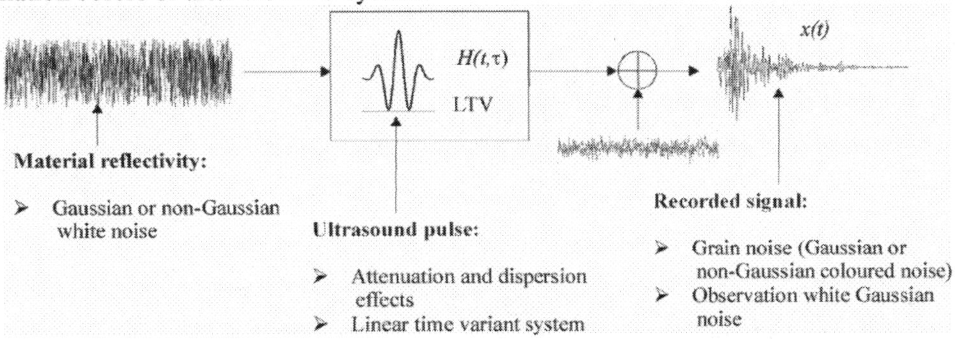

Figure 2. Ultrasound through-transmission LTV model.

In any case, what it is important from the perspective of the work described in this paper is that these models (apart from simulation use) may inspire about possible features for classifying the material. It is clear that those features should be related to the input (reflectivity) or to the system itself (impulse response or zero-memory nonlinear function, for example). Reflectivity may be considered a stochastic process and so statistical features directly obtained from first, second or higher order statistics are reasonable candidates. These features may be stationary or non-stationary, depending on the nature of the material. For example in the restoration framework, the obtained features should change in areas where the restoration material is present in comparison with areas where it is not: the form that a given statistics varies with depth suggests a signature for classification purposes between different qualities of restoration.

Regarding the impulse response or other models of the systems, we can try to estimate them in time or frequency domain from the recorded signals. Thus, we can make spectral analysis to smooth spectral estimates of the system frequency response. This can be made assuming an LTI model or a LTV model (time-frequency spectral analysis). Then some signatures and parameters can be computed from the estimated spectra to be used as features at the classifier input.

3. Conclusion

The standardization of an efficient and non-destructive method for ceramic characterization based on ultrasonic testing will be an important contribution for archaeologists. This method could complement or replace destructive, costly, and time-consuming techniques, which are currently being used by archaeologists in the area of ceramic characterization. We have discussed here the possibility of devising NDT by ultrasounds for archaeological ceramic classification from the perspective of linear time variant modelling. The differences in physical properties, composition and processing of the materials allow the extraction of signatures by using ultrasound signals.

Acknowledgments

This work has been supported by the *Generalitat Valenciana* under grant PROMETEO/2010/040, and the Spanish Administration and the FEDER Programme of the European Union under grant TEC2011-23403 01/01/2012.

References

[1] Krautkrämer J 1990 *Ultrasonic Testing of Materials 4th edition* (New York : Springer)
[2] Hahn D A, Williams G N, Wilcox M and Wilcox P 1985 *IEEE Oceans Conference Rec.* (San Diego, California, 12-14 November 1985) pp 104-109
[3] Hergli M, Bouslama F, Baili J, Lansari A, Al-Rawi A and Besbes K 2004 *International Symposium on Control, Communications and Signal Processing ISCCSP* (Hammamet, Tunesia, 1-24 March 2004) pp 229-232
[4] Sellers W I and Chamberlain A T 1998 *Journal of archaeological science* 25 867-873
[5] Matsuo Y, Tao L, Watanabe T and Motooka S 2002 Japan J. *Applied Physics* **41** (5B) 3539-44
[6] Jiménez A R and Seco F 2005 *Sensors and Actuators* A **123** (124) 224-233
[7] Zucciatti A, Gutiérrez Neira P C, Climent-Font A, Escudero C and Barrera M 2011 *Nuclear Instruments and Methods in Physics Research* B **269** 3115-3119
[8] Marzaioli F, Lubritto C, Nonni S, Passariello I, Capano M and Terrasi F 2011 *Analytical Chemistry* **83** (6) 2038- 2045
[9] Mitchell P D 2009 *American Journal of Physical Anthropology* **139** (2) 274
[10] Kanda S, Shioya K, Yanagiya Y, Tamura Y and Adachi K 1998 *Proc IEEE Ultras Symp* (Sendai, Japan, 5-8 October 1998) pp 743-746
[11] Siddiolo A M and Maev R M 2005 *Proceedings of IEEE Ultrasonic Symposium* (Rotterdam, 18-21 September 2005) pp 890-893
[12] Badde A and Illerhaus B 2008 Scanning 30 16-26
[13] Capizzi P and Cosentino P L 2011 *Journal of Geophysics and Engineering* **8** S117-S125
[14] Kamh GME 2007 *Environ Geol* **52** 1539-1547
[15] Tsobgou RA 2009 *Journal of Archaeological Science* **36** 1615-1625
[16] Salazar A and Vergara L 2010 *Eurasip Journal on Advances in Signal Processing 2010* 125201
[17] Salazar A, Vergara L, Serrano A and Igual J 2010 *Pattern Recognition* **43** (1) 69-85
[18] Salazar A, Vergara L and Llinares R 2010 *Mechan. Systems and Signal Proc.* **24** (6) 1870-1886
[19] Cheeke J D 2002 *Fundamentals and Applications of Ultrasonic Waves* (Florida: CRC Press)

Ultrasonic variable-focus optical lens using transparent gel

D Koyama, R Isago and K Nakamura

Tokyo Institute of Technology, 4259-R2-26, Nagatsutacho, Midori-ku, Yokohama, 226-8503 JAPAN

E-mail: dkoyama@sonic.pi.titech.ac.jp

Abstract. Varifocal imaging using an optical lens that employs acoustic radiation force and a viscoelastic material and that has no mechanical moving parts is investigated. The lens has a simple and thin structure that consists of an annular ultrasonic transducer and silicone gel. An axially symmetric acoustic standing wave can be generated in the gel by exciting a vibration mode in the radial direction on the transducer. The lens profile can be altered by varying the acoustic radiation force of the transducer. The focal length can be controlled by varying the transducer input voltage so that the lens functions as a variable-focus lens.

1. Introduction

Crystalline lenses in the human eye change their shape to control the focal point. Most camera modules require actuators and a gearing system to move the position of the lens in order to focus on an object. Several studies have reported variable-focus liquid lenses that do not contain any moving mechanical parts. Most liquid lenses utilize electrowetting to vary the lens shape, but this mechanism gives relatively long response times of the order of tens of milliseconds [1, 2]. Several studies have reported variable-focus liquid lenses that utilize sound pressure. L□pez and Hirsa have developed an oscillating liquid lens that changes its focal length by changing its shape when excited by an audible sound, allowing it to be used as a variable-focus lens [3]. We have developed a liquid lens that utilizes acoustic radiation force; it has a shortest response time of 6.7 ms [4, 5, 6, 7]. However, these conventional liquid lenses are sensitive to the ambient temperature because their optical properties depend on the temperature dependences of the interfacial tension and the liquid viscosities. Since the lens profile is determined by the contact angle between the two liquids, which depends on the interfacial tension between the two liquids, the temperature dependence of the interfacial tension directly affects the lens profile. The optical properties and the dynamic response of the lens also depend on the kinematic viscosity of the liquids because the propagation loss of sound waves increases with increasing viscosity [4]. Furthermore, microbubble generation and erosion of the liquids reduce the operating life of liquid lenses. The present study investigates a variable-focus lens that has a simple and thin structure and that uses a silicone gel, which is stable with temperature and does not generate bubbles.

2. Configuration

Figure 1 shows the configuration of the lens. The center of a piezoelectric lead zirconate titanate (PZT) ring (inner diameter: 15 mm; outer diameter: 30 mm; thickness: 2 mm) polarized in the thickness direction was filled with silicone gel (KE-1052(A/B), Shin-Etsu Silicone; refractive index: 1.4; specific gravity: 0.97; complex elastic modulus: 2×10^4 N/m^2; speed of sound: 1030 m/s), which is both transparent and viscoelastic. The complex elastic modulus of silicone gel varies by less than

±10% in the range 0-100 °C. The viscoelastic properties of the silicone gel can be controlled by controlling the mixture ratio of KE-1502-A and -B; this ratio was 1:0.7 in the prototype. A silicone gel layer was formed by attaching a 0.1-mm-thick polyethylene terephthalate (PET) film to one side of the PZT ring to prevent the silicone gel from flowing out. After molding, the shape of the gel layer changed little when the ambient temperature was varied in the range 0-100 °C. An acoustic standing wave is efficiently generated in the silicone gel at the center of the ring by exciting a vibration mode in the radial direction (r direction in figure 2(a)) in the PZT ring.

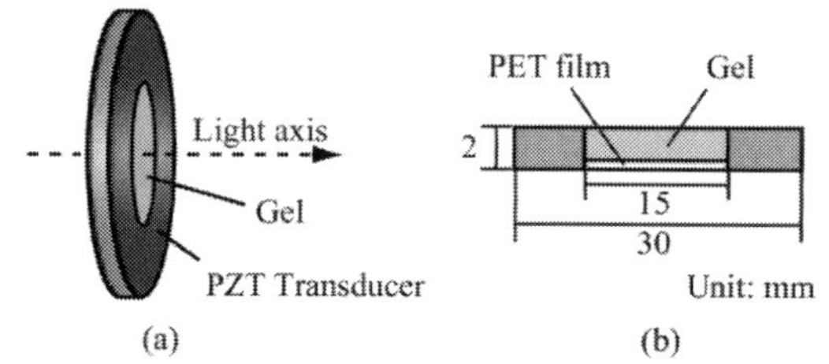

Figure 1. Configuration of the lens.

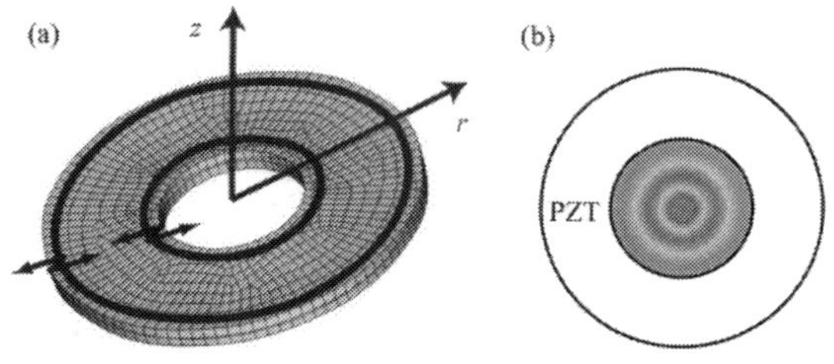

Figure 2. (a) Vibration mode of the transducer and (b) sound pressure distribution in the gel at 222 kHz predicted by FEA.

A finite element analysis (FEA) simulation using commercial FEA software ANSYS 11.0 (ANSYS, Inc., Canonsburg, PA, US) predicts that a vibration mode in the radial direction will be generated in the PZT ring at 222 kHz (see figure 2(a)). The simulation model includes the gel and the PET film; a continuous sinusoidal voltage was applied to the electrodes of the PZT ring as the boundary condition. A half-wavelength standing wave is generated between the inner and outer diameters of the lens. The standing wave has a nodal circle at r = 11.25 mm, which is suitable to be supported. Figure 2(b) shows the sound pressure amplitude distribution in the gel at 222 kHz calculated by FEA. FEA predicts that an acoustic standing wave with two concentric nodal circles will be generated in the gel and that the sound pressure amplitude will be largest on the central axis. The sound pressure distribution in the gel obtained by FEA was similar to a Bessel function, which is the theoretical solution for a disk cavity [8]. By utilizing the resonance frequency of the vibration mode in the radial direction of the PZT ring, which is almost entirely determined by the inner and outer diameters and is almost independent of the ring thickness, thin and compact lenses can be developed for use in MEMS.

3. Results and discussions

When an ultrasound standing wave at 222 kHz is excited in the silicone gel, a difference in the acoustic energy densities of the gel and air is generated at the lens surface. This is because most of the acoustic energy generated by the PZT ring vibrations remains in the gel because it is reflected at the lens surface due to the large acoustic impedance difference between the gel and air. This discontinuity in the acoustic energy density generates a static force on the lens surface, which is called the acoustic radiation force [9]. The acoustic radiation force acts from the higher energy density medium (the gel) to the lower energy density medium (air) so that the lens surface is statically deformed toward air. Figure 3 shows radial profiles of the lens obtained by optical coherence tomography (OCT) when the lens is excited by voltages in the range 0-21 V_{pp} at 226 kHz. Figure 3 also shows the distribution of the

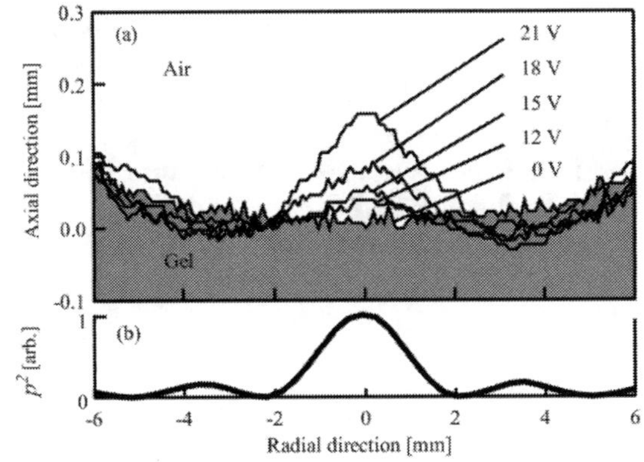

Figure 3. (a) Radial profiles of the lens excited by five different input voltages at 226 kHz and (b) radial distribution of square of the sound pressure amplitude p^2 calculated by FEA.

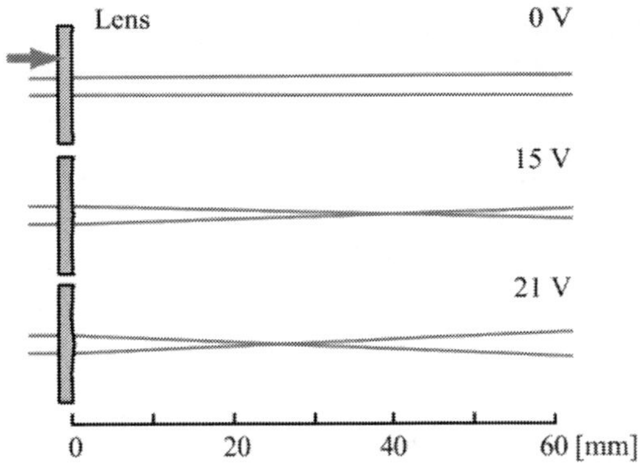

Figure 4. Ray-tracing results for the lens excited by five different input voltages at 226 kHz.

square of the sound pressure amplitude p^2 calculated by FEA (see figure 3(b)) since the acoustic radiation force is a function of p^2. The center of the lens surface was statically deformed outward. The displacement amplitude increased with increasing input voltage since the larger input voltage generates a larger acoustic radiation force; the displacement amplitude was approximately 150 μm for an input of 21 V_{pp}. The vibrations at 226 kHz overlapped on the statically deformed lens surface. This vibrational displacement amplitude was considerably smaller than the static displacement amplitude (it was tens of nanometers for 21 V_{pp}). The lens surface was deformed only at the center axis, which corresponds to the predicted peak position of p^2 in the gel by FEA. The lens profile is determined by the balance between the acoustic radiation force, the surface tension, and the elastic restoring force of the gel. The PET film on the reverse side was deformed much less than the gel. The displacement amplitude should be less than 7.7 μm, which is the spatial resolution of OCT. The effect of gravity on the lens profile is considered to be small since no change in the gel surface was observed by OCT when the lens was inverted.

Figure 4 shows the light paths calculated by ray tracing for a lens excited by voltages between 0 and 21 V_{pp}. The lens aperture is considered to act as a variable-focus lens and incident light with a beam width of 2 mm passed through the lens from left to right. The lens profiles obtained by OCT (see figure 3) were used in the simulation. When no input voltage is applied, the lens functions as a concave lens with a radius of curvature of 84 mm. By controlling the input voltage, the transmitted light was focused and the focal point of the lens could be varied. Increasing the input voltage reduces the focal length; the focal length was approximately 24 mm from the lens surface when the input voltage was 21 V_{pp}.

The dynamic response of the lens was investigated by changing the focal point. A resolution test target (1951 USAF) and an image of a swallow on a glass plate were positioned parallel to the lens at distances of 10 and 23 mm from the lens surface, respectively. A high-speed camera with a commercial lens was employed and optical images were captured through the prototype lens so that the focal point of the camera corresponded to the position of the swallow (i.e., 23 mm) when no voltage was applied to the lens. The transient motion of the lens surface along the central axis was observed by M-mode OCT. Figure 5 shows sequential optical images and the transient response of the lens surface. The lens was excited by 20 V_{pp} at $t < 0$ and the input voltage was switched off at $t = 0$. When the excitation voltage was switched off, the lens surface began to move to its default position and the focal point moved from 10 to 23 mm (i.e., from the resolution test target to the image of the swallow). The response time of the lens, which corresponds to the duration of the transient state, was

0.3 s. The lens response time depends on the viscoelastic properties of the gel. We intend to reduce the response time in future research.

4. Conclusions

A variable-focus optical lens that employs a viscoelastic material and acoustic radiation force was proposed. This lens has a simple and thin profile and consists of a PZT ring and silicone gel. The optical profile of the lens was investigated. The focal point could be controlled by varying the input voltage so that the lens acted as a variable-focus lens.

Acknowledgments

This work was partially supported by a research grant from the Ministry of Education, Science, Sports and Culture, a Grant-in-Aid for Young Scientists (B) (23760220, 2011), the Kurata Memorial Hitachi Science and Technology Foundation, the Japan Science Society, Ozawa and Yoshikawa Memorial Electronics Research Foundation, the Noguchi Institute, and the SNK-FAM.

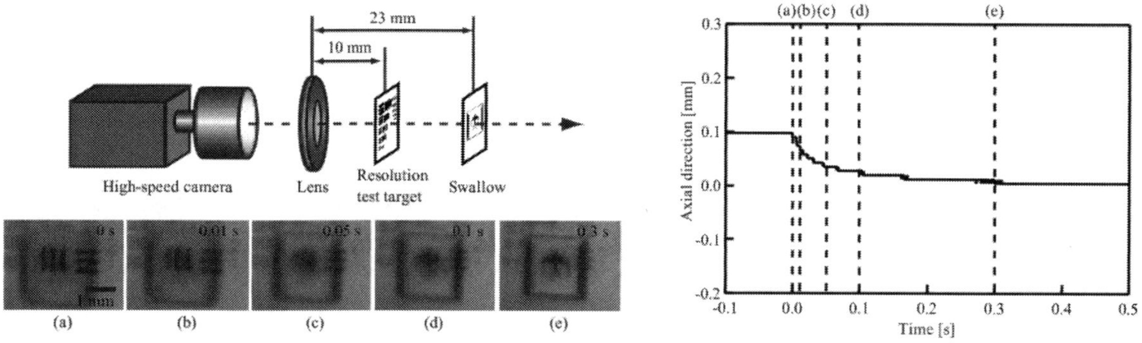

Figure 5. Transient response of the lens and images captured by a high-speed camera through the lens when the input voltage was switched off at $t = 0$.

References

[1] Gorman C B and Biebuyck H A and Whitesides G M 1995 *Langmuir* **11** 2242
[2] Berge B and Peseux J 2000 *Eur. Phys. J. E.* **3** 159
[3] L□pez C A and Hirsa A H 2008 *Nature Photonics* **2** 610
[4] Koyama D and Isago R and Nakamura K 2010 *Opt. Express* **18** 25158
[5] Koyama D and Isago R and Nakamura K 2011 *IEEE Trans. Ultrason., Ferroelect., Freq. Contr.* **58** 596
[6] Koyama D and Isago R and Nakamura K 2011 *Jpn. J. Appl. Phys.* **50** 07HE26
[7] Koyama D and Isago R and Nakamura K 2011 *IEEE Trans. Ultrason., Ferroelect., Freq. Contr.* **58** 2720
[8] Rayleigh J W S 1945 *The Theory of Sound volume two* (New York : Dover) p 297
[9] Chu B and Apfel E 1982 *J. Acoust. Soc. Am.***72** 1673

Fabrication of polylactic acid (PLA) microcapsules and release of the internal hydrophilic substance under ultrasound irradiation

Hiraku Tabata[1], Yukako Kato[1], Kenji Yoshida[1], Daisuke Koyama[2], Kentaro Nakamura[3] and Yoshiaki Watanabe[1]

[1] Faculty of Life and Medical Sciences, Doshisha University, Kyoto, JAPAN
[2] Faculty of Science and Engineering, Doshisha University, Kyoto, JAPAN
[3] Precision and Intelligence Laboratory, Tokyo Institute of Technology, Yokohama, JAPAN

E-mail: dml0132@mail4.doshisha.ac.jp

Abstract. Hollow microcapsules have been considered for potential application as drug or gene carriers. Targeted drug or gene release to tumor vascular can be achieved by destroying the capsules by high-intensity ultrasound. This report investigates fabrication of microcapsules having biocompatible polylactic acid (PLA) shell and the mechanical property for the destruction under ultrasound irradiation. The PLA hollow microcapsules with the diameter of 1 to 35 μm were fabricated by double emulsion method. Red fluorescent dye which is substitute for drug was enclosed into the capsules. The capsules have trilaminar structure. Inner air is surrounded by red fluorescent, and the PLA shell encloses them. Shell thickness of capsules was distributed from 100 nm to 3 μm. Destruction of the fabricated capsules was investigated by using ultrasound pulse with the center frequencies from 500 kHz to 2 MHz. The behavior of capsule was observed by using an optical microscope under ultrasound irradiation. The frequency dependence of the capsule destruction was investigated and 35% of the capsules with the resonance size was destroyed with 400 kPa at 700 kHz. The capsule shell was ruptured by the ultrasound, and then the inner dye was released; the dye was completely released for 120 seconds under ultrasound irradiation.

1. Introduction

Hollow microcapsules such as contrast agents in ultrasound imaging are expected to be employed as drug and gene carriers in ultrasound drug delivery system (UDDS). In UDDS, the capsules having drug are injected and flow through blood vessels, and then they are destroyed by high-intensity ultrasound so that the internal drug can be released locally to the diseased part. Comparing with contrast agents such as Sonazoid® or Levovist® [1], microcapsule with harder shell is required in UDDS to deliver drug safety to the diseased part though blood vessel. PLA, which is biocompatible material, is suitable for UDDS to fabricate microcapsules with harder surrounding shell than Sonazoid® or Levovist®. The vibration and destruction characteristics of the PLA capsule depend on the mechanical property of capsule including the shell elasticity. Lu et al. [2], Daiguji et al. [3] and Makuta et al.[4] reported the fabrication of PLA capsules. Our group has also fabricated PLA capsules and observed the capsule vibration under ultrasound irradiation by using a laser Doppler vibrometer [5]. In this report, we investigated the fabrication method of PLA capsule with red fluorescent dye which is substitute for drug, and the destruction characteristics of the capsules.

2. Fabrication procedure

PLA capsules were fabricated by double emulsion method. The material of surrounding shell of the capsule is biodegradable poly-L-lactic acid (PLLA). Fabrication procedure is indicated below (see figure 1).

1).Prepare 100 ml of 0.2% polyvinyl aqueous solution as dispersion aqueous solution of the bubble (Solution A).

2).Dissolve 0.5 g of PLLA by 20 ml of methylene chloride (Solution B). Add 0.2 g of red fluorescent to 4 ml water (Solution C).

3).Mix two of the solutions above (Solutions B and C) and emulsify by using homogenizer at 10000 rpm. Then add the solution A which is high molecular solution (Solution D).

4).Pour hot water of 200 ml (100 °C) into solution D. Stir at 500 rpm to make precursors of capsule. At this time, PLLA membrane surrounding capsule is formed as a solid shell but the capsule is not hollow.

5).Freeze-dry the capsules to volatilize inner water.

Fabricated capsules by this method are shown in figure 2. The radius of the fabricated capsules mainly ranged from 1 to 35 μm, and average diameter is 15 μm and standard deviation is 7 μm. The cross section of capsule is shown in figure 2(b). The thickness of the capsule shell ranged from 500 nm to 1.5 μm. Several tens of thousands of capsules were fabricated through one procedure.

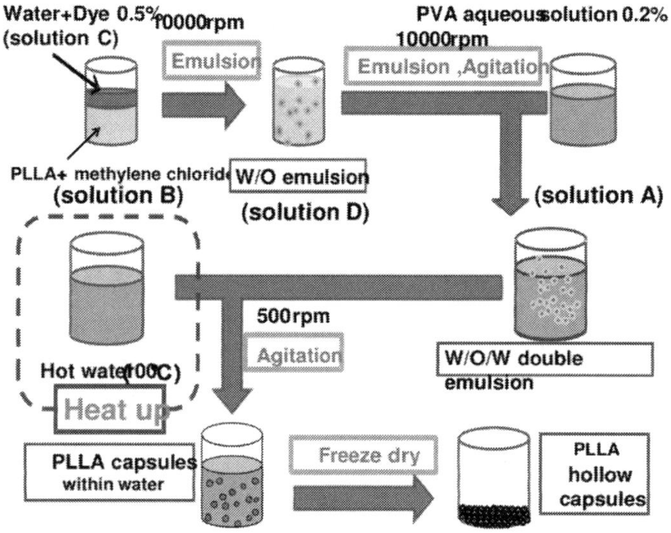

Figure 1. Fabrication procedure of PLLA capsules.

Figure 2. SEM image: (a) top sand (b) cross-sectional view.

3.Experimental method

The destruction characteristics of the fabricated capsule were investigated. Figure 3 shows the observational system. The capsule destruction was observed by an optical microscope (IX71 , OLYMPUS) with a high-speed camera. The radius distribution of the destroyed capsule was counted. Ultrasound transducers with the resonance frequencies of 700 kHz, 1 and 2 MHz were attached on an observational cell (30×30×30 mm^3) filled with degassed water. The capsules were attached on a cover glass at bottom of the cell. Figure 4 shows the sound pressure amplitude distribution at bottom of the cell at the frequency of 2 MHz. An ultrasound standing-wave field was generated and maximum sound pressure amplitude was approximately 400 kPa.

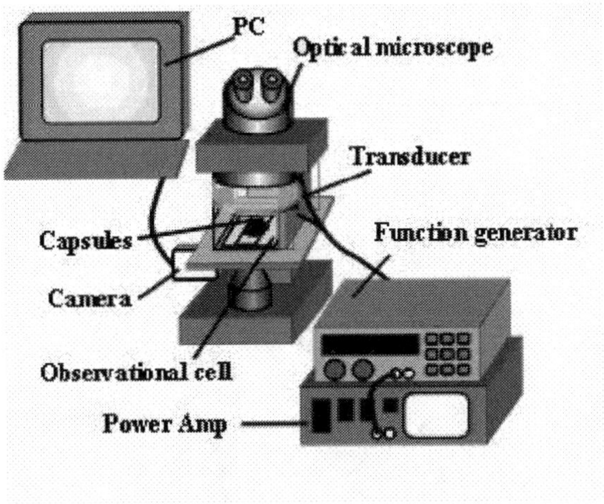

Figure 3. Experimental system

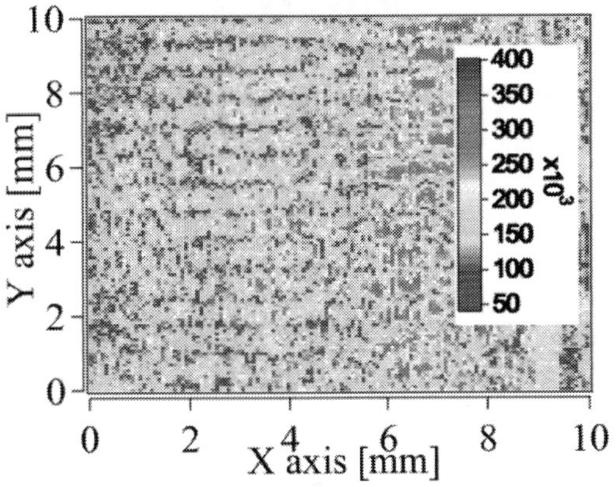

Figure 4. Sound pressure amplitude distribution at the bottom of the cell at 2 MHz

4.Results

Most of capsules appear black in figure 5(a), since the back light of the optical microscope refracts at the internal air of capsule. Comparing figure 5(a) which is the image of capsules before ultrasound irradiation, figure 5(b) clearly shows the capsules were broken by the ultrasound. Several capsules in figure 5(b) pass the light of optical microscope, since the broken capsules filled with water. In this report, we defined this change as capsule destruction. It should be noted that when the capsules have broken, there is little change in the shape of capsule surface because of the high rigidity of PLA. Figure 6 shows the destruction rate of the capsules at 700 kHz, 1 and 2 MHz. There is peak diameter for the rate of capsule destruction at each frequency: they are 15, 20, and 25 μm at 700 kHz, 1 and 2 MHz, respectively, and these results imply the destruction rate of the capsule is high at the resonance condition. Figure 7 shows the relationship between the capsule size and the resonance frequency. The smaller capsule has the higher resonance frequency. Considering the capsule is injected in human blood vessel with the size of approximately 10 μm, the resonance frequency of the capsule of 10 μm is estimated to be approximately 5 MHz. Figure 8 shows the sequential images of the capsule destruction. In the process of the capsule destruction, first, the surrounding shell of capsule was ruptured and the surrounding water flow into the capsule. And then the internal red fluorescent dye was solved and gradually released. Most of the dye was released from 75 to 120 s (see figure 8).

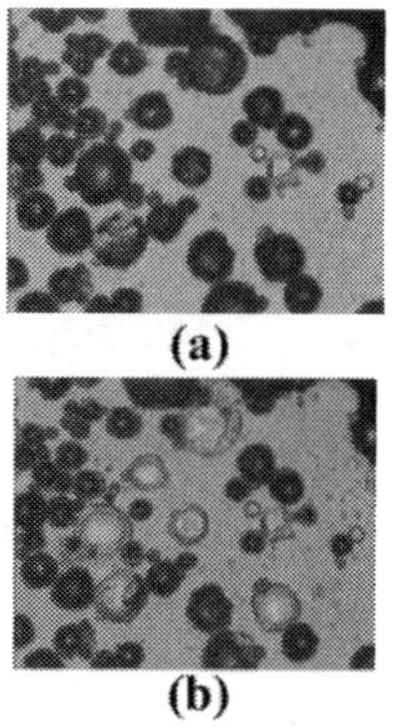

Figure 5. Optical microscope images of hollowing capsules. (a) before ultrasound irradiation (b) after ultrasound irradiation.

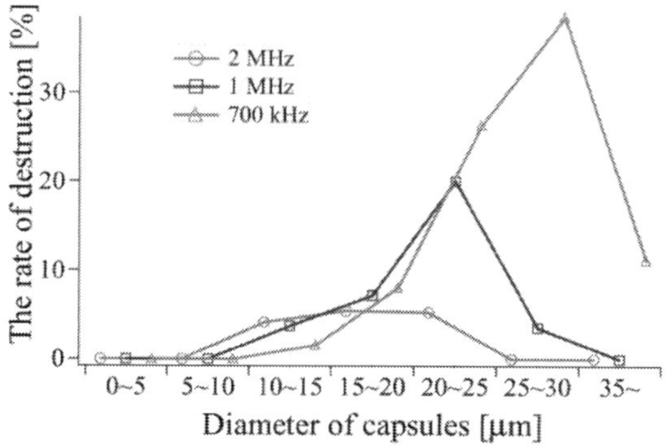

Figure 6. Destruction rate of capsules by the ultrasound irradiation of 400 KPa.

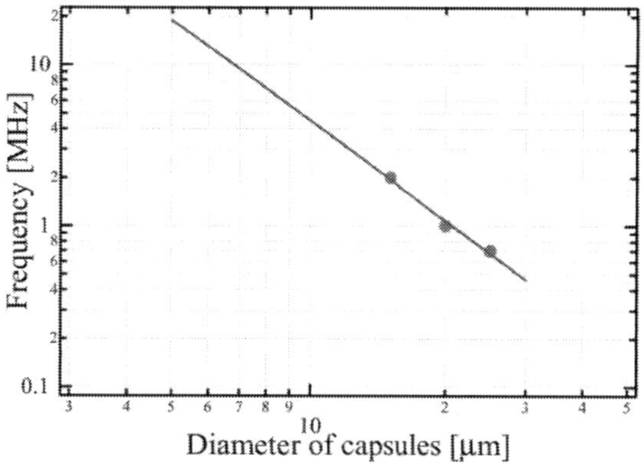

Figure 7. Relationship between diameter of capsule and the resonance Frequency.

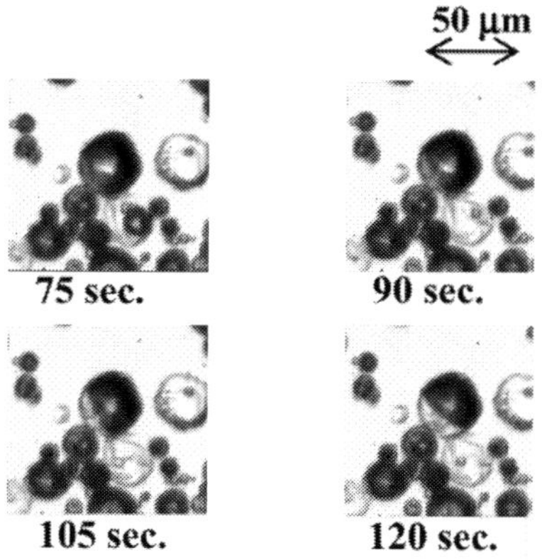

Figure 8. Optical image of capsule destruction and drug release.

5. Conclusion

We have succeeded in fabricating the capsules including hydrophilic substance. And frequency-destruction property become clear. The frequency dependence of the capsule destruction was investigated and 35% of the capsules with the resonance size was destroyed with 400 kPa at 700 kHz. The capsule shell was ruptured by the ultrasound, and then the inner dye was released; the dye was completely released for 120 seconds under ultrasound irradiation.

Acknowledgement

This work was supported by MEXT KAKENHI (22700517), Casio Sci. Promotion Foundation, JGC-S Scholarship Foundation.

References

[1] Maruyama K, Suzuki R *et al* 2007 *Yakugakuzasshi* **127** (5) 781-787
[2] Rong L, Hongjing D Yuanyuan Q, Dong Z, Kang S, Yuqi Z and Kun S 2009 Colloid Polym. Sci. **287** 683–693
[3] Daiguji H, Takada S, Jay J, Molino C and Fumio T 2009 J. Phys. Chem. B 113(45) 15002–09
[4] Makuta T, Takada S, Daiguji H and Takemura F *Materials Letters* **63** 703–705
[5] Kotera H, Kitazawa N *et al* 2010 *Proc. IEEE Ultrasonics Symp.* (San Diego,CA, 11-14 October 2010) pp 1708-11

International Symposium on Ultrasound in the Control of Industrial Processes (UCIP 2012) IOP Publishing
IOP Conf. Series: Materials Science and Engineering **42** (2012) 012013 doi:10.1088/1757-899X/42/1/012013

Characterisation of a multi-frequency cavitation vessel

L. Wang, G. Memoli, I. Butterworth, M. Hodnett and B. Zeqiri

National Physical Laboratory, Hampton road, Teddington, TW11 0LW, UK

E-mail: Lian.wang@npl.co.uk

Abstract. A multi-frequency reference vessel (RV-multi) has been established at NPL, with the ultimate aim of establishing a standard source of acoustic cavitation in water. The vessel is a cylindrical cavity and can work at 6 frequencies, from 21 kHz up to 140 kHz, in controlled temperature conditions. RV-multi is capable of self-tuning its driving frequency to lock onto a fixed modal pattern, based on the maximised phase derivative of the driving voltage signal. The ability of system self-locking was tested under various power levels. As part of the validation process for the new reference vessel, its performance has been investigated using various methods. A small hydrophone was used to determine the acoustic pressure distribution along selected planes of symmetry inside the vessel. The generated cavitation distribution was also monitored using sonoluminescence, erosion and acoustic emission. In a novel step, broadband laser vibrometry was used to measure the motion of the vessel wall to investigate the feasibility of a non-invasive method of cavitation detection.

1. Introduction

In order to produce a measurement standard to quantify acoustic cavitation process, a multi-frequency reference vessel system (RV-multi) has been installed at NPL, extending the work of previous investigations on a single frequency vessel [1],[2],[3]. The system was designed by NPL, manufactured by Sonic Systems, and covers most of the frequency range commonly used by industry and research organisation. The system can work over six frequency windows, with nominal centre frequencies of 20, 30, 40, 60, 90 and 120 kHz respectively. Once its metrological potential has been assessed, in terms of stability and reproducibility at the different frequencies, RV-multi will be used to evaluate and calibrate different types of sensing techniques, and vice versa. The protocols and results of this research may eventually be incorporated into a measurement standard.

This paper presents some preliminary results of the characterisation of RV-multi, obtained from different measurements, using established techniques and methods. A laser vibrometer was used to acquire information on the modal behaviour in the sonoreactor. The initial investigation was mainly concentrated on the acoustic field in the vessel over the entire frequency range at low or very low power levels before inception of cavitation: in these conditions, since the field is linear, stability and repeatability can be verified easily, both in terms of temporal and spatial variations. This process provided a good map of the field distribution within the vessel and highlighted the regions where it is suitable to deploy sensors for tests. The cavitation field, although highly non-linear and random, will develop with a very high degree of correlation from the linear acoustic field at its driving frequency.

IOP Conf. Series: Materials Science and Engineering **42** (2012) 012013 doi:10.1088/1757-899X/42/1/012013

2. The multi-frequency reference vessel system

The RV-multi as show in figure 1 is a cylindrical vessel made of stainless steel plate of 2 mm thickness, with internal diameter 269 mm and maximum water level 297 mm. The maximum capacity is about 18 litres of water.

Figure 1. The multi frequency reference vessel described in this work

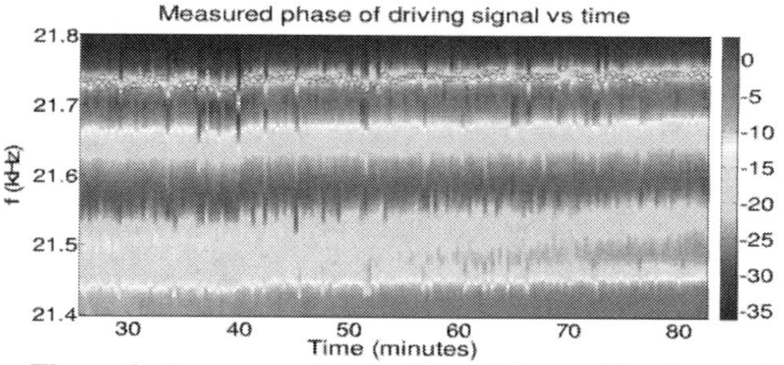

Figure 2. Frequency drift in RV-multi over 80 minutes, in absence of temperature control. Once a selected field distribution is chosen (one of the red areas in the above plot), the feedback loop tracks it (dotted points plot).

The bottom of the vessel is of a conical shape with a very small angle such that it can be considered flat. There are a total of 21 Tonpilz transducers aligned in three rows, with each row being made of 7 equally spaced transducers of equal size. The transducer size varies between the three rows and each row can work with up to 150 W of nominal power. The three rows are arranged with an equal depth increment between the bottom of the vessel and the water surface at the maximum level. The coordinate system is chosen here so that the origin is at the centre of the water surface at its maximum level and the X-axis is in the same direction of the acoustic axis of the transducer (T1) as shown on the top right side of Figure 1.

The driving voltage is generated by a lock-in amplifier (AMETEK 7230). A matching and tuning circuit is used for each of the six frequencies to provide a properly matched resistive load to an E&I 1040L power amplifier before the signal is delivered to one of the bank of the transducers. The lock-in amplifier receives current and voltage feedback from the impedance matching circuitry: these or other signals (such as the reading from a pressure sensor mounted at the bottom of the vessel or the phase derivative of the signal applied to the transducer), can be used to build a control loop setting the operating frequency. In this way, once a particular distribution of field is achieved in the vessel, this can be maintained over time (i.e. "tracked"), even when environmental factors may change.

One such factor is the temperature of the water inside RV-multi, which has been demonstrated to affect the field distribution significantly in previous studies [3]. Figure 2 shows how the tracking system works. The plot presents the phase (colour scale) between the current and voltage of the driving signal, as acquired by the lock-in amplifier. The signal was a frequency sweep from 21.4 to 21.8 kHz. It is noticed that the phase of the signal drifts over the 80 minutes of this test due to a temperature change, in the absence of a temperature control loop (0.5 °C increase with 3W operating power). The lock-in amplifier measures the phase-derivative and selects the frequency of operation based on the maximum of the phase-derivative, which follows the drift (see dots in figure 2). Under these conditions, the field distribution inside the vessel should remain constant.

In addition to the tracking system, a forced convection, air-based temperature control loop was incorporated in the system, to maintain temperature within ±1 °C during operation, even with high power levels. This should ensure a stable and reproducible field inside RV-multi and, consequently, a known cavitation distribution.

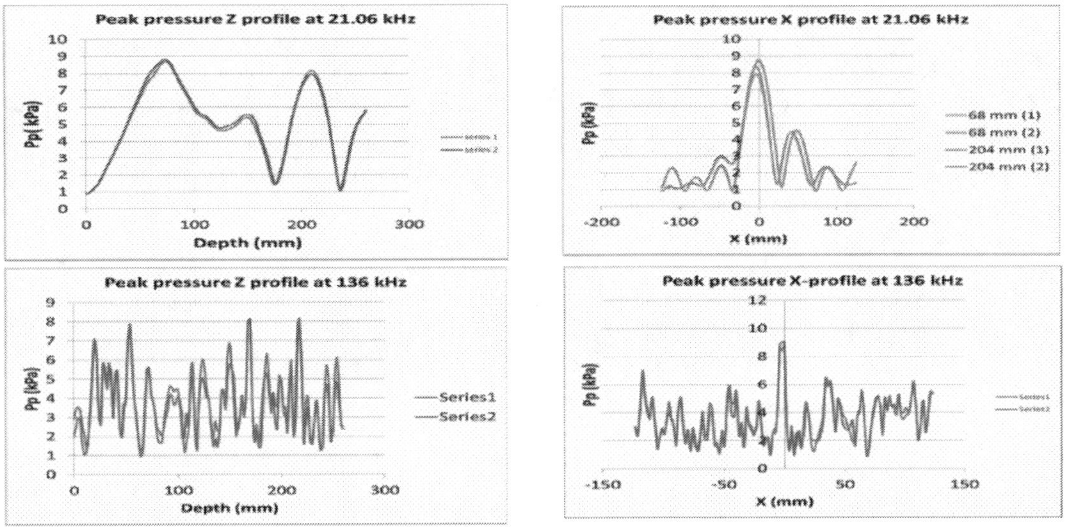

Figure 3. Measured sound pressure field profiles: axial profiles (left), X-profiles (right) along the acoustic axis direction of the left most transducer at various depths for two frequencies, 21 and 136 kHz.

3. Experimental results

A calibrated Reson 4038 hydrophone was used to measure the field distribution along pre-selected tracks. The acoustic measurements were made along the Z-axis (Z-scan) first to locate the peaks of the pressure profile in the vertical direction. Then X-scans in the direction of the acoustic axis of T1 transducer were carried out to obtain the pressure profile at a selected depth. Two consecutive scans were taken to illustrate the quality of the measurements as shown in figure 3. The plots in figure 3 represent the Z-scans (left) and the X-scans (right) at 21 and 136 kHz, with the X-scans taken at different depths (2 depths for 21 kHz and 1 for 136 kHz).

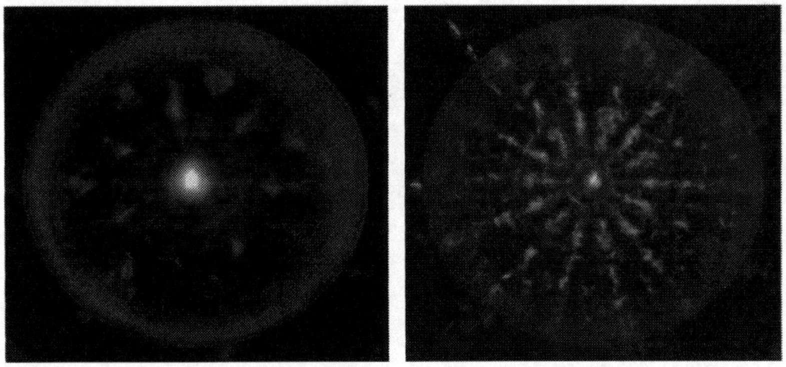

Figure 4. Sonoluminescence inside the RV-multi at 21 (left) and 44 kHz (right).

Sonoluminescence images in figure 4 were taken from water-filled RV-multi, with 20 g of luminol and 400 g of sodium carbonate, at a power level of 100W (well above the power needed for cavitation) and with an exposure time of 1 minute. They show the top-view of the generated cavitation fields at two frequencies, 21 and 44 kHz. The presence of high cavitation activity can be revealed in the areas where blue light spots appear, the higher the cavitation intensity, the brighter the light.

A scanning laser vibrometer (Polytec PSV400-M) was used in the frequency domain to examine different portions of the outer surface of the cell during operation, monitoring the local vibrational velocity in mm/s. Figure 5 shows the velocity distribution on some areas of the vessel at 21 and 136 kHz, confirming their modal component.

IOP Conf. Series: Materials Science and Engineering **42** (2012) 012013 doi:10.1088/1757-899X/42/1/012013

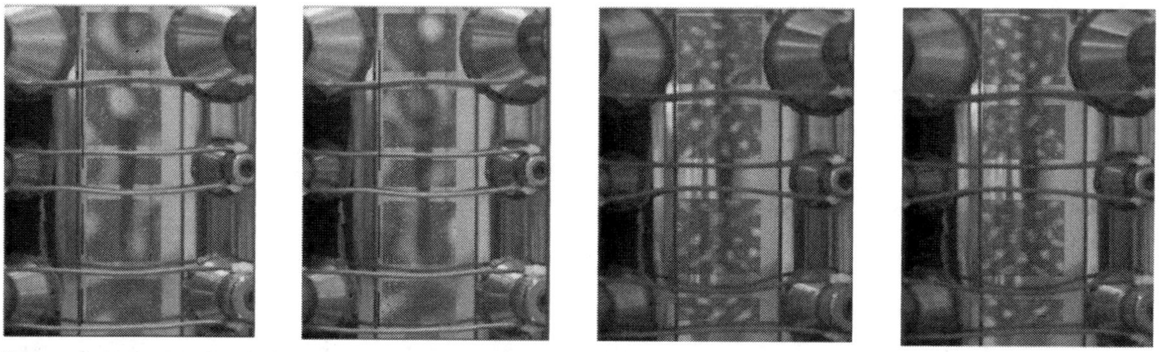

Figure 5. Velocity distribution at 21 and 136 kHz on the wall of RV-multi vessel: the two subsequent images report a difference in phase of 180°.

4. Conclusions

Various measurements were performed with the NPL's RV-multi vessel in order to determine its characteristics in terms of acoustic field distribution and over the entire available range of frequency. The results presented here illustrate different aspects of the field inside the vessel. Further work will be carried out in order to achieve the aim of realising fully stable, controllable and repeatable experimental conditions for cavitation process.

References

[1] Hodnett M, Choi MJ and Zeqiri B 2007 Toward a Reference Ultrasonic Cavitation Vessel: Part 1—Preliminary Investigation of the acoustic field distribution in a 25 kHz cylindrical cell," *Ultrasonics Sonochem.* **14** 29-40

[2] Hodnett M and Zeqiri B 2008 Toward a Reference Ultrasonic Cavitation Vessel: Part 2—Investigating the Spatial Variation and Acoustic Pressure Threshold of Inertial Cavitation in a 25 kHz Ultrasound Field *IEEE transactions on ultrasonics, ferroeelectrics, and frequency control* **55** 1809-22

[3] Memoli G, Gelat PN, Hodnett M and Zeqiri B 2012 Characterisation and improvement of a reference cylindrical sonoreactor *Ultrasonics Sonochemistry* **19** 939–952

IOP Conf. Series: Materials Science and Engineering **42** (2012) 012014 doi:10.1088/1757-899X/42/1/012014

Non-contact transportation system of small objects using Ultrasonic Waveguides

K Nakamura and D Koyama[1]
Precision and Intelligence Laboratory, Tokyo Institute of Technology,
R2-26, 4259 Nagatsuta, Midori-ku, Yokohama 226-8503, JAPAN

E-mail: knakamur@sonic.pi.titech.ac.jp

Abstract. A transportation system for small object or fluid without contact is investigated being based on ultrasonic levitation. Small objects are suspended against gravity at the nodal points in ultrasonic pressure field due to the sound radiation force generated as the gradient of the energy density of the field. In this study, the trapped object is transported in the horizontal plane by introducing the spatial shift of the standing waves by the switching the lateral modes or travelling waves. The goal of the study is to establish a technology which can provide a total system with the flexibility in composing various transportation paths. Methods for linear/rotary stepping motions and continuous linear transportation are explained in this report. All the transportation tracks are composed of a bending vibrator and a reflector. The design for these acoustic cavity/waveguide is discussed.

1. Introduction

Non contact technology is highly demanded in the next generation process for chemical, pharmacy and food industries as well as production process in microelectronics and precision machinery. Small objects or droplet are to be transported without contact to avoid contamination. Conventional magnetic levitation is not applicable for non-magnetic materials. Suspension technique by air flow is expensive method and easy to induce dust. Authors are focusing on ultrasonic levitation technology, as an alternative method, where small objects or droplet [1] can be trapped at the nodal points of the sound pressure field in air due to acoustic radiation force. To compose a total transportation system in practical use, various kinds of transportation components are required as illustrated in figure 1. The most basic one is linear transportation track. Joint or switch between two transportation tracks are essential for practical use. Elevation of position shall be needed in addition to the horizontal movement. Some of the concepts to realize these variations of the transportation paths using ultrasonic levitation are explained in this report. In order to transport the objects without contact in horizontal direction, the authors have introduced the spatial shift in standing wave field between a vibrating plate and a reflector. The position of the trapped object is moved step by step if the lateral mode pattern is shifted by changing the phase difference between two driving points. The use of travelling wave is also investigated for continuous linear transportation. Design for the travelling wave acoustic waveguide is discussed.

---

[1] Currently, he is at Doshisha University, Kyoto, Japan.

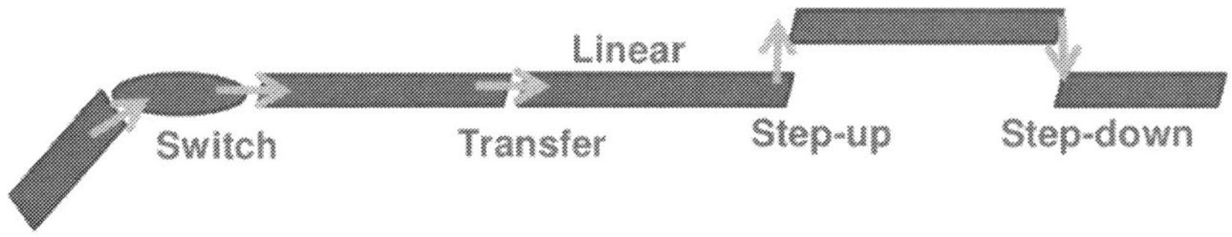

Figure 1. Concept of non contact transportation system for small objects.

2. Stepping motion using standing waves

2.1. Basic levitation phenomena

Basic levitation setup is composed of a vibrating plate and a reflecting plate as shown in figure 2. The bottom plate is vibrating in the bending mode and standing wave of pressure field is excited between the two plates. Small objects such as a polystyrene ball and a piece of tablet can be suspended at the nodal position due to the radiation force. The radiation force is acting on the object in accordance with the gradient of the energy density of the sound field.

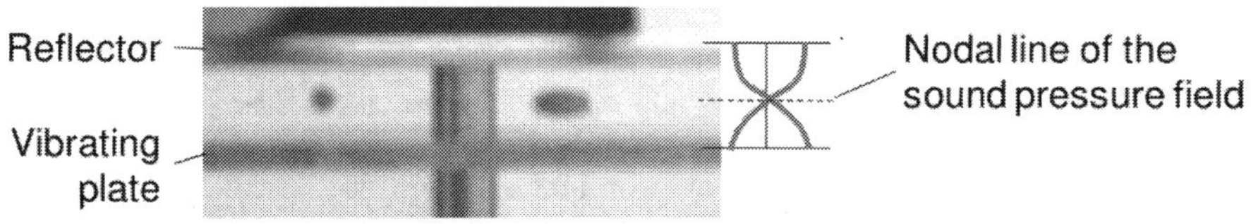

Figure 2. Levitation of small objects between the vibrating plate (lower) and the reflector (upper). A small polystyrene ball (left) and a tablet (right) are trapped at the nodal point of the sound pressure field against the gravity.

2.2. Linear transportation with stepping motion

The horizontal position of trapping objects is determined by the sound field distribution. Then, the position is moved if the excitation location of the standing waves is shifted. In the linear transportation system, this can be practically achieved by changing the phase difference between two driving point at the vibrating plate [2] as shown in figure 3. White circles indicate the polystyrene balls levitated in the standing wave field of sound pressure. If the phase difference between the voltages applied to Langevin transducers connected at the left and right ends of the plate is changed from 0 to 60 degrees, the sound pressure pattern is moved rightward by the distance corresponding to the phase shift. As a result, the position of the objects is also shifted.

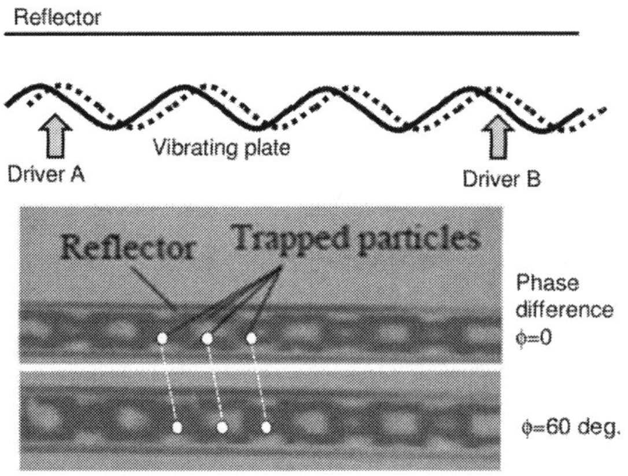

Figure 3. Principle for moving the suspended objects along the horizontal plane being based on the shift in the position of standing waves.

2.3. Rotary transportation with stepping motion

Rotary motion can be achieved by utilizing the same principle. Sound field pattern is rotated by switching the driving position instead of the phase shift. Figure 4 shows an example using a disk vibrator and a disk reflector [3]. Higher order mode of bending vibration is excited using a annular PZT element bonded on the back surface. The electrode on the PZT element is divided into many parts. Angular position of the mode pattern is determined by selecting the electrode to be activated. Small object trapped between the disk vibrator and the reflector is transported along angular direction if the stimulation position is switched one after another.

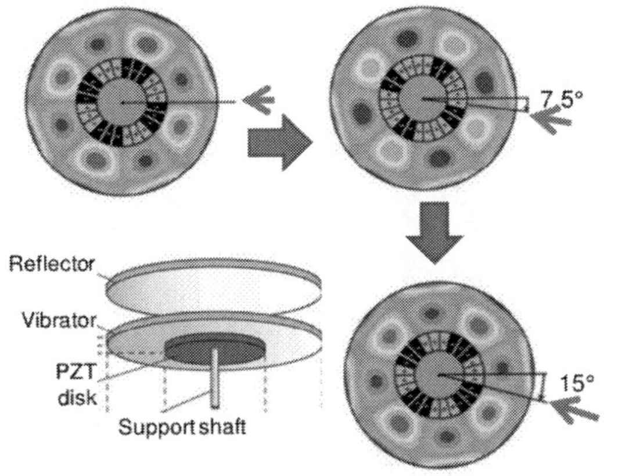

Figure 4. Method for rotary motion. Vibration mode pattern is rotated by switching the activated electrode.

3. Continuous linear transportation using travelling waves

In this section we discuss about continuous linear transportation of object at high speed using travelling waves. Let us consider sound pressure field between parallel rigid reflectors using figure 5. Here, the case with two nodal lines is investigated. The spacing between the reflectors W is 11 mm and the frequency is 37 kHz in our experiments. The relationship among the spacing W, the free-space wavelength λ_0 and the angle θ for propagation is denoted as

$$\lambda_0 = W \sin\theta \tag{1}$$

where λ_0=9.2 mm. The angle of the incidence was set to 60 degrees in accordance with equation (1) as shown in figure 6, where a commercial airborne ultrasonic radiator (Murata co.) was utilized. In the figure, the sound pressure distribution measured using a probe microphone is overlapped. Red indicates higher pressure, while blue shows zero. Horizontal distribution was not uniform, but since the phase changed almost linearly with the distance as shown in figure 7, travelling wave field was obtained. Along the horizontal nodal lines, polystyrene particle were transported. The distance of transportation was limited under 20-40 mm because of the irregular pressure distribution and the low pressure level. The sound field was attenuated as the distance as can be seen from figure 6.

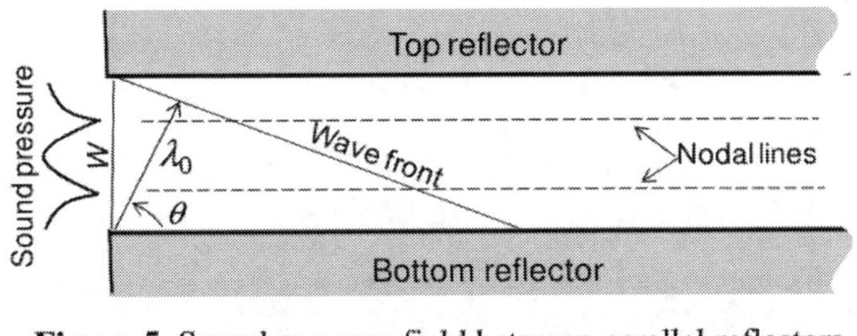

Figure 5. Sound pressure field between parallel reflectors.

In order to obtain higher sound pressure, the radiator was replaced by a combination of a bolt-clamped transducer and a vibration plate of 65 mm x 30 mm. The frequency was 25.7 kHz. The moving speed of the transported object is plotted as a function of the distance in figure 8, which is calculated from the trajectory recorded using a video camera. The object was accelerated and reached to the velocity of approximately 0.5 m/s. The transportation distance was successfully extended. The traction force acting on the polystyrene ball of 2 mm in diameter and 0.3 mg in weight is calculated from the acceleration to be 3.5 μN.

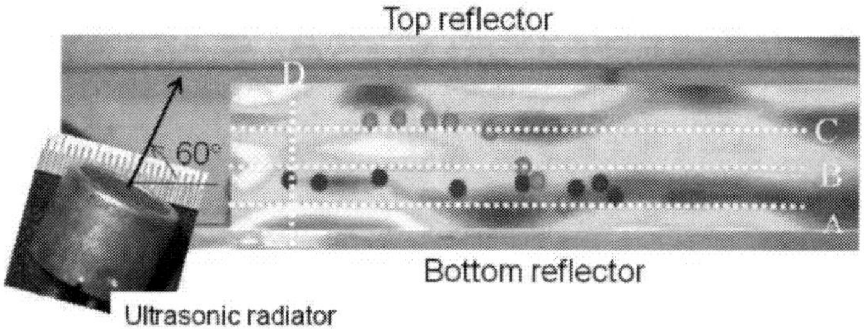

Figure 6. Measured sound pressure field and the trajectory of small object.

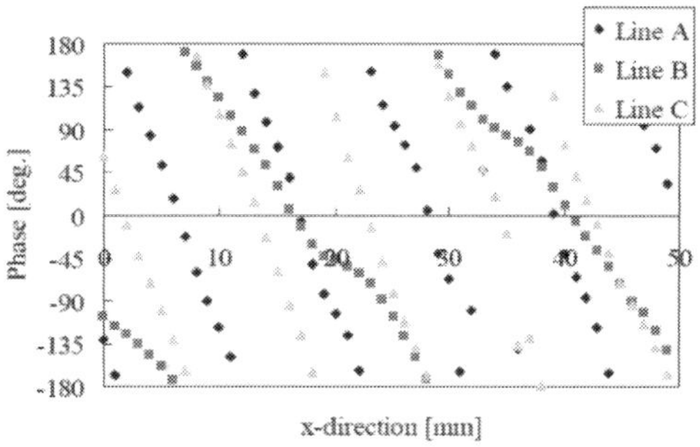

Figure 7. Phase distribution along the lines A, B and C in figure 6.

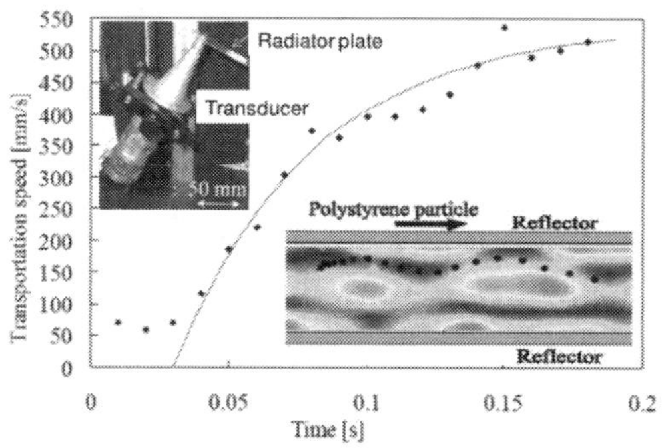

Figure 8. Temporal change in the transportation speed.

4. Conclusions

Non-contact transportation technique was discussed being based on ultrasonic levitation in standing wave field. Stepping motion in horizontal and rotational directions was achieved with the phase shift or switching the excitation position. Continuous linear transportation was realized using travelling sound field, which was obtained by oblique incidence to the waveguide between the parallel reflectors. Using the same setup, droplet can be transported without contact.

Acknowledgments

A part of this study has been supported by Grants-in-Aid for Scientific Research (Challenging Exploratory Research, 23656163), JSPS.

References

[1] Koyama D and Nakamura K 2010 *IEEE Trans. Ferroelect., Freq. Contr.* **57** 1152
[2] Koyama D and Nakamura K 2010 *IEEE Trans. Ferroelect., Freq. Contr.* **57** 1434
[3] Otsuka T and Nakane T 2002 *Japan. J. Appl. Phys.* **41** 3259

A numerical and experimental study of ultrasonic metal welding

Z Al-Sarraf, M Lucas and P Harkness
School of Engineering, University of Glasgow, Glasgow, G12 8QQ, UK

E-mail: z.sarraf.1@research.gla.ac.uk, Margaret.Lucas@glasgow.ac.uk

Abstract. Ultrasonic metal welding has been the subject of ongoing research and development, most recently concentrating on metal joining in miniature devices, for example to allow solder-free wire bonding. As well as at the small scale, there are also opportunities to research the joining of thicker sheet metals and to widen the range of similar and dissimilar materials that can be successfully joined using this technology. This study presents the design, characterisation and test of a lateral-drive ultrasonic metal spot welding device. The ultrasonic metal spot welding horn is modelled using finite element analysis (FEA) and its vibration behaviour is characterised experimentally to ensure ultrasonic energy is delivered effectively to the weld coupon. The welding stack and fixtures are then designed and mounted on a test machine to allow a series of experiments to be conducted for various welding and ultrasonic parameters. Weld strength is subsequently analysed using tensile-shear tests. The results show how the weld strength is particularly sensitive to the combination of clamping force and ultrasonic vibration amplitude of the welding tip, but there are optimal combinations of these and also limits that must be clearly identified.

1. Introduction

Ultrasonic welding can be defined as a solid-state joining process in which materials are held together by a normal force whilst a high-frequency shear vibration is applied. During this process the vibration deforms, shears, and flattens surface asperities, scatters interstitial oxides and contaminants, and increases the contact area of the parts being welded [1]. The first demonstration of ultrasonic welding was in the early 1950s and was limited to grain refinement and soldering [2], but now the technique can be applied to various softer metals, such as copper and aluminium, as well as harder metals [3]. Ultrasonic welding has become an efficient joining technique for many industrial and scientific applications, using lighter and more versatile equipment to produce a stronger, smaller, and more precise weld [2]. Furthermore, ultrasonic welding does not require any solder or filler and therefore has some associated environmental and economic benefits [4].

Ultrasonic welding systems consist of a power supply, transducer, booster, horn and anvil. The horn is tuned to operate in the longitudinal mode but imprecise design can affect the dynamic characteristics of the device, reducing both vibration amplitude and weld quality. In this study, two investigations have been carried out to improve weld strength; numerical design and subsequent experimental characterisation of an integrated ultrasonic spot welding horn to enhance the vibration characteristics at the welding surface, and secondly an experimental study of the effects of process parameters on the weld itself, considering issues such as tool/workpiece adhesion and weld quality. The weld strength is characterised experimentally in terms of the results of repeated tensile shear tests.

2. Design and characterisation of the welding system

2.1 Numerical design of the ultrasonic spot welding horn

A lateral-drive system has been used to investigate ultrasonic metal spot welding because it can produce high amplitude vibration but yet apply the low forces necessary to weld thinner parts [5]. Several factors are considered in the design of such an ultrasonic horn including resonant frequency, frequency separation, amplitude amplification, amplitude uniformity and stress concentration. High amplification and low stress are obtained through the use of a catenoidal horn, the horn being modelled as steel to take advantage of its high fatigue strength and good wear qualities. An integrated welding tip with knurled welding flats 6 mm in diameter is employed, the horn being clamped at the nodal plane as shown in figure 1. This horn is modelled in the FEA package, Abaqus, and a harmonic analysis is carried out through the application of an excitation vibration to the base, to extract the shape of the operating vibration mode and those of some of the surrounding modes.

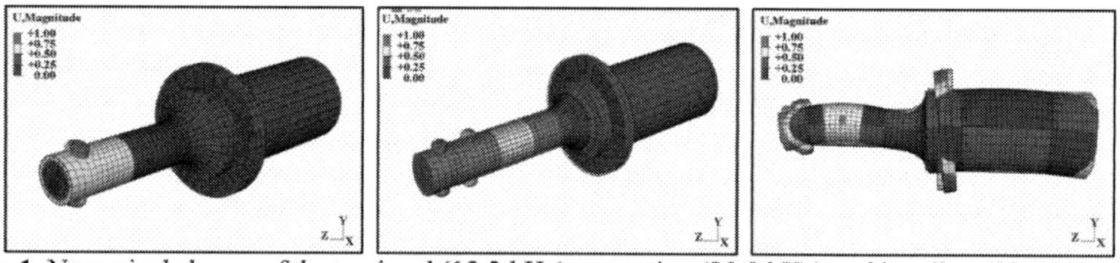

Figure 1. Numerical shapes of the torsional (13.2 kHz), operating (20.8 kHz) and bending (22.6 kHz) mode.

2.2 Experimental characterisation of the ultrasonic spot welding horn

The numerical results are compared to the results of an Experimental Modal Analysis (EMA), shown in figure 2, carried out using a Polytec 3D-Laser Vibrometer and random excitation over a frequency range from 0 to 40 kHz. The amplifications, resonant frequencies and mode shapes of the horn are found to be in agreement and the system is therefore considered to operate as designed.

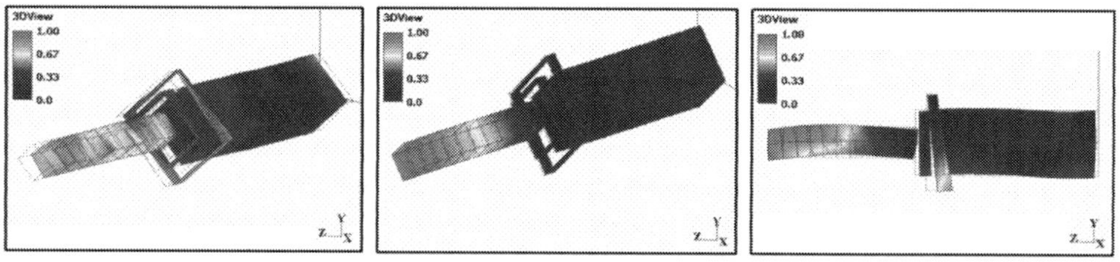

Figure 2. EMA shapes of the torsional (13.7 kHz), operating (20.8 kHz) and bending (23.5 kHz) mode.

3. Welding Experiments

3.1 Experimental set up

A 1 kW ultrasonic generator (Sonic Systems L500/3-20) is used to drive the transducer and hence the welding horn, known together as the welding stack. A tensile test machine (Zwick Roell) is employed to secure the stack by holding it at the nodal plane, as shown at left in figure 3 (a), and also to measure the clamping force and weld time. The 6 mm knurled flat contacts the upper specimen during welding, while the lower specimen is secured by a knurled anvil as shown in figure 3 (b). The specimens themselves are 0.1 mm and 0.5 mm thick aluminium and copper strips, as shown in figure 3 (c), and their properties are obtained from ASTM [6] and BSI [7] codes. Five tensile tests at each set of weld conditions are carried out, the results are averaged, and the standard deviation for the tests is obtained. Tests of both similar and dissimilar metals and of different thicknesses, are performed with different arrangement or stacking order. Several tests are also carried out to ameliorate the problem of horn sticking.

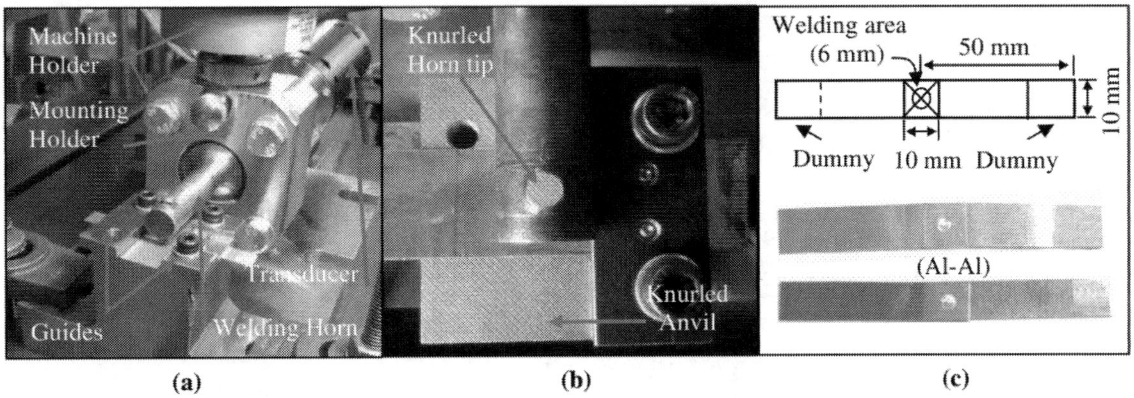

Figure 3. Experimental set up of welding tools, knurled surfaces, specimen layout and welded coupons.

3.2 The effect of process parameters on weld strength in similar materials

Figure 4 shows the average measured values of weld strength against clamping force applied for various welds with different vibrational amplitudes and specimen thicknesses. The error bars represent one standard deviation of the five tests for each parameter set. Welding strength is seen to increase with clamping force although it is noted that excessive clamping force may generate high friction and suppress the relative motion of the surfaces, resulting in reduced weld strength [8]. Al-Al welds appear to be slightly stronger than Cu-Cu welds created under identical process parameters, regardless of sample thickness. Scattering of weld strength across the five tests tends to become proportionally smaller as clamping force is increased.

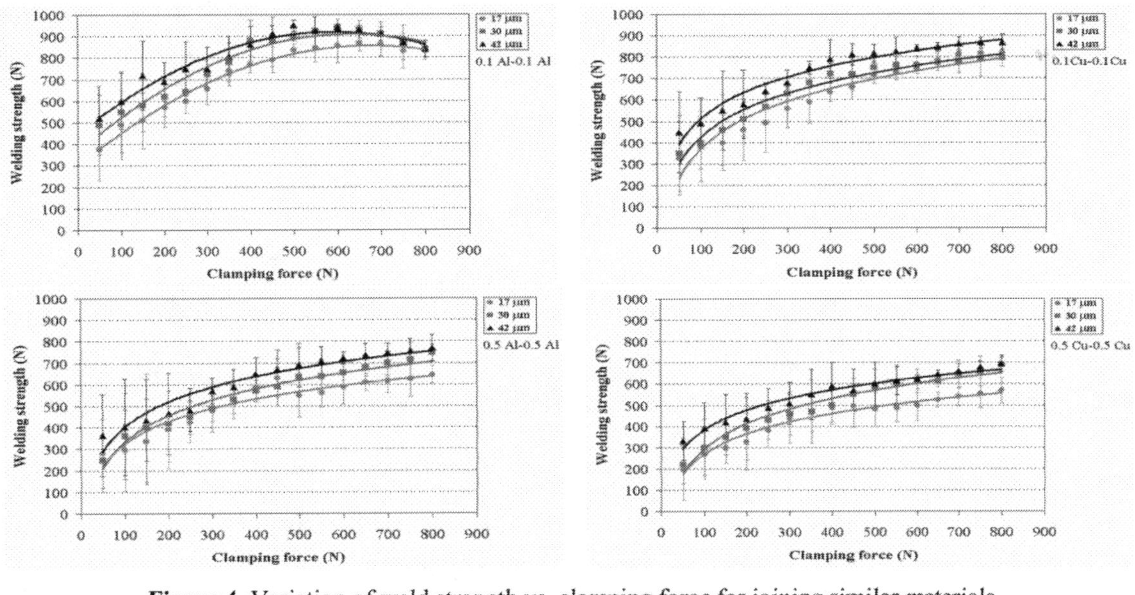

Figure 4. Variation of weld strength vs. clamping force for joining similar materials.

3.3 The effect of process parameters on weld strength in dissimilar materials

In welding dissimilar materials it is important to determine which strip is placed on top, as the results in figure 5 suggest a superior bond is created when the horn bears against aluminium. It is also noted that sticking becomes more prevalent at higher clamping forces, particularly where thinner samples are employed.

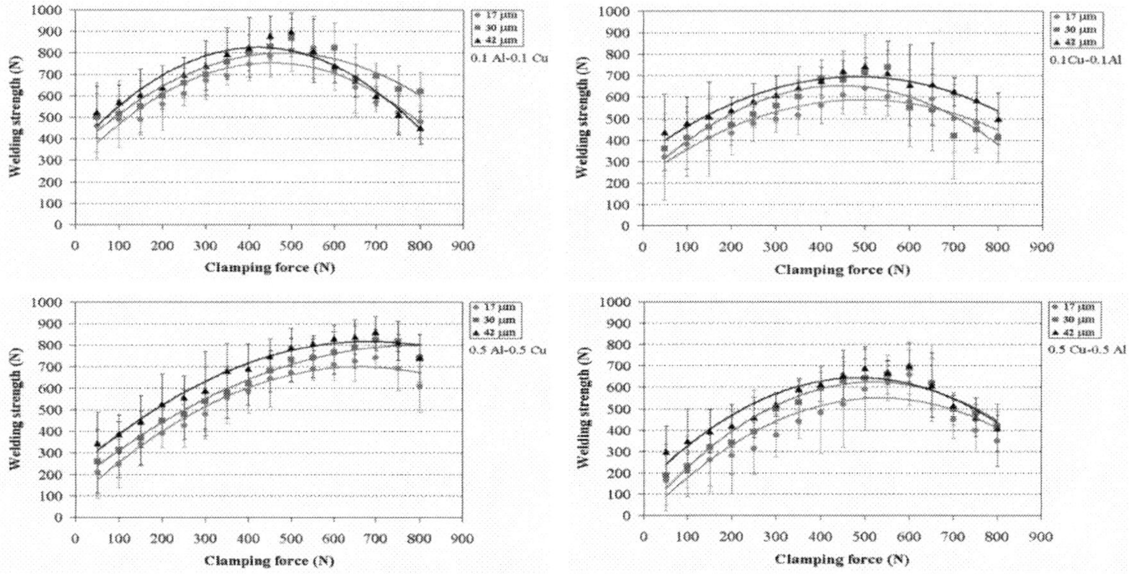

Figure 5. Variation of weld strength vs. clamping force for joining dissimilar materials.

4. Conclusions

Design and fabrication of a lateral-drive ultrasonic spot welding system has been carried out to investigate the welding of thin metal strips. It has been observed that vibration amplitude, clamping force and, in some circumstances, material arrangement order, have a significant effect on weld strength. Al-Al welds are stronger than Cu-Cu welds and weld strength in both cases tends to increase with clamping force within the range of forces examined. However, when dissimilar metals are considered, slightly stronger welds are obtained when the aluminium layer is placed on top and in direct contact with the ultrasonic horn and, in general, there is a decrease in weld strength when clamping forces above approximately 500 N are applied.

References

[1] Graff K and Bloss M 2008 Understanding ultrasonic welding http://www.thefabricator.com/article/arcwelding/understanding-ultrasonic-welding

[2] Ahmed N 2005 *New Developmets in Advanced Welding* (Florida:CRC Press) p 298

[3] Bloss M and Graff K 2009 Ultrasonic metal welding of advanced alloys: the weldability of stainless steel, titanium, and nickel-based superalloys *Trends in Welding Research. Proceedings of the 8th International Conference* (ASM International)

[4] Kim S, Jang H, Kim E and Park D 2010 Vibration analysis of ultrasonic metal welding horn for optimal design. *Proc. Int. Conf. Mechanical, Industrial and Manufacturing Technologies (MIMT)* (Sanya, China, 22-24 January 2010)

[5] Staff PDL 2008 *Handbook of Plastics Joining* (William Andrew Publishing Plastics Design Library)

[6] ASTM International Codes 2009 *Standard Test Methods for Tension Testing of Metallic Materials* pp 1-24

[7] British Standard Codes 2009 *Test Pieces and Test Methods for Metallic Materials for Aircraft. Metric Units* pp 1-7

[8] Hiraishi M and Watanabe T 2003 Improvement of ultrasonic weld strength for Al-Mg alloy by adhesion of alcohol *Quarterly Journal of the Japan Welding Society* **21**(2) 295-301

IOP Conf. Series: Materials Science and Engineering **42** (2012) 012016 doi:10.1088/1757-899X/42/1/012016

Bubble size distribution analysis and control in high frequency ultrasonic cleaning processes

M Hauptmann[1,2], H Struyf[1], P Mertens[1], M Heyns[1,3], S De Gendt[1,4], C Glorieux[2] and S Brems[1]

[1] IMEC vzw, Kapeldreef 75, B-3001 Leuven (Belgium)
[2] Laboratorium voor Akoestiek en Thermische Fysica, Katholieke Universiteit Leuven, Celestijnenlaan 200D, B-3001 Leuven (Belgium)
[3] Departement Metaalkunde en Toegepaste Materiaalkunde, Katholieke Universiteit Leuven, Kasteelpark Arenberg 44, B-3001 Leuven (Belgium)
[4] Afdeling Moleculair Design en Synthese, Katholieke Universiteit Leuven, Celestijnenlaan 200F, B-3001 Leuven (Belgium)

E-mail: Marc.Hauptmann@web.de

Abstract. In the semiconductor industry, the ongoing down-scaling of nanoelectronic elements has lead to an increasing complexity of their fabrication. Hence, the individual fabrication processes become increasingly difficult to handle. To minimize cross-contamination, intermediate surface cleaning and preparation steps are inevitable parts of the semiconductor process chain. Here, one major challenge is the removal of residual nano-particulate contamination resulting from abrasive processes such as polishing and etching. In the past, physical cleaning techniques such as megasonic cleaning have been proposed as suitable solutions. However, the soaring fragility of the smallest structures is constraining the forces of the involved physical removal mechanisms. In the case of "megasonic" cleaning –cleaning with ultrasound in the MHz-domain – the main cleaning action arises from strongly oscillating microbubbles which emerge from the periodically changing tensile strain in the cleaning liquid during sonication. These bubbles grow, oscillate and collapse due to a complex interplay of rectified diffusion, bubble coalescence, non-linear pulsation and the onset of shape instabilities. Hence, the resulting bubble size distribution does not remain static but alternates continuously. Only microbubbles in this distribution that show a high oscillatory response are responsible for the cleaning action. Therefore, the cleaning process efficiency can be improved by keeping the majority of bubbles around their resonance size. In this paper, we propose a method to control and characterize the bubble size distribution by means of "pulsed" sonication and measurements of acoustic cavitation spectra, respectively. We show that the so-obtained bubble size distributions can be related to theoretical predictions of the oscillatory responses of and the onset of shape instabilities for the respective bubbles. We also propose a mechanism to explain the enhancement of both acoustic and cleaning activity observed when turning the ultrasound periodically on and off in comparison to the case of continuous sonication.

1. Introduction

The ongoing downscaling of nanoelectronic devices leads to more, stringent requirements for the processes involved in their fabrication. For instance, there is a need for the removal of nano-particulate contamination that results from various abrasive fabrication processes and resides at the

IOP Conf. Series: Materials Science and Engineering **42** (2012) 012016 doi:10.1088/1757-899X/42/1/012016

semiconductor substrate surface. [1] Conventional wet-chemical removal is not applicable anymore, as the underetching that is needed for particle lift-off exceeds the allowed tolerances for substrate losses by far [2]. However, when particle removal is assisted by nano-mechanical forces acting on the particles, substrate losses can be reduced significantly.

One way to introduce those nano-mechnical forces is through the application of high-frequency ultrasound to the cleaning liquid ("Megasonic Cleaning"). The cleaning action accumulating during sonication is strongly linked to the occurrence of acoustic cavitation, which is the existence of oscillating microbubbles in the cleaning liquid [3, 4, 5]. However, the pressure amplitude of the driving sound field has to be kept below a certain limit in order to prevent damage to the fragile nano-structures.

Unfortunately, at lower acoustic intensities the inception of acoustic cavitation in the cleaning liquid is cumbersome. Also the size distribution of the emerging bubbles has to be kept around the active bubble radii (typical μm), where bubbles respond efficiently to the acoustic driving. Pulsing the acoustic field, meaning turning the acoustic driving periodically on and off, can lead to a dramatic enhancement of cavitation activity as characterized by acoustic cavitation noise measurements [6]. The pulse parameters (Pulse-on time *POnT* and Pulse-off time *POffT*) at which this enhancement is observed, suggest that it is driven by the partial dissolution and ensuing reactivation of active bubbles between subsequent pulses.

The correlation between the dissolution behavior of a bubble and the deterioration of cavitation activity measured by means of sonoluminescence has already been used to map out active bubble size distributions [7]. However, this analysis is limited to the case of very low POnTs and very high acoustic pressure amplitudes, where the phenomenon of sonoluminescence occurs and no signal enhancement with the respect to continuous sonication can be observed.

In this paper, we present a modified approach that is adapted to our process conditions (longer pulse durations and lower acoustic intensity). Cavitation activity is measured by means of ultraharmonic cavitation noise (UHCN) [6, 8] and particle removal efficiency (PRE) [9]. A model distribution is fitted to the experimental cavitation activity data which is obtained as a function of the POffT. An upper dissolution threshold is introduced as an additional fit parameter to represent the necessary partial dissolution of the active bubbles. Finally we present a phenomenological model to explain and quantify the enhancement mechanism resulting from the recycling of active cavitation bubbles in between subsequent acoustic wavetrains.

2. Experimental approach and results

2.1. Setup and Experimental Details

A detailed description of the setup, the experimental approach and the mathematical methodology will be given in [10]. All experiments were carried out in a megasonic cleaning tank with a volume of 13 l. The liquid was sonicated using an ultrasonic transducer (Prosys, 928 kHz driving frequency, 60 cm^2 surface area) with (cleaning experiments) and without (cavitation noise measurements) a Silicon wafer under traveling wave conditions [11]. The liquid was degassed and subsequently regassed to 105% oxygen saturation using a set of degas/regas phasors (Entegris, Phasor II). The liquid recirculation flow was set to 6.5 l/min. UHCN was measured with a hydrophone (HNR-500, Onda), subsequently amplified (homebuilt amplifier) and captured using an oscilloscope (Hameg, HNO2524) [6,8]. Silicon wafers with a diameter of 300 mm were contaminated with 78 nm silica particles (10^6 particles / cm^2). Particle concentrations were measured before and after cleaning using a light scattering technique [12] to yield PREs. To improve uniformity, the wafers were scanned vertically during sonication.

2.2 Active and "cleaning" bubble size distributions

The determination of the active bubble size distributions involves the measurement of the UHCN as a function of the POffT at a constant POnT of 50 ms. Next, a model distribution is fitted to the data as described in [10]. For the determination of the "cleaning" bubble size distribution a similar approach is used. The wafers are reused multiple times, as cleaning is confined to small areas corresponding to the size of the transducer (+scanning distance).

Figure 1 displays PRE maps for different POffTs. There is a clear enhancement of cleaning activity with respect to continuous sonication (POffT 0 ms) for certain POffTs. The average PRE over the (projected) transducer area as well as the corresponding UHCN signal are given in figure 2(a) as functions of the PoffT. The bubble size distributions displayed in figure 2(b) correspond to the model fits in figure 2(a). Both distributions are centered around the linear bubble resonance size R_{res} (3.4 µm). The upper dissolution thresholds R_{thres} are located at the left side of the distributions, indicating that bubbles have to dissolve at least partially out of the active bubble size distributions during the POffT before they can be reactivated again. For long POffTs (corresponding to large bubble radii), a very good correspondence between cleaning and cavitation noise can be observed. However, cleaning action extends to shorter POffTs (corresponding to smaller bubble radii) than acoustic activity due to changed acoustic boundary conditions for bubbles close to the wafer surface [10].

2.3 *Enhancement mechanism*

The enhancement mechanisms can be modeled by summarizing the different mechanism that contribute to cavitation activity as described in [10]. The first mechanism involves the reactivation of partially dissolved bubbles due to the growth of bubbles towards resonance size. The second mechanism involves the deactivation of bubbles due to growth of the active bubbles "out of resonance". When the POffT is kept constant (e.g. to 150 ms) and cavitation activity is measured as a function of the POnT (figure 3), the enhancement model can be fitted to the experimental data.

The interdependency between reactivation and loss of active bubbles leads to an optimal POnT around 50 ms. For POnTs below that value, not enough partially dissolved bubbles can be reactivated, above that the loss mechanisms dominate.

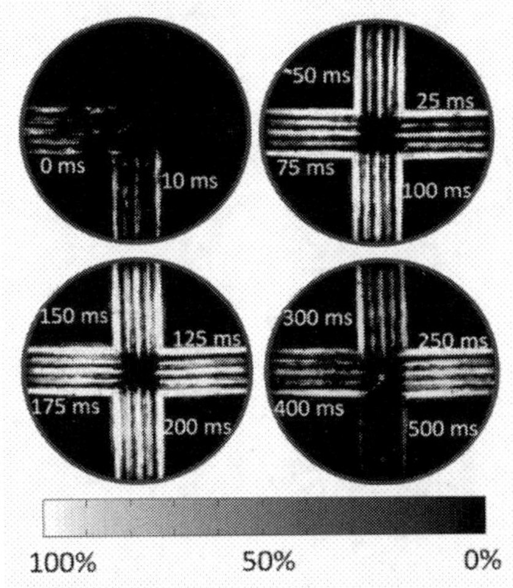

Figure 1. (up) PRE maps for different POffTs as indicated. A nominal acoustic intensity of 0.42 W/cm^2 has been used for all cleaning tests.

Figure 2. (right) (a) PRE (squares) and UHCN (circles) as functions of the POffT for 0.42 W/cm^2. (b) Bubble size distribu-tions calculated from the fits in (a).

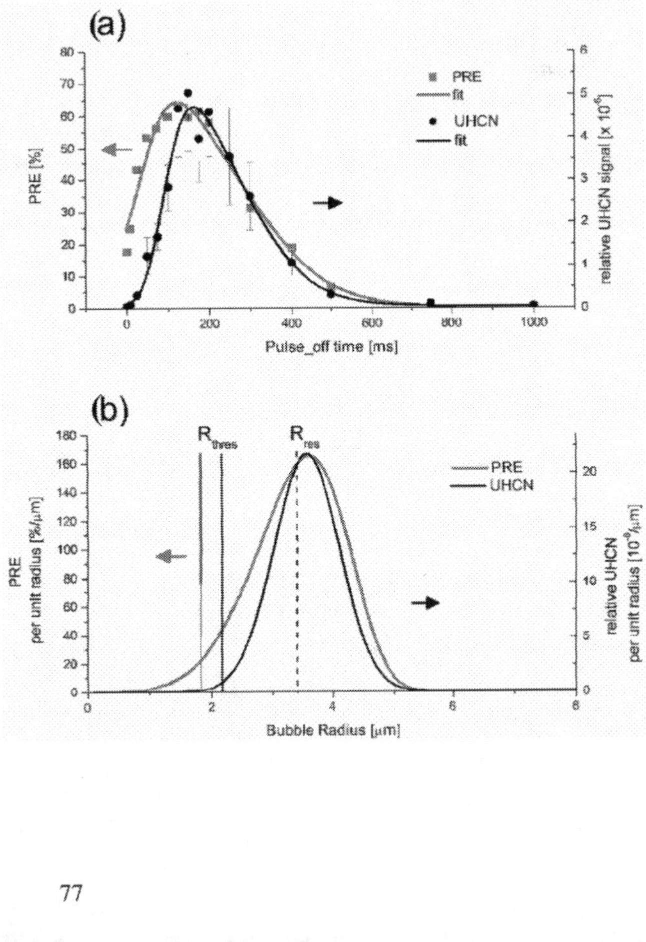

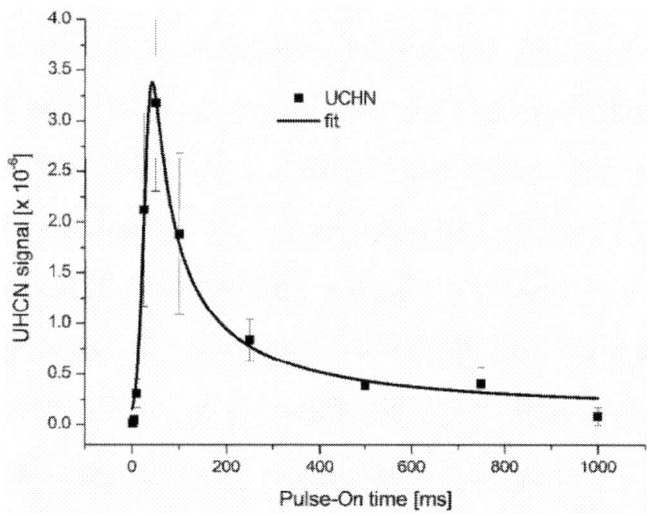

Figure 3. Cavitation activity measured by means of UHCN as a function of the POnT for a constant POffT of 150 ms and a nominal acoustic intensity of 0.33 W/cm^2. The enhancement model has been fitted to the experimental data (black solid line).

3. Summary & conclusion

We demonstrated that the active and "cleaning" bubble size distribution can be determined by combining ultrasonic pulsing with Cavitation Noise and Particle Removal measurements. Enhancement of both cavitation as well as cleaning activity can be attributed to the partial dissolution of active bubbles in between subsequent pulses. The optimal pulse parameters strongly depend on the involved enhancement mechanisms.

Acknowledgment

We would like to acknowledge ProSys Inc. and Entegris. C.G. would like to thank FWO (project G.0492.10) and KU Leuven Research Council (project OT/11/064) for financial support.

References

[1] Semiconductor Industry Association 2011 Front End Processes (FEP) Table FEP1 *The International Roadmap for Semiconductors (ITRS) 2011 Edition*
[2] Hattori T 1998 *Ultraclean Surface Processing of Silicon Wafers* ed T Hattori (Springer)
[3] Holsteyns F *et al* 2003 *ECS Proc.* **26**, 161
[4] Marmottant P and Hilgenfeldt S 2003 *Nature* **423** 153
[5] Ohl CD, Arora M, Dijkink R, Janve V and Lohse D 2006 *Appl. Phys. Lett.* **89** 074102
[6] Hauptmann M, Frederickx F, Struyf H, Mertens P, Heyns M, De Gendt S, Glorieux C and Brems S 2012 Enhancement of cavitation activity and particle removal with pulsed high frequency ultrasound and supersaturation *Ultrason. Sonochem.* **20**(1) 69
[7] Brotchie A, Grieser F and Ashokkumar M *2010, J. Phys. Chem. B* **114** 11010
[8] Hauptmann M, Struyf H, Mertens P, Heyns M, De Gendt S, Glorieux C and Brems S 2012 Towards an understanding and control of cavitation activity in 1 Mhz ultrasound fields *Ultrason. Sonochem.* **20**(1) 77
[9] Christenson KK 1997 *J. Inst. Environ. Sci.* **41** 45
[10] Hauptmann M *et al* (submitted) Enhancement of cavitation activity in pulsed megasonic cleaning *Ultrason. Sonochem.*
[11] Brems S *et al* (submitted) Physical forces exerted by microbubbles on a surface in a traveling wavefield *Ultrasonics*
[12] Xu K *et al* 2003 *Sol. Stat. Phenom.* **92** 161

IOP Conf. Series: Materials Science and Engineering **42** (2012) 012017 doi:10.1088/1757-899X/42/1/012017

Characterising the acoustoplastic effect in an ultrasonically assisted metal forming process

S A Aziz and M Lucas

School of Engineering, University of Glasgow, Glasgow G12 8QQ, UK

Email: s.abdul-aziz.1@research.gla.ac.uk, Margaret.Lucas@glasgow.ac.uk

Abstract. An investigation through experiments and finite element analysis (FEA) has been carried out to study the effects of applying ultrasonic oscillations to the lower platen in forming tests for two different metals. Previous research has shown that by applying ultrasonic vibrations to the lower platen in compression tests on pure aluminium specimens, the resulting stress-strain relationship can be characterised by a temporary effective softening of the material properties during intervals of ultrasonic excitation. The current research demonstrates this effect in two different metal specimens and additionally shows that finite element simulations can be used to model the behaviour in terms of both volume and surface effects. In this study, the ultrasonic excitation was introduced both prior to and post yield and the process simulations were developed in the FEA software Abaqus. The data recorded from experiments and predicted by the FEA illustrate how ultrasonically assisted metal forming can result in a lowering of the static flow stress, consistent with the effective material softening proposed previously in studies of the acoustoplastic effect.

1. Introduction

Previous studies [1,2] have shown how the influence of ultrasonic vibrations on metal forming processes can be considered by analysis of both volume effects and surface effects. These studies all demonstrated that superimposing ultrasonic excitation on a static forming process resulted in a reduction in flow stress. In the early studies [3] the interpretation of the results of stress reduction relied on experimental data from a test machine load cell which could not provide measurement of the oscillatory stress and therefore only the mean flow stress was considered. As with more recent research [4], this present study incorporates a piezoelectric force transducer into the test configuration to allow simultaneous measurement of the static and oscillatory stress response. The study aims to characterise the metal work hardening behaviour of two different metal specimens under superimposed ultrasonic excitation applied on a static compression test under dry surface conditions. Consequently, the results of this study can provide interpretation of the material flow behaviour from measurement of the stress-strain relationship. In parallel, finite element models are developed to simulate the ultrasonic compression test, allowing investigation of the result of modifying the material properties and contact condition.

2. Ultrasonic system and experimental procedure

A 20 kHz axially resonant ultrasonic system, consisting of a tuned piezoelectric transducer and half-wavelength ultrasonic horn, was positioned to form the lower platen on a Zwick Roell test machine as shown in figure 1.In these experiments, cylindrical metal specimens were compressed between a flat die and punch. Two types of metal specimen were used in the tests, with all specimens machined to a cylindrical bar of 8 mm in diameter and 8 mm in height, giving an aspect ratio of 1. Specimen surfaces were sanded and polished to provide a uniform surface roughness and the interfaces between the specimens, die horn and punch tool were not lubricated. Compression tests were performed on specimens of commercial grade metals of aluminium and magnesium.

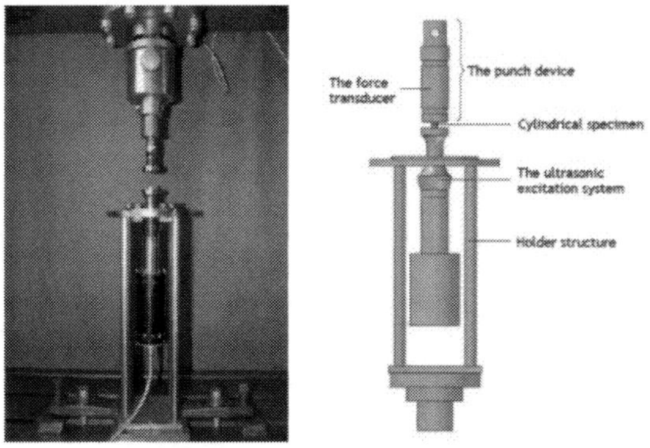

Figure 1. Schematic of the experimental set-up.

The ultrasonic horn provides a uniform nominal vibration amplitude in the range 5 to 20 μm, which is controlled by the ultrasonic generator. The flat punch was connected to the cross-head of the test machine which provided a constant cross-head speed of 5 mm/min for these experiments. For measuring the static-oscillatory force response, a Kistler piezoelectric transducer was mounted between the punch and machine cross-head. For each test, ultrasonic excitation was applied after 1 mm deformation of the specimen height. A finite element model was developed using axisymmetric four node elements, taking advantage of symmetry. The model material properties and boundary conditions applied aim to simulate the experiments for static and ultrasonically assisted deformation.

3. Experimental and numerical results

Measured and predicted data from the experimental and numerical tests respectively, are shown in figure 2. Figure 2(a) shows the static and oscillatory stress derived form measurements using the piezoelectric force transducer. Figure 2(b) shows the static and oscillatory stress results from the finite element analysis. Comparison of (a) and (b) shows that measurement of the flow stress without reference to the oscillatory stress behaviour does not provide a very meaningful measure of the effects of ultrasonic excitation, especially since the deviation between the results of the FE model and measurements is best illustrated by the change in the path of the maximum oscillatory stress. It can be observed for both the experimental and FE data that the path of the mean oscillatory stress is significantly reduced from the path of the static flow stress when ultrasonic excitation is introduced, but the path of the maximum oscillatory stress is also significantly reduced in the measurements but not in the FE results. This change in behaviour, characteristic of an effective material softening, is a result of the acoustoplastic effect [1,5]. This volume effect is incorporated into the FE model by applying an adjustment to the material properties at the onset of ultrasonic excitation. This involves calculation of a material dependent constant which can be estimated from the measured oscillatory stress as the decrease of flow stress per unit amplitude as described by Izumi [6]. However, it has also been shown previously that ultrasonic excitation additionally alters the friction contact condition between the specimen and upper and lower test machine platens [7]. This surface effect is included in the FE model as a 40% reduction in the coefficient of friction, consistent with the maximum achievable friction reduction generally agreed in the literature [1]. The resulting stress-strain relationship is shown in figure 2(c). The results show that a close match between the experiment and the FE model data is achievable by modelling the effects of superimposed ultrasonic excitation as a combination of an adjustment of material properties and interfacial friction. Figure 2(d) and (e) show the effects of superimposed ultrasonic excitation predicted by the FE model by adjustment of (d) material properties only and (e) interfacial friction only.

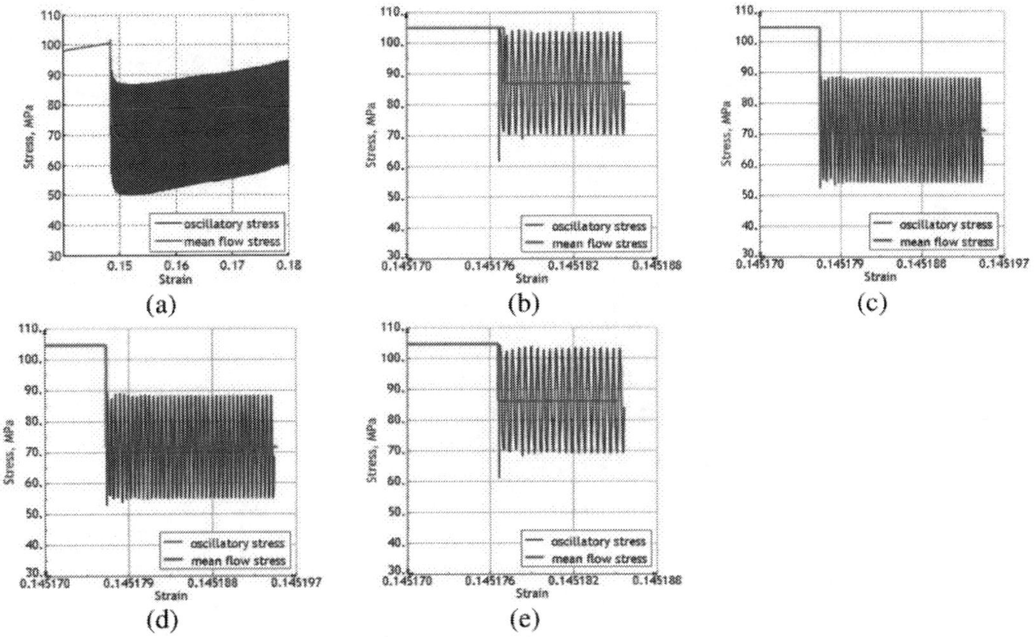

Figure 2. Oscillatory stress for aluminium specimen and 20 µm ultrasonic amplitude, (a) measured by piezoelectric force transducer, (b) from FE analysis, (c) from FE analysis with adjustments to material properties and contact friction (d) from FE analysis with adjustments to material properties only and (e) from FE analysis with adjustments to contact friction only.

Similar measurement and FE modelling procedures were employed for studying the effects of ultrasonic excitation on compression tests for magnesium metal specimens for the range of ultrasonic amplitudes. Similar measurement results to those in figure 2(a), illustrating both a reduction in the mean flow stress and a reduction in the path of the maximum oscillatory stress from the path of the static flow stress, were obtained in all tests. Therefore, the FE models were developed to match the experimental results by combining an adjustment in the material properties with an adjustment in the friction contact condition during ultrasonic excitation. The results are summarised in figure 3.

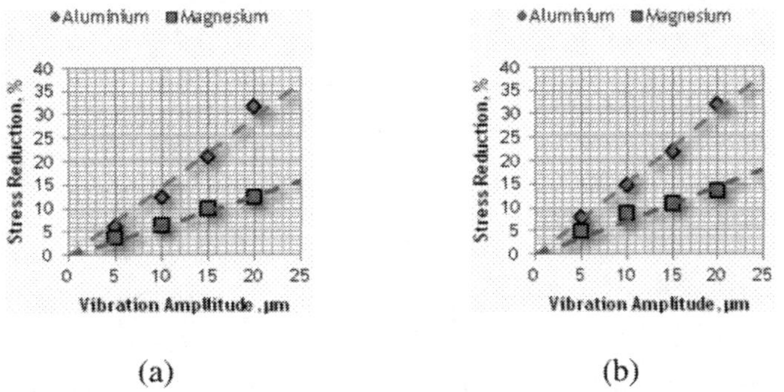

Figure 3. Reductions in mean flow stress during ultrasonic excitation from (a) experiment and (b) FE analysis.

The results in figure 3 show how the percentage stress reductions for the same ultrasonic amplitude are highly dependant on the metal of the specimen, with softer metals allowing larger stress reductions to be achieved, and that stress reduction increases linearly with ultrasonic amplitude in the range of this study (5 μm to 20 μm). The ratio of stress reduction to ultrasonic amplitude ($\Delta\sigma$-USA) is adopted as a measure of the separate and combined contribution of volume effects, in terms of adjustment in material properties, and surface effects, in terms of adjustment to the friction contact condition, data extracted from figure 2(c), (d) and (e). The results are presented as a bar chart in figure 4. The bar chart shows that for plastic deformation, the volume effect of effective material softening dominates the stress-strain behaviour during superimposed ultrasonic excitation.

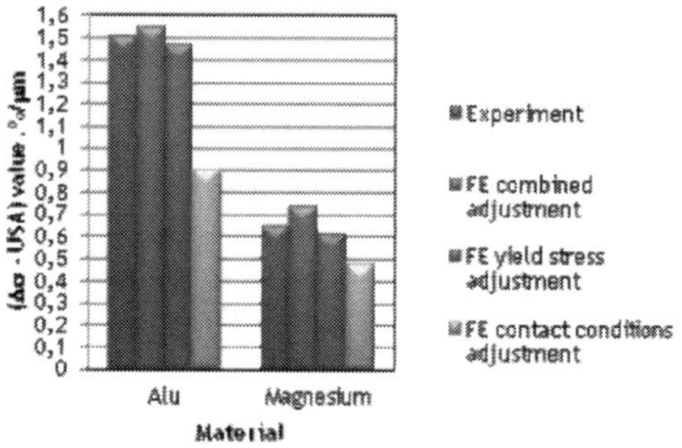

Figure 4. The contribution to stress reduction of volume and surface adjustments in the FE model.

4. Conclusion

A study of the effects of superimposed ultrasonic excitation on compression of aluminium and magnesium cylindrical specimens has been carried out. The numerical model was adjusted at onset of and during ultrasonic excitation by a combination of adjustment to the material properties and the contact friction. This allowed close agreement to be obtained between experimental and FE data illustrating the effects of reduction in the mean flow stress and, additionally, a reduction in the path of the maximum oscillatory stress from the path of the static flow stress due to ultrasonic excitation of the die. The influence of volume and surface effects were investigated separately in the FE model and it was shown that the reduction in mean flow stress is dominated by the effective material softening during ultrasonic excitation.

References

[1] Daud Y, Lucas M and Huang Z 2007 *J. Mat. Process. Tech.* **186** 179-190
[2] Winsper CE and Sansome E 1967 *8th International MTDR Conference* 1359-60
[3] Brehl DE and Dow TA 2008 *Precision Engineering* **32** 153–172
[4] Yaoa Z, Kim GY, Faidley L, Zou Q, Mei D and Chen Z 2012 *J. of Mat. Process. Tech.* **212** 640– 646
[5] Lucas M, Gachagan A and Cardoni A 2009 *Proc. Inst. Mech. Eng. Part C-J. Mech. Eng. Sci.* **223**(12) 2949-65
[6] Izumi O, Oyama K and Suzuki Y 1966 *Trans. JIM* **7** 162-167
[7] Siddiq A and Sayed T 2012 *Ultrasonics* **52** 521-529

IOP Conf. Series: Materials Science and Engineering **42** (2012) 012018 doi:10.1088/1757-899X/42/1/012018

Effect of ultrasonic pre-treatment of thermomechanical pulp on hydrogen peroxide bleaching

E Loranger[1], **A Charles and C Daneault**[2]

Lignocellulosic Materials Research Center, Université du Québec à Trois-Rivières, 3351 boulevard des Forges, C.P. 500, Trois-Rivières, Québec, Canada, G9A 5H7

E-mail: Eric.loranger1@uqtr.ca

Abstract. Ultrasound pre-treatments of softwood TMP had been carried to evaluate its impact on the efficiency of hydrogen peroxide bleaching. The trials were performed after a factorial design of experiment using frequency, power and time as variables. The experiments were conducted in an ultrasonic bath and then bleached with hydrogen peroxide. Measurements such as brightness, L*A*B* color system coordinate, residual hydrogen peroxide and metal content were evaluated on bleached pulp. The results indicate that the effect of ultrasonic treatment on brightness was dependent on the ultrasound frequency used; the brightness increased slightly at 68 kHz and decreased at 40 and 170 kHz. These results were correlated to the ultrasound effect on the generation of transition metals (copper, iron and manganese) which are responsible for catalytic decomposition of hydrogen peroxide. The influence of metal interference was minimized by using a chelating agent such as diethylene triamine pentaacetic acid (DTPA). With the results obtained in this study we have identified a set of option conditions, e.g. 1000 W, 40 kHz, 1.5 % consistency and 0.2% addition of DTPA prior to the bleaching stage (after ultrasonic pre-treatment) who improve brightness by 2.5 %ISO.

1. Introduction

Thermomechanical pulp (TMP) is used in a wide array of paper products including some value-added papers grades. Since TMP is a high yield pulp ($\approx$ 95 %) its inclusion in papermaking can greatly reduce the production cost. However, TMP pulp has relatively low brightness (50-60 %ISO) related to its high lignin content and must be greatly enhanced [1]. The brightness enhancement stage, by alkaline hydrogen peroxide oxidation [2], is then an important part of profitability and must therefore, be improved. In Ilker Mistik et al. [3], cotton fibres have been bleached successfully by hydrogen peroxide under ultrasound. Their study has shown a significant bleaching time gain which was correlated to the improved statistical probability of contact between reagents resulting from transient cavitation (jet of matter and energy) but yet very small brightness improvement was observed. A recent study by Xing et al. [4] have shown that a Chinese white poplar chemimechanical pulp (CMP) pre-treated by ultrasonic energy will better respond to hydrogen bleaching. They have reported enhanced brightness values by up to 3.5 %ISO and directly shown that the improvement was attributed to morphologic changes in fibres structure. However, their study [4] was explored using very limited ultrasonic conditions (20 kHz, 750 W, 5 – 40 min). The objective of our study was to further

1 To whom any correspondence should be addressed.
2 Canada Research Chair in Value-added Paper.

investigate the effect of ultrasonic pre-treatment, at various powers and frequencies, on the hydrogen peroxide bleaching of softwood fibres.

2. Materials and methods

A thermomechanical pulp (TMP) made of softwood and diethylene triamine pentaacetic acid (DTPA), a commercial sequestration agent, was obtained from an Eastern Canadian pulp mill. All other chemicals used were of A.C.S. reagent grade and chemical solutions were prepared with deionised water ($\Omega < 0.8$ µS). The ultrasonic treatments were conducted in a stainless steel bath (model BT90, Ultrasonic Power Corporation, USA) equipped with a 1000 W temperature controlled heater and an external heat exchanger as seen in figure 1. Commercial ultrasound generators (Ultrasonic Power Corporation, USA) at frequencies of 40, 68 and 170 kHz were used to produce up to 1000 W of nominal ultrasonic energy. In this ultrasonic bath reactor, a potential ultrasonic density of about 4350 W/m² can be achieved with 12 Vibra-bar® transducers.

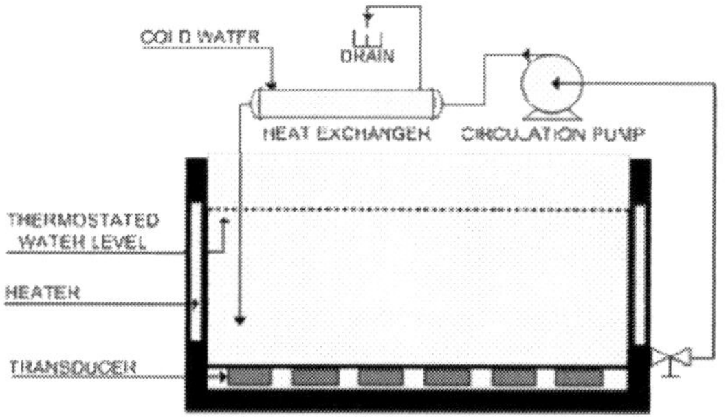

Figure 1. Schematic diagram of the ultrasonic bath.

The ultrasound treatment conditions were the variables of a fractional design of experiments (DOE) built on 22 conditions including 6 repetitions for the purpose of accessing the experimental error (table 1). The ultrasound conditions explored covered power level ranging from 0 to 1000 W, an irradiation time between 10 and 40 min and the frequencies were fixed at 40, 68 or 170 kHz. Ultrasonic pre-treatment was performed in a polyethylene bag which had the same size as the bottom of the ultrasonic bath in conditions similar to [4] (1.5% consistency, 50 °C). After the required ultrasound exposition time, the pulp sample was filtered (Reeve Angel grade 202 filter paper) and washed by re-soaking in 1 L of deionized water and filtered again. On subsequent experiments, a supplement of DTPA (0.2 %) was added immediately before the first filtration (DTPA/US). Conventional alkaline hydrogen peroxide bleaching (3% charge) of pulp and measurements were conducted as in [5]. Metals content in the pulp were measured by ICP-OES (Perkin Elmer 4300DV).

Table 1. Experiment design for ultrasonic pre-treatment.

Trials	Power (W)	Time (min)	Frequency (kHz)
1	250	10	40
2	1000	10	40
3	1000	40	40
4	625	25	40
5	250	40	40
6/7	625	25	40
8	250	40	68
9/13	625	25	68
10	1000	10	68
11	250	10	68
12	1000	40	68
14	250	10	170
15/19	625	25	170
16	250	40	170
17	1000	40	170
18	1000	10	170
20-22	---	---	---

3. Results and discussion

In comparison to the blank brightness (71.2 %ISO), two ultrasonic conditions emerged as potentially optimal in figure 2: trial 10 at 71.7 %ISO and trial 8 at 71.6 %ISO but since the brightness gains were similar to the standard deviation (0.5 %ISO), the observed improvements were not significant. The majority of the other trials yielded lower brightness by ≈ 1.7 %ISO when compared to the blank. High power level (greater than 625 W) combined with long exposition time (longer than 25 min) was detrimental to brightness regardless of the frequency levels. From the conclusion found in Xing et al. [4], the results from figure 2 should have been maximal at 40 kHz. Indeed, a low frequency (e.g. 40 kHz) will promote the mechanical effect of the cavitation, a high frequency (e.g. 170 kHz) promote water sonolysis while a mid-frequency (e.g. 68 kHz) has intermediate effects [6]. As reported in [7], water sonolysis could lead to in-situ hydrogen peroxide production, which could have beneficial effect. On the other hand, water sonolysis also generates various radical species such as HO•, HO_2• and ROO•, which could potentially lower the final pulp brightness. The brightness values obtained in both cases reported by [7] were not identified as maxima in figure 2.

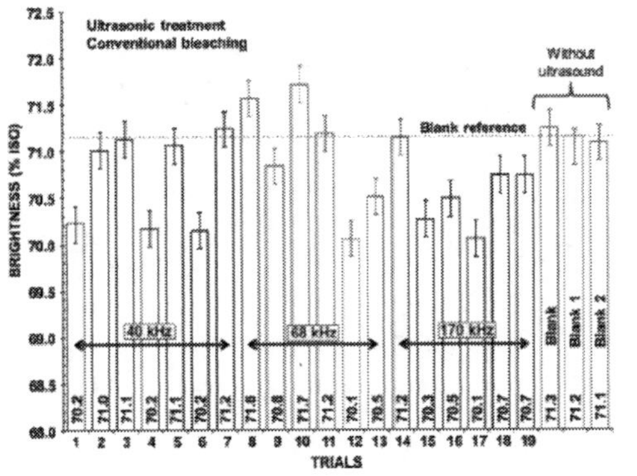

Figure 2. Final brightness for ultrasonic treatment without additional DTPA.

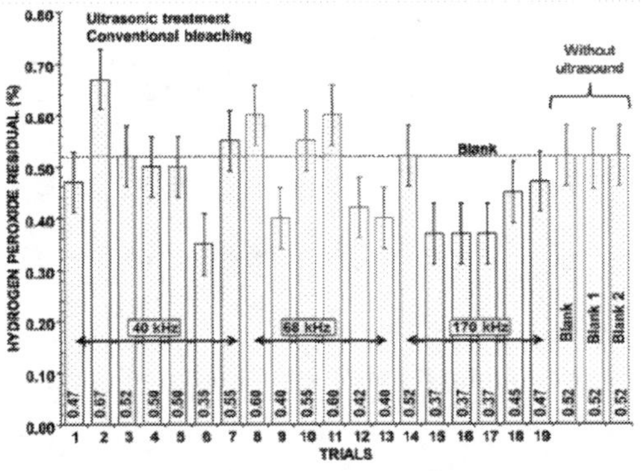

Figure 3. Hydrogen peroxide residual for ultrasonic treatment without additional DTPA.

As seen in figure 3, residual hydrogen peroxide obtained at 40 kHz was similar to that for conventional bleaching (blank). When the experiments were conducted at 68 kHz, half of the trials showed residual hydrogen peroxide lower than that of the blank, and all the trials performed at 170 kHz gave lower residual hydrogen peroxide. Low hydrogen peroxide concentration can be related to high radical content which promotes catalytic degradation of H_2O_2. The low brightness found at 170 kHz (figure 2) was, therefore, possibly attributable to high peroxide degradation, meaning less peroxide was available to bleach. High transition metal content (copper, iron and manganese) could be another potential cause of hydrogen peroxide degradation. These elements were chosen because they are mainly responsible for the catalytic decomposition of peroxide [8]. With respect to standard deviation, the mean Cu content was similar for the bleached (2.8 ppm) and unbleached (2.0 ppm) pulps while the mean Fe content was also similar for the bleached (16 ppm) an unbleached (13 ppm) pulp (data not shown). There was a substantial difference in Mn content between the bleached pulp mean (17 ppm) and the unbleached (42 ppm) pulp (data not shown) but this effect was attributable to the bleaching mechanism and not to ultrasound treatment. Cu, Fe and Mn contents were independent of the ultrasonic treatment conditions. Even if in terms of the standard deviation all results are not significant, a pattern seems to emerge when the metal contents were related to the final brightness.

Considering that pattern, we repeated the experiments according to the same DOE with a 0.2 % DTPA charge in post-addition. We compared the final brightness (figure 4) with the manganese content (figure 5) obtained from ultrasonic treatment with or without DTPA post-addition. As seen in figure 4, the DTPA post-addition had a very positive effect on final brightness for all the frequencies studied. A mean increase in brightness of 1.9 %ISO was observed. From all the trials, one optimal experimental condition emerged and yields a brightness of 73.6 %ISO. A statistical analysis with a confidence level of 95% as confirmed that trial 3 (1000 W, 40 kHz, 40 min time) is statistically significant. This optimal condition is now in agreement with conclusions reported by Xing et al. [4]. The manganese content (figure 5) was lower for every trial made with a DTPA post-addition but the standard deviation do not conclusively allow us to identify an optimal set of ultrasonic conditions. Nevertheless, when the blank value of manganese (16.1 ppm) with DTPA addition was compared to the mean value (4.6 ppm) of ultrasound trials, it is clear that ultrasounds are responsible for a decrease in manganese content by a factor of nearly 4. Since the manganese has a higher valence than iron or copper, the former will decompose more peroxide [9]. Our results suggest that less free manganese in the pulp will increase the efficiency of hydrogen peroxide bleaching.

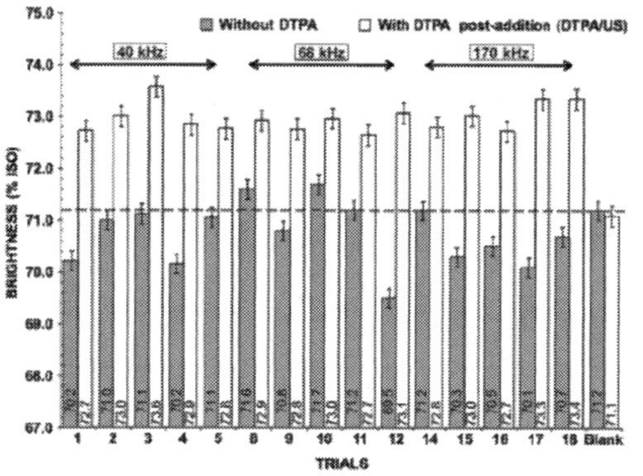

Figure 4. Final brightness for ultrasonic treatments with and without a 0.2% DTPA post-addition.

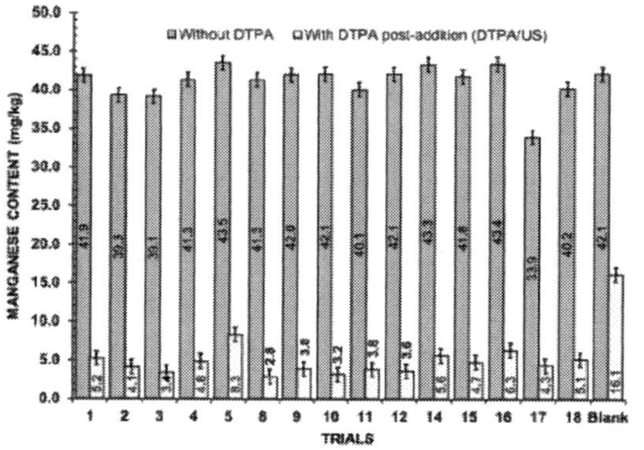

Figure 5. Manganese (Mn) content for ultrasonic treatments with and without a 0.2% DTPA post-addition.

4. Conclusions

This study was conducted to determine the impact of ultrasonic pre-treatment of thermomechanical pulp (TMP) on the efficiency of hydrogen peroxide bleaching. Ultrasonic treatment of TMP without DTPA post-addition has little positive impact on final pulp brightness. Ultrasound treatments improve manganese content sequestration by DTPA but have no improvement relative to iron and copper. High manganese sequestration was directly correlated to the final increase in brightness. A statistical analysis permit us to locate, with a confidence level of 95 %, an optimal condition (1000 W, 40 kHz 40 min exposition, 1.5 % consistency) that gives a brightness of 73.6 %ISO (2.5 %ISO increase).

Acknowledgements

The authors gratefully acknowledge the financial supports from the Canada Research Chair in Value-added Paper and the Natural Sciences and Engineering Research Council of Canada.

References

[1] Liu Z, Ni Y, Li Z and Court G 2005 *Pulp Paper Can.* **106** 34-7
[2] Liu S 2003 *Chem. Eng. Sci.* **58** 2229-44
[3] Ilker Mistik S and Müge Yükseloglu S 2005 *Ultrasonics* **43** 811-4
[4] Xing M, Yao S, Zhou S.-K, Zhao Q, Lin J.-H and Pu J.-W 2010 *Bioressources* **5** 1353-65
[5] Leduc C, Lanteigne-Roch L.-M. and Daneault C 2011 *Cellul. Chem. Technol.* **45** 657-63
[6] Mason T.J, Cobley A.J, Graves J.E and Morgan D 2011 *Ultrason. Sonochem.* **18** 226-30
[7] Santos H.M, Lodeiro C and Capelo-Martinez J.L 2009 The power of ultrasound *Ultrasound in Chemistry: Analytical Applications* (Weinheim, Germany: Wiley-VCH)
[8] Lapierre L, Bouchard J, Berry R.M and Lierop B.V 1995 *J. Pulp Pap. Sci.* **21**(8) 268-73
[9] Sun Y, Fenster M, Yu A, Berry R.M and Agyropoulos D.S 1999 *Can. J. Chem.* 77 667-75

International Symposium on Ultrasound in the Control of Industrial Processes (UCIP 2012) IOP Publishing
IOP Conf. Series: Materials Science and Engineering **42** (2012) 012019 doi:10.1088/1757-899X/42/1/012019

Effect of Sonotrode Design on the Performance of the Ultrasonic Patterning Process

Keun Park[1], Young-Soo Seo and Woosin Jung
Dept. Mechanical System Design Engineering, Seoul National University of Science & Technology, Seoul, 139-743, Republic of Korea

E-mail: kpark@seoultech.ac.kr

Abstract. The present study examines the use of ultrasonic waves in micro-pattern replication on polymer substrates. The ultrasonic patterning process was developed in order to use ultrasonic vibration energy to replicate the micro-patterns engraved on a sonotrode into the surface of a polymer substrate. To replicate the micro-patterns on the polymer substrate with high accuracy, the sonotrode should be designed to allow only longitudinal vibration, not transverse or twisting vibrations. In addition, the vibration magnitude should be maintained uniformly on the tip of the sonotrode in order to obtain uniform pattern replication. For this purpose, the sonotrode design is investigated through finite element analyses, from which the resulting natural frequency of the sonotrode can be adjusted to be within the desirable range. These FE analyses are performed for two types of sonotrodes, and the validity of the developed method is discussed from the ultrasonic patterning experiments using the optimally-designed sonotrodes.

1. Introduction

The micro-molding of thermoplastic polymers is one of the most popular fabrication methods for micro-devices. Various molding techniques have been developed to enable the micro-molding of thermoplastic polymers including hot embossing and micro-injection molding [1]. In recent years, ultrasonic vibration energy has been used to facilitate these micro-molding processes, and several researchers have used ultrasonic vibration to improve moldability in the conventional hot embossing [2] and micro-injection molding [3]. The ultrasonic patterning process using a micro-patterned sonotrode has also been developed to replicate micro-patterns without use of thermal energy [4].

This study investigates the effect of the sonotrode design on the replication characteristics of the micro-pattern. In the ultrasonic patterning process, a sonotrode acts not only as a transmitter but also as an amplifier of the ultrasonic wave. Thus, the vibration characteristics of the sonotrode should be analyzed and optimized to replicate micro-patterns with a high, uniform replication ratio. Finite element (FE) analyses were performed as a method of predicting the vibration characteristics of two types of sonotrodes: a small circular sonotrode and a large rectangular sonotrode.

2. Micro-patterning on a small area using a circular sonotrode

The ultrasonic patterning process uses ultrasonic waves to generate frictional heat between the sonotrode and the polymer substrate. Thus, the surface region of the polymer substrate is repeatedly

[1] To whom any correspondence should be addressed.

heated as a result of this friction and is softened sufficiently for the embossing using the pressure of the sonotrode. Figure 1 illustrates the schematic configuration of the ultrasonic patterning system, which consists of a power supply, a pressing unit, and an ultrasonic stack. The ultrasonic stack is composed of a piezoelectric transducer, a booster, and a sonotrode.

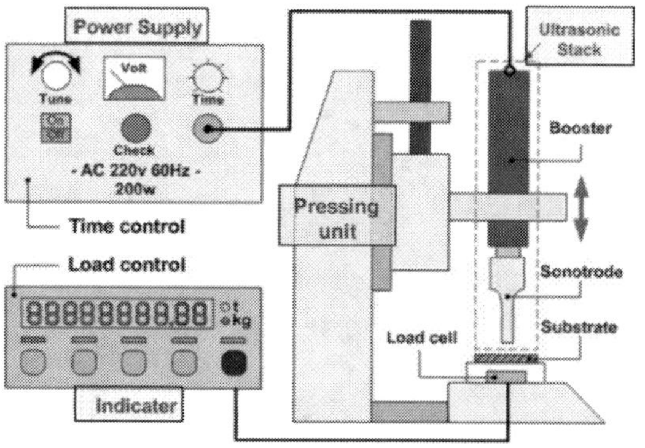

Figure 1. Configuration of micro-patterning system using ultrasonic vibration.

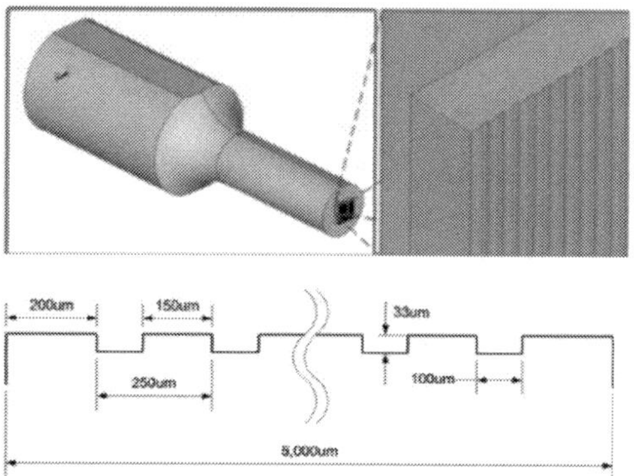

Figure 2. Geometry of the circular sono-trode with micro-channels.

2.1. Design and analysis of the circular sonotrode

Because the sonotrode must resonate at a specified ultrasonic frequency, its design parameters must be determined to satisfy the desirable vibration characteristics. In this section, a stepped sonotrode with a circular cross section is discussed. Figure 2 presents the geometry of the circular sonotrode containing 14 micro-channels with a width of 100 μm and a depth of 33 μm. These micro-channels are engraved on the outlet of the sonotrode with a base area of 5 × 5 mm. The inlet and outlet diameters of the sonotrode are 28 mm and 13 mm, respectively.

On the assumption that the sonotrode is rotationally symmetric, its dynamic equilibrium equation is given using the one-dimensional wave equation [4] as follows:

$$\rho A(x)\frac{\partial^2 u(x,t)}{\partial t^2} = E\frac{\partial}{\partial t}\left(A(x)\frac{\partial u(x,t)}{\partial t}\right) \tag{1}$$

where $u(x,t)$ is the amount of vibration in the axial direction and $A(x)$ is the cross sectional area at an axial position x. Equation (1) can be rewritten using the separation of variables, $u(x, t) = U(x)G(t)$:

$$U''(x) + \frac{S'(x)}{S(x)}U'(x) + \omega_n^2 U(x) = 0 \tag{2}$$

where ω_n is the natural frequency. The lengths of regions 1, 2, and 3 (l_1, l_2, and l_3) can then be determined by solving the following equation:

$$\tan\omega_d l_2 = \frac{(\tan\omega_n l_1 + \tan\omega_n l_3)\sqrt{\xi^2 - 1}}{\tan\omega_n l_1 \tan\omega_n l_3 + (\tan\omega_n l_3 - \tan\omega_n l_1)\xi - 2\xi^2 + 1} \tag{3}$$

where ω_d and ξ are defined using equation (4):

$$\omega_d = \omega_n\sqrt{\xi^2 - 1}, \qquad \xi = -\frac{1}{2\omega_n}\frac{S'(x)}{S(x)} \tag{4}$$

Thus, the regional lengths (l_1, l_2, and l_3) were determined to be 45.1, 14.0, and 33.0 mm, by solving equation (3). To investigate the vibration characteristics of the sonotrode, FE modal analyses were performed. The natural frequency for the longitudinal vibration mode was 28.608 kHz, which was available to resonate with the excitation frequency of the ultrasonic wave (28±1 kHz).

2.2. Replication of the micro-channels using the designed sonotrode
A circular sonotrode was fabricated using the determined design parameters. An experiment for the ultrasonic excitation was performed in order to measure the amount of vibration of the sonotrode. Figure 3 compares the longitudinal displacement of the inlet and outlet sections of the sonotrode. The peak displacements were 1.8 μm for the inlet and 6.9 μm for the outlet. This corresponds to an amplification ratio of 3.83, which indicates that successful resonation was obtained.

The ultrasonic patterning experiments were performed with variations in the embossing load. An ultrasonic excitation of 28 kHz was applied for 2.5 s. The workpiece was a polymethylmethacrylate (PMMA) substrate with a thickness of 500 μm. Figure 4 plots the sectional channel profiles with the various embossing loads. It can be seen that the micro-pattern was not filled ertirely, while the replication ratio increased as the embossing load increased. The area-based replication ratio was calculated to be 83.6% when the embossing load was 5 kgf, showing an improvement of 72% in comparison with that of the 4 kgf load (48.7%). Figures 5(a) through 5(c) show the replicated micro-patterns for the 3.0, 4.0, and 5.0 kgf of embossing loads, respectively. It can be seen that the increase of the embossing load improves the replication characteristics of the micro-channels.

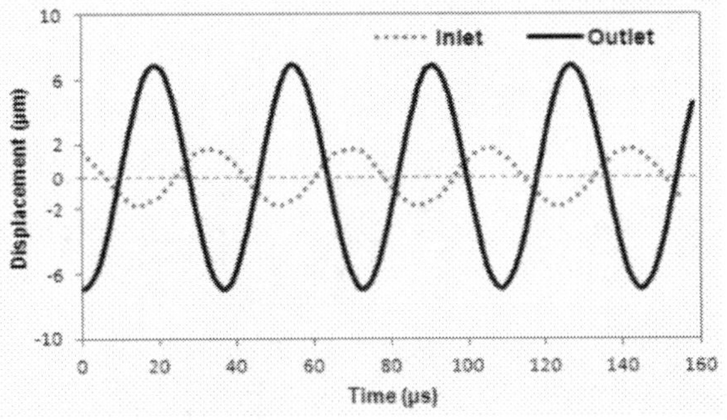

Figure 3. Measured longitudinal displacement profiles of the sonotrode.

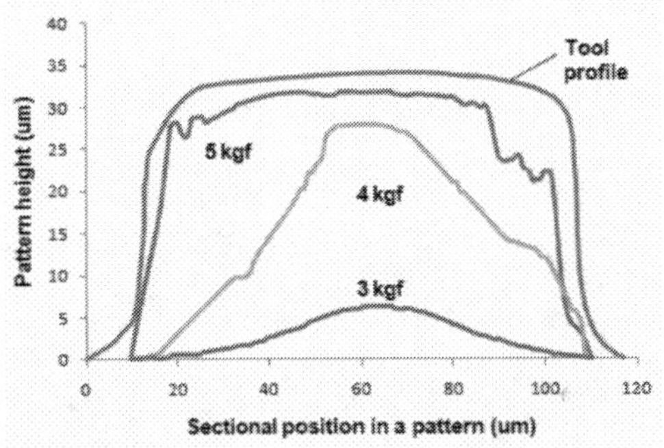

Figure 4. Sectional profiles of the replicated channel with various embossing loads.

(a) 3.0 kgf (b) 4.0 kgf (c) 5.0 kgf

Figure 5. Photographs of the replicated patterns with variations in the forming load.

3. Micro-patterning on a large area using a rectangular sonotrode

3.1. Design and analysis of a rectangular sonotrode containing side slots
In this section, a rectangular sonotrode is considered for the ultrasonic patterning of large area (63 × 49 mm). In order to replicate micro-patterns on a large area, the sonotrode should be designed to ensure not only the vibration-amplification characteristics but also uniformity in the vibration. Figure 6 is the geometry of the rectangular sonotrode containing two side slots [5]. FE modal analyses were performed for two types of sonotrodes: a regular sonotrode without a slot and the modified sonotrode with side slots. Figure 7 presents the FE analysis results for the two sonotrodes, demonstrating that the

non-uniformity in the vibration was reduced to 6.5% using the modified design including side slots from 20.1% for the regular horn design.

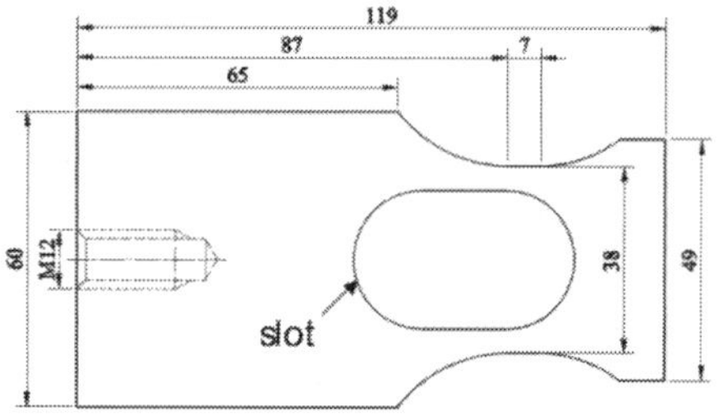

Figure 6. Geometry of the modified rectangular sonotrode with side slots.

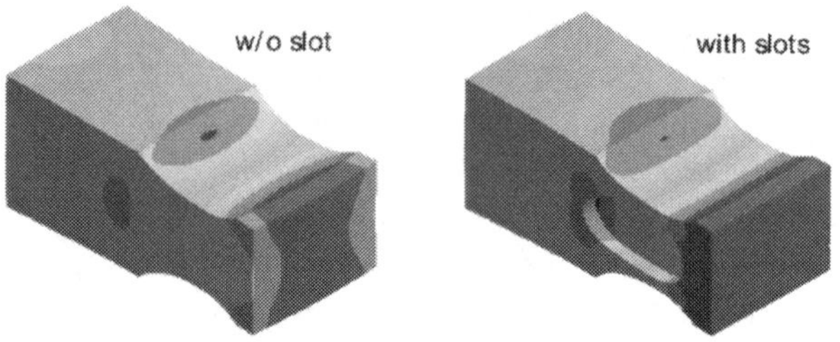

Figure 7. FE analysis results for the longitudinal displacement for the regular and modified sonotrodes.

3.2. Replication of the micro-patterns on a large area using the designed sonotrode

The ultrasonic horn was fabricated reflecting the optimized slot design parameters. Micro-prism patterns with a pitch of 490 μm and a depth of 200 μm were grooved on the sonotrode outlet face. The ultrasonic patterning experiments were performed using the modified sonotrode. An ultrasonic excitation of 20 kHz was applied for 4.0 s under a pressure of 0.5 MPa. The workpiece was a polystyrene (PS) film with a thickness of 300 μm. Figure 8 presents sectional photographs of the replicated patterns at various locations. It can be seen that the micro-patterns were uniformly replicated for all regions when the proposed horn design was used.

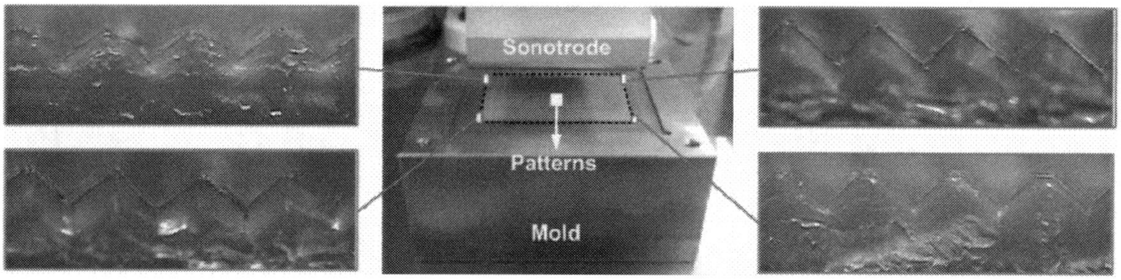

Figure 8. Replicated micro-prism patterns at various locations.

4. Conclusion

In this study, the design of ultrasonic sonotrodes was investigated in order to improve the replication characteristics of micro-patterns. FE analyses were performed to obtain uniform vibration at the specified frequency for various sonotrode designs, from which micro-patterns were successfully replicated on polymer substrates. Based on these results, it is expected that the proposed ultrasonic patterning process can be applied in the manufacturing of high quality micro and nano components.

Acknowledgements

This research was financially supported through the Basic Science Research Program of the National Research Foundation (NRF) funded by the Ministry of Education, Science & Technology (Grant number: 2010-0008435).

References

[1] Heckel M and Schomburg W K 2004 *J. Micromech. Microeng.* **14** R1
[2] Mekaru H, Goto H and Takahashi M 2007 *Microelectron Engng.* **84** 1282
[3] Michaeli W, Kamps T and Hopmann C 2011 *Microsyst Technol.* **17** 243
[4] Seo Y S and Park K 2012 *Microsyst Technol.* **18** DOI: 10.1007/s00542-012-1524-y.
[5] Jung W, Ra J and Park K 2012 *Int. J. Prec. Engng. Manuf.* **13** (accepted for publication)

IOP Conf. Series: Materials Science and Engineering **42** (2012) 012020 doi:10.1088/1757-899X/42/1/012020

A piezoelectric vibration-based syringe for reducing insertion force

Y C Huang, M C Tsai[1] and C H Lin

Department of Mechanical Engineering, National Cheng Kung University, Tainan 701, Taiwan

E-mail: mctsai@mail.ncku.edu.tw

Abstract. Puncturing of the human skin with a needle is perhaps the most common invasive medical procedure. Clinical studies have revealed that tissue deformation and needle deflection are the primary problem for needle misplacement in percutaneous procedures. To avoid this, various techniques for reducing insertion forces during needle insertion have been considered. This paper presents a piezoelectric vibration-based syringe to reduce insertion force. AC power was applied to the piezoelectric elements to vibrate the needle with high frequency and thereby reduce the friction and cutting forces between the needle and tissue. Vibration mode shapes of the needle were observed by finite element analysis and verified by experimental results. Effects of reducing insertion force via the vibrating needle were also confirmed by inserting the needle into the porcine tissues. The proposed syringe, which minimizes the insertion force and overcomes limitations of needle materials, can be widely utilized in robot-assisted needle insertion systems.

1. Introduction

Numerous modern medical diagnoses and therapies involve percutaneous intervention in which thin tubular devices are deeply inserted into inhomogeneous tissue to reach the target. Percutaneous needle insertion can be utilized in several medical procedures, such as biopsy, regional anesthesia, blood sampling, neurosurgery, and brachytherapy [1]. In these procedures, the accuracy of the process carries great importance; however, it is difficult to insert the needle accurately in inhomogeneous tissue because of several reasons, such as elastic stiffness, tissue heterogeneity, needle bending, and tissue deformation and movement [2]. Reducing the insertion forces is helpful for improving the accuracy and reducing the pain for patients during the procedure. Therefore, this paper proposed a syringe with an ultrasonic vibration and verified its effect on insertion force during the process of puncture.

2. Structure of the Piezoelectric Vibration Based Syringe

In this study, a Langevin type transducer was used to generate extension vibration and actuate a commercial needle (TERUMO NEEDLE 27G, 13mm, Bevelled tip). The scheme of the proposed piezoelectric syringe is shown in figure 1, where the needle is connected to the Langevin transducer. This transducer is designed based on a sandwich structure, which is composed of two metallic pieces and stacks of piezoelectric annular wafers (PZT-4). Bolts clamp all elements together and provide a

[1] To whom any correspondence should be addressed.

pre-load on the piezoelectric wafer [3]. Notice that a central hollow channel is built in the syringe and is able to inject medicine or aspirate tissue. The transducer is fixed at the flange, which experiences no displacement during vibration.

Extension vibration mode occurs when the transducer is driven by an electrical input to the piezoelectric wafers to generate the needle with extension and compression displacements. In order to design the dimension of the device for achieving acoustic impedance matching, the resonance frequency and vibration mode shape of the needle is evaluated by using finite element analysis. The resonance frequency of piezoelectric annular wafers must be chosen at around the resonance frequency of needle with extension vibration mode. In the illustrated design of figure 1, the outer diameter and the inner diameter of the piezoelectric annular wafer are 15mm and 6mm, and the thickness is 3mm, respectively. The dimension of the metallic pieces is designated by the wavelength to achieve higher amplitude. Based on the resonance frequency of the needle and the piezoelectric annular wafer, the dimension of the metallic pieces can be determined. The diameter of this transducer is 15mm, and the length without the needle is 76mm. The resonance frequency of the transducer is reconfirmed by the finite element analysis.

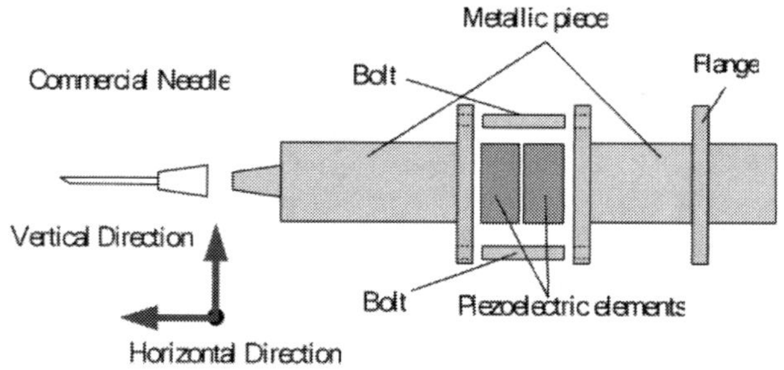

Figure 1. Scheme of Piezoelectric Vibration Based Syringe.

The transducer's impedance was measured to confirm the resonance frequency by using the impedance analyzer (Wayne Kerr 6520A). Note that the electrical impedance is the ratio of input voltage to output current. The magnitude and phase of impedance, in terms of frequency responses, is shown in figure 2. From the results, the resonance frequencies were 84 kHz, 114 kHz, and 121 kHz. In order to confirm the vibration mode shape, the vibration amplitude can be measured at the needle tip by the laser interferometer (Graphtec AT0023). Since it is difficult to spot the laser beam at the top of bevel needle tip, we measured the horizontal and vertical displacements at the side of the needle tip via two laser interferometers. Hence, the projection of displacement was a combination vector composed of vertical and horizontal displacement vectors [4]. The amplitudes at the needle tip at three different resonance frequencies were measured when the applied voltage was 5 Vp–p. The measured results of the vibration mode shape are indicated in figure 3, and it demonstrated that the amplitude at the first resonance frequency, 0.52 μm, is more than 10 times larger than others and also proved that the extension vibration mode occurred at 84 kHz. Therefore, the operating frequency of the transducer is chosen at 84 kHz.

(a) (b)

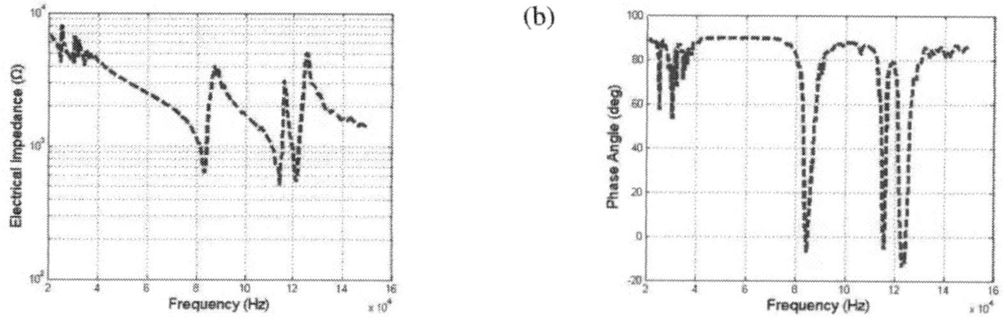

Figure 2. Frequency responses of electrical impedance (a) magnitude plot and (b) phase plot.

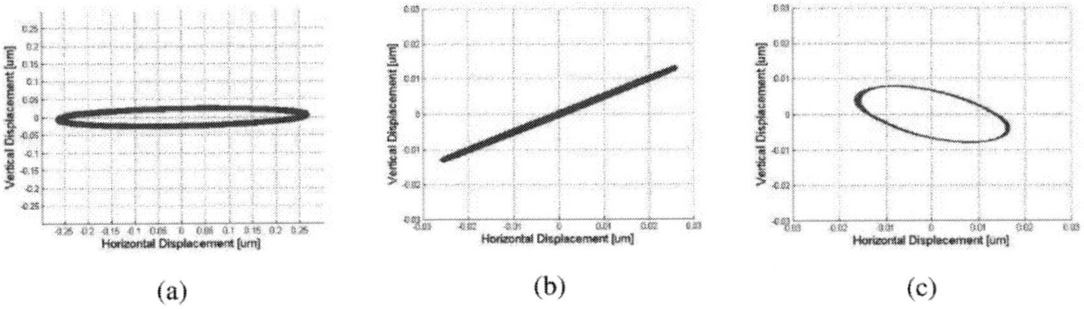

Figure 3. Trajectories of projection at 84 kHz, 114 kHz, and 121 kHz (left to right).

3. Experimental Setup

An insertion system was set up for studying needle insertion force into porcine tenderloin tissues without its skin. The experimental system is shown in figure 4, and the speed of insertion can be controlled by the stepping motor. A load cell is attached to the needle holder in order to measure the forces acting on the tissue. The proposed insertion method, with 84 kHz extension vibration, was compared to normal insertion procedure. In the proposed method, the devices were driven with voltages as high as 300 Vp-p in order to generate a larger vibration displacement. Each experiment of insertions was made in a close neighbourhood to better ensure consistent tissue behaviour and was under the constant insertion speed, 1.5 mm/s.

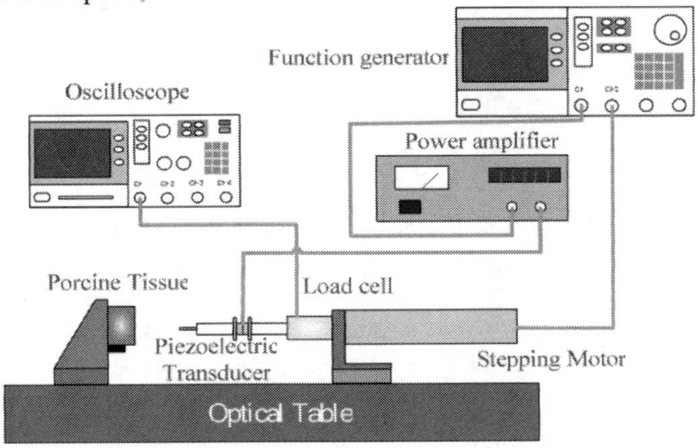

Figure 4. Scheme of insertion system.

4. Result and Discussion

The experimental results of the insertion force are shown in figure 5. The force increases with further needle displacement until the tissue is punctured by the needle, where a drastic reduction in force can be seen. The force rises again as the needle penetrates into the underlying tissue until the insertion is stopped. The relationship between the reduced force and insertion time is shown in figure 6. It is evident that the insertion force with ultrasonic vibration was reduced about 0.3 N (28%) after the puncture of the capsule.

According to the model of needle insertion into soft tissue [5], the insertion force is a summation of stiffness, friction, and cutting force. The force for puncturing the tissue surfaces, which is referred as stiffness force, is almost the same between the two methods. Therefore, the reduced forces are friction force and cutting force. The friction force occurs along the length of the needle inside the tissue and is due to Coulomb friction, tissue adhesion, and damping. Cutting force is defined as the force required to puncturing the skin and is generally constant. As can be seen from figure 6, the reduction of insertion force increased after needle punctured the skin until insertion cessation. It demonstrated that when the deeper needle was inserted, the larger force was reduced, which is the sign of friction force reduction. The main reason of the effects for reducing friction force is that the high frequency vibration changes the friction coefficient between the needle and skin. In our result, there are no significant signs of cutting force reduction during vibration insertion. However, extension vibration provides energy in the insertion direction which may reduce the cutting force.

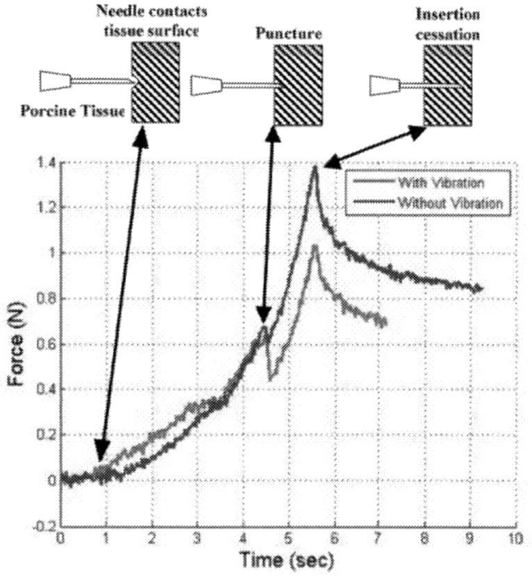

Figure 5. Result of forces during inserting needle into porcine tissue.

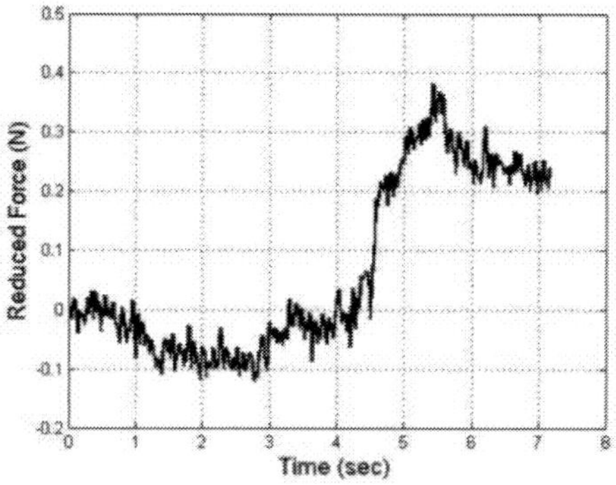

Figure 6. Reduced force by using ultrasonic vibration insertion.

5. Conclusions

In this paper, a piezoelectric vibration-based syringe and its fundamental design procedures were proposed. The proposed device provides an alternative method to insert a needle with ultrasonic extension vibration, whose amplitude was 0.52 μm when the applied voltage was 5Vp–p. The experimental results indicated that the high frequency vibration has significant influences on friction and cutting forces, and it showed 28% force reduction by using ultrasonic vibration. It also demonstrated that the proposed method can reduce insertion force further when the insertion depth is increased. These results suggest that insertion with ultrasonic vibration has potential for safer, faster, and more precise insertion, such as vertebra insertion, bone biopsy, and other medical procedures. Moreover, minimizing the insertion force and overcoming limitations of needle materials are also benefits of the proposed needle insertion system, which can be widely utilized in robot-assisted insertion systems.

References

[1] Abolhassani N, Patel R and Moallem M 2007 *Med. Eng. Phy.* **29** 413-31
[2] Podder T K, Clark D P, Fuller D, Sherman J, Ng W S, Liao L, Rubens D J, Strang J G, Messing E M, Zhang Y D and Yu Y 2005 *Engineering in Medicine and Biology 27th Annual Conf. (Shanghai)* vol 1 (New Jersey: IEEE EMBS) pp 5766-70
[3] Wang S H and Tsai M C 2011 *Ultrasonics* **51** 617-24
[4] Shigematsu T and Kurosawa M 2008 *IEEE T. Ultrason. Ferr.* **55** 2005-15
[5] Okamura A M, Simone C and O'Leary M D 2004 *IEEE T. Bio-Med. Eng.* **51** 1701-16

IOP Conf. Series: Materials Science and Engineering **42** (2012) 012021 doi:10.1088/1757-899X/42/1/012021

Atmospheric freeze drying assisted by power ultrasound

J V Santacatalina[1], J A Cárcel[1], S Simal[2], J V Garcia-Perez[1] and A Mulet[1]

[1]Grupo de Análisis y Simulación de Procesos Agroalimentarios (ASPA), Departamento de Tecnología de Alimentos, Universidad Politécnica de Valencia, Camí de Vera s/n, Valencia, E46022, Spain, Tel.:+34 96 387 93 76
[2]Grupo de Ingeniería de Alimentos, Departamento de Química, Universitat de les Illes Balears. Crta. Valldemosa, km. 7.5, E07122, Palma de Mallorca, Spain.

E-mail: jogarpe4@tal.upv.es

Abstract. Atmospheric freeze drying (AFD) is considered an alternative to vacuum freeze drying to keep the quality of fresh product. AFD allows continuous drying reducing fix and operating costs, but presents, as main disadvantage, a long drying time required. The application of power ultrasound (US) can accelerate AFD process. The main objective of the present study was to evaluate the application of power ultrasound to improve atmospheric freeze drying of carrot. For that purpose, AFD experiments were carried out with carrot cubes (10 mm side) at constant air velocity (2 ms^{-1}), temperature (-10°C) and relative humidity (10%) with (20.5 kWm^{-3}, USAFD) and without (AFD) ultrasonic application. A diffusion model was used in order to quantify the influence of US in drying kinetics. To evaluate the quality of dry products, rehydration capacity and textural properties were determined. The US application during AFD of carrot involved the increase of drying rate. The effective moisture diffusivity identified in USAFD was 73% higher than in AFD experiments. On the other hand, the rehydration capacity was higher in USAFD than in AFD and the hardness of dried samples did not show significant ($p<0.05$) differences. Therefore, US application during AFD significantly ($p<0.05$) sped-up the drying process preserving the quality properties of the dry product.

1. Introduction

The increasing need to achieve an equilibrium between high quality of dried products and low-operating cost has induced researchers to investigate the use of new techniques to intensify drying processes. From a quality point of view, freeze drying is considered to be one of the best food drying methods, but it is quite expensive due to freezing and the low pressures applied [1]. Atmospheric freeze drying (AFD) is an alternative consisting in the use of cold gas with low water vapor pressure as drying medium to cause direct moisture sublimation from frozen material at atmospheric pressure [2]. This technique could combine the advantages of both freeze drying (high product quality) and convective drying (low cost and continuous process). However, it presents as main drawback longer drying times than in traditional freeze drying processes.

Power ultrasound (US) has been recently applied in hot air drying of different products reducing both the drying times and energy costs. The potential of power ultrasound to improve mass transfer phenomena during drying of several fruits and vegetables has been showed in previous works [3] [4]. Ultrasonic energy could be applied during AFD to accelerate the drying process due to the mechanical energy introduced in the medium, which could contribute to reduce both the external and the internal mass transfer resistance.

Mathematical modeling constitutes an approach for analyzing drying and drier's operation [5]. Modeling of drying kinetics allows obtaining the effective moisture diffusivity of the product, essential parameter in simulation and optimization of drying process [6].

On the other hand, most of dried products should be rehydrated before consumption. Thus, it is interesting the knowledge of the behavior of the rehydrated product, being softening and loss of texture the main problems. The texture profile analysis is frequently used to evaluate the textural properties of rehydrated products indicating the tissue damage underwent during the operation.

The main aim of this work was to evaluate the application of power ultrasound to improve atmospheric freeze drying of carrot, quantifying its influence on drying kinetics and quality of dried products.

2. Materials and methods

2.1. Drying experiments

Carrot (*Daucus carota* var. Nantesa) cubes (10 mm side) were obtained using a houseware tool. Samples were sealed with film, frozen and stored at -18°C during 24 hours before drying. Initial moisture content was measured placing samples at 70°C and 200 mmHg until constant weight according to AOAC standards [7].

Atmospheric freeze drying experiments were conducted in a convective drier with air recirculation, temperature and air velocity control and an ultrasonically activated drying chamber, already described in literature [8]. Kinetics were carried out at constant air velocity (2 ms-1), temperature (-10°C) and relative humidity of the air (10%) with (20.5 kWm-3, USAFD) and without ultrasonic application (AFD). AFD and USAFD kinetics were carried out, at least, in triplicate and extended until samples lost 82% of the initial weight.

2.2. Rehydration experiments

The dried cubes (AFD and USAFD) were rehydrated until constant weight in distilled water at 25 °C using a thermostatic bath. Experiments were conducted, at least, in triplicate and samples were weighed at preset times.

2.3. Texture

Textural properties of dried/rehydrated and fresh carrot cubes were measured with a TA-XT2 texturometer (SMS, Godalming, UK) with a load cell of 25 kg. Texture profile analysis (TPA) was carried out by two compression cycles between parallel plates, at 25% strain, using a flat 75 mm diameter aluminium plunger (SMS P/75) and with a 5 s set period of time between cycles. Hardness was calculated from force/deformation profiles. At least, 10 samples were analyzed for each set of samples (AFD and USAFD rehydrated and fresh carrot cubes).

Analysis of variance (ANOVA) ($p<0.05$) was carried out and LSD (Least Significant Difference) intervals were identified using the statistical package Statgraphics Plus 5.1. (Statistical Graphics Corp., Warrenton, USA) in order to estimate significant differences.

2.4. Modeling drying kinetics

A diffusion model based on the Fick's law was used to mathematically describe the drying kinetics (AFD and USAFD) of carrot cubes. The differential equation of diffusion can be obtained combining Fick's law and the microscopic mass balance. For cubic geometry, the diffusion equation was written (1) considering constant effective moisture diffusivity and isotropic solid.

$$\frac{\partial W_p(x,y,z,t)}{\partial t} = D_e\left(\frac{\partial^2 W_p(x,y,z,t)}{\partial x^2} + \frac{\partial^2 W_p(x,y,z,t)}{\partial y^2} + \frac{\partial^2 W_p(x,y,z,t)}{\partial z^2}\right) \quad (1)$$

where W_p is the local moisture (kg w/kg d.m.), t is the time (s), D_e is the effective moisture diffusivity (m^2s^{-1}) and x, y and z represent the characteristic coordinates in cubic geometry (m).

In order to solve equation (1), some assumptions were considered: solid symmetry, uniform initial moisture content and temperature, constant shape during drying and negligible external resistance to water transfer. Taking into account these assumptions, the analytical solution of the diffusion equation is expressed in terms of the average moisture content in equation (2) [9].

$$W(t) = W_e + (W_e - W_0)\left[\sum_{n=0}^{\infty} \frac{8}{(2n+1)^2 \pi^2} e^{\left(-\frac{D_e (2n+1)^2 \pi^2 t}{4L^2}\right)}\right]^3 \quad (2)$$

where W is the average moisture content (kg w/kg d.m.), L the half-length of the cube side (m) and subscripts 0 and e represent the initial and equilibrium state, respectively.

2.5. *Model fitting*

For drying kinetics, the effective moisture diffusivity was identified by minimizing the sum of the squared difference between experimental and calculated average moisture content of samples. Identification was carried out by using the Generalized Reduced Gradient (GRG) method, available in Microsoft Excel TM spreadsheet (Microsoft Corporation, Seattle, WA, USA). The goodness of the fit was determined by the percentage of explained variance (%VAR, (3)) [10].

$$\%VAR = \left[1 - \frac{S_{xy}^2}{S_y^2}\right] \cdot 100 \quad (3)$$

where S_{xy} and S_y are the standard deviation of the estimation and the sample, respectively.

3. Results and discussion

3.1. *Drying kinetics*

AFD and USAFD kinetics of carrot cubes carried out at -10°C and 2 ms^{-1} are shown in figure 1. It can be observed that the application of US accelerated the AFD of carrot obtaining a reduction of drying time by 60% to achieve a weight loss of 82%. The kinetic improvement by US application was also observed in the values of the effective moisture diffusivity identified, which was 73% higher in USAFD ($7.98 \cdot 10^{-11}$ m^2s^{-1}) than in AFD ($4.62 \cdot 10^{-11}$ m^2s^{-1}) experiments. The percentage of explained variance obtained was 92.45 and 97.64% for USAFD and AFD, respectively, showing the better fit of the model to the AFD experimental data (figure 1). This fact could be explained considering that in the AFD experiments the internal resistance to mass transfer was much higher than the external one, which was nearly negligible. In the case of USAFD experiments, US diminished the internal resistance to mass transfer, which leads to a more significant external resistance [8]. This fact explains also the low %VAR provided by the diffusion model in the USAFD experiments.

3.2. *Rehydration kinetics*

The USAFD samples rehydrated faster than AFD ones (figure 2). The equilibrium moisture (W_{eq}) reached by the USAFD and the AFD samples was 5.05±0.06 and 4.06±0.21 kg w/kg d.m., respectively. The long drying times in AFD could involve a high degradation of carrot structure, which could explain the lower value of W_{eq} obtained in AFD than in USAFD experiments.

3.3. *Texture*

The hardness, determined from force/deformation profiles obtained by texture profile analysis of the rehydrated samples, was 3.34±0.63 N and 3.76±0.43 N for AFD and USAFD, respectively. Therefore, the hardness of the AFD and USAFD dried samples did not show significant ($p<0.05$) differences. The value for the fresh cubes (100.12±5.91 N) was much higher than for dried samples because dehydration involved a high degradation of carrot structure. Thereby, rehydrated samples do not recover the initial texture and dried/rehydrated samples were much softer than fresh ones.

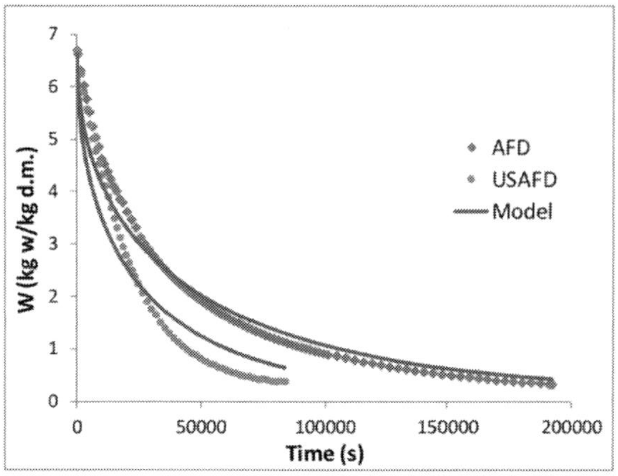

Figure 1. Atmospheric freeze drying kinetics of carrot cubes with (USAFD, -10°C, 2 ms^{-1} and 20.5 kWm^{-3}) and without (AFD, -10°C and 2 ms^{-1}) power ultrasound application.

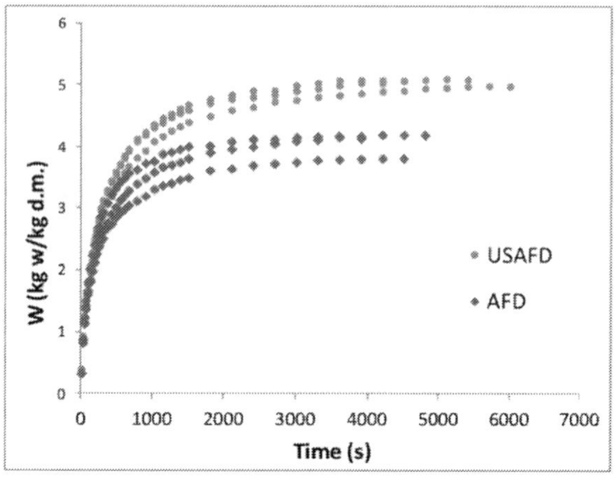

Figure 2. Rehydration kinetics (25°C) of AFD and USAFD dried carrot cubes.

4. Conclusion

The ultrasonic application during AFD kinetics of carrot involved the increase of drying rate achieving a reduction of drying time by 60%. The diffusion model proposed was adequate for describing the AFD kinetics but not fitted well to the USAFD experimental data. The effective moisture diffusivity identified in USAFD was 73% higher than in AFD experiments. On the other hand, the rehydration capacity was higher in USAFD than in AFD and the hardness of dried samples did not show significant ($p<0.05$) differences. Therefore, USAFD could represent an interesting alternative to vacuum freeze drying to achieve high quality dried products with lower cost.

Acknowledgments

The authors acknowledge the financial support of the Ministerio de Economía y Competitividad (MINECO) of Spain from the project DPI2009-14549-C04-04.

References

[1] Stawczyk J and Li S 2007 *Transp. Porous Med.* **66** 159-172
[2] Wolff E and Gibert H 1990 *Drying. Technol.* **8** 385-404
[3] Garcia-Perez J V, Cárcel J A, Benedito J and Mulet A 2007 *Food Bioprod. Process.* **85** 247-254
[4] Gallego-Juarez J A, Riera E, de la Fuente Blanco S, Rodriguez-Corral G, Acosta-Aparicio V M and Blanco A 2007 *Drying. Technol.* **25** 1893-1901
[5] Mulet A, Cárcel J A, Sanjuan N and Garcia-Perez J V 2010 Food dehydration under forced convection conditions *Recent Progress in Chemical Engineering* ed J Delgado (Houston, TX, USA: Studium Press LLC) pp 153-177
[6] Hassini L, Azzouz S, Peczalski R and Belghith A 2007 *Journal of Food Engin.* **79** 47-56
[7] AOAC 1997 *Official Methods of Analysis* (Arlington, Virginia, USA: Association of Official Analytical Chemist)
[8] Garcia-Perez J V, Cárcel J A, Riera E, Rosselló C and Mulet A 2012 *Drying. Technol.* DOI: 10.1080/07373937.2012.675533
[9] Crank J 1975 *The Mathematics of Diffusion* (London: Oxford University Press)
[10] Berthouex P M and Brown L C 1994 *Statistics for Environmental Engineers* (New York, USA: Lewis Publishers Inc)

IOP Conf. Series: Materials Science and Engineering **42** (2012) 012022 doi:10.1088/1757-899X/42/1/012022

The acoustic spectroscopy of asphaltene aggregation in petroleum

G D Abbott[1] and M J W Povey[2]

[1]School of Civil Engineering and Geosciences, Drummond Building, Newcastle University, Newcastle upon Tyne NE1 7RU, UK
[2] School of Food Science and Nutrition, University of Leeds, Leeds LS2 9JT, UK

E-mail: geoff.abbott@newcastle.ac.uk

Abstract. Asphaltenes are well known to be important in causing blockages during production, transportation and refining. They are also responsible for the formation of tar mats in petroleum reservoirs, which can impede crude oil recovery. Acoustic spectroscopy in the frequency range 2 – 120 MHz was used to detect the aggregation of asphaltenes in a North Sea crude oil (UK continental shelf). In the absence of aggregation, both the toluene solvent and the maltenes isolated from the oil exhibit a power law dependence which is a function of the continuous phase. In contrast the crude oil itself showed clear signs of a changing size of the scattering entities. With sufficient information, it may in future be possible to determine the size of the scattering particles from the acoustic spectrum including the changing size of the asphaltene particles during aggregation.

1. Introduction

Asphaltenes are known to be important components causing a number of problems during production, transportation and refining e.g. deposition in flowlines [1, 2]. They are also responsible for the formation of tar mats in petroleum reservoirs, which can impede crude oil recovery [3]. Despite the amount of effort that has been expended on studying these nanoparticles over a number of decades, it is really only in the last few years that the connection between molecular architecture and aggregation has been clarified leading to the development of the modified Yen model (also known as the Yen-Mullins model) where the asphaltenes form nanoaggregates with aggregation numbers of ~ 6 [4]. This model has recently been validated using centrifugation and DC-electrical conductivity methods [5]. Complementary neutron and X-ray small angle scattering has revealed that each nanoaggregate is best described by a disk of total radius 32 Å and a height of 6.7 Å with an aromatic core and a highly aliphatic shell [6]. Furthermore the ultrasonic velocity has been plotted as a function of asphaltene concentration in order to measure critical nanoaggregate concentrations (CNACs) in toluene for asphaltenes isolated from two Kuwaiti crude oils [7]. In this study we report the application of acoustic spectroscopy (frequency range 2 – 120 MHz) for detecting asphaltene aggregation in a North Sea oil (UK continental shelf).

2. Materials and methods

2.1 Samples

A selection of samples were analysed including a fresh non-biodegraded North Sea oil (UK continental shelf), a biodegraded North Sea oil (UK continental shelf), a source rock bitumen from the blackstone band in the Kimmeridge Clay Formation (Upper Jurassic, Dorset, UK) as well as isolated asphaltene and maltene fractions from these oils and the Kimmeridge Clay source rock. To obtain the bitumen, the rock sample was extracted for 72 hours with a dichloromethane (DCM)/methanol mixture (93:7, v/v) using the Soxhlet extraction method. The solvent was removed from the extract by rotary evaporation (30 °C, 25 mmHg) followed by drying under a stream of nitrogen until a constant weight of bitumen was obtained. The asphaltene and maltene fractions were isolated both from the bitumen and the crude oils by dissolving each of them in 1 mL of DCM which was treated with 40 mL excess of *n*-hexane. The mixture was stirred for 2 hours and then allowed to equilibrate for 24 hours. The asphaltenes were recovered by centrifugation (3500 rev/minute, 15 minutes) and re-dissolved in 1 mL DCM followed by re-precipitation with 40 mL of *n*-hexane. After about 30 minutes of stirring, the asphaltenes were recovered which also yielded a maltenes fraction. This was repeated two more times. The asphaltenes were finally extracted with *n*-hexane for 10 days using the Soxhlet method to remove co-precipitated maltene [8].

2.2 Experimental

Acoustic attenuation measurements were carried out using an Ultrasizer MSV by Malvern Ltd [Malvern, UK] (figure 1), an instrument which performs acoustic spectroscopy of liquids and emulsions in the frequency range 2 – 120 MHz. In brief, the instrument operates in transmission mode where one transducer emits sound into the sample while a second transducer receives the sound waves. Two such pairs, one operating at low frequency and one at high frequency, are employed. Sub-bands are 2 MHz to 15 MHz (low frequency pair), 10 MHz to 150 MHz (high frequency pair). We did not record the actual set used, since this is set during optimization. Due to their broad-band response, the acoustic frequency of each pair can be swept over a wide range. The attenuation is measured from the decay of the received voltage with propagation distance through the sample. The diffraction of the beam, which also causes variations in the received voltage, is calibrated automatically and its effects factored out. Furthermore, by comparing the received signal at two propagation distances the non-trivial response of the transducer and electronic circuitry and any other systematic errors can also be accounted for. Hence, an absolute measure of the acoustic attenuation is determined. Finally, the transducer separation is calculated for a given fluid so as to exploit the full dynamic range of the measurement system and maximise the signal-to-noise ratio.

Figure 1. The Malvern Ultrasizer with its associated temperature control unit. The sample vessel is shown foremost in the picture.

The Ultrasizer requires 500 mL of sample and is connected to an external Huber Ministat temperature control unit which enables the sample temperature to be varied across a range of 0 - 50 °C. During the measurement process the sample under test is agitated by a stirrer whose angular speed can be varied in the range 0 - 500 rpm. The agitator is employed to reduce thermal fluctuations in the sample (for emulsions it also suppresses sedimentation and creaming). This combination maintains the sample at the selected temperature to within 0.5 °C, as measured by the internal thermometer.

The Ultrasizer therefore provides a stable temperature to determine the associated acoustic attenuation. Note that the acoustic power levels are of the order of milliwatts and so the associated heating of the sample is negligible.

The instrument implements automated processes to conduct measurements at 50 different frequencies selected within the range 2 – 120 MHz. At each selected frequency of the acoustic spectra, 30 repeated attenuation measurements were made and, from these, an average attenuation value was calculated.

3. Results and discussion

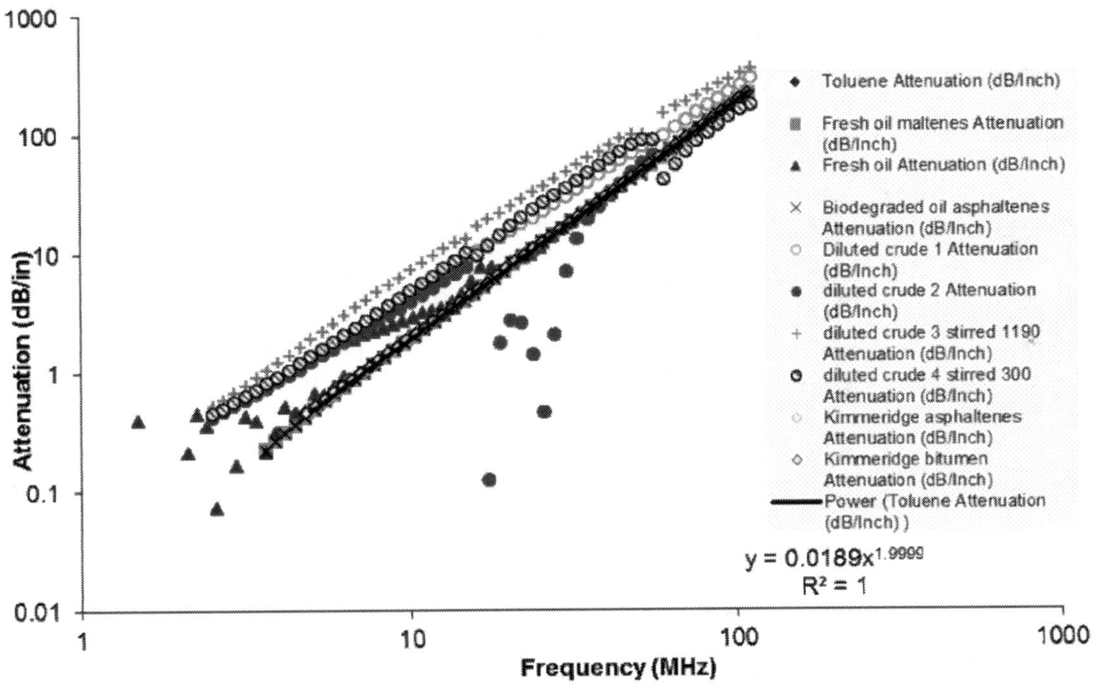

Figure 2. Attenuation vs frequency for toluene (closed diamond), fresh oil maltenes (closed square), fresh North Sea oil (closed triangle), biodegraded oil asphaltenes (cross), diluted North Sea crude (open circle), diluted North Sea crude (closed circle), diluted and stirred North Sea crude (plus), diluted North Sea crude (oblique line in circle), Kimmeridge asphaltenes (open diamond), Kimmeridge bitumen (closed diamond), linear fit to toluene attenuation (line).

In the absence of aggregation, both the toluene (closed diamonds) and fresh oil maltenes (closed squares) exhibit a power law dependence which is a function of the continuous phase (figure 2). The attenuation spectrum for the whole fresh oil contained breaks and departures from the normal power law dependence (see ▲in figure 2). These breaks and jumps are occurring because the measurement frequency sweep takes around 7 minutes during which the aggregation process which, as is evident in figure 2, takes place changing the sample during the measurement and leading to significant departures from power law dependence.

The fresh crude oil therefore shows clear signs of a changing size of the scattering entities. This is also the case when the crude oil is diluted with toluene. We might expect the toluene to cause the aggregates to disperse, however our data suggest that nanoaggregates (detectable by ultrasound) remain since the data departs from power law dependence. Once the diluted fresh crude is stirred sufficiently (+ in figure 2) the re-suspended asphaltene's spectrum approaches a power law once again. In contrast, the toluene diluted bitumen (including the isolated asphaltenes) behaves exactly like the toluene in which the asphaltenes are suspended, indicating that in this case no measurable excess scattering was occurring.

4. Conclusions

Ultrasound spectroscopy shows clearly that aggregation of asphaltenes is taking place. In the absence of aggregation, the crude oil exhibits a power law dependence which is a function of the continuous phase together with the suspended particles which will scatter sound. With sufficient information, it may in future be possible to determine the size of the scattering particles from the acoustic spectrum [9, 10], including the changing size of the asphaltene particles during aggregation.

These results demonstrate the potential that lies in bringing together the normally disparate disciplines of organic geochemistry and ultrasonic spectroscopy in order to improve our fundamental understanding of the asphaltene fraction.

References

[1] Sheu E Y 2002 *Energy Fuels* **16** 74
[2] Creek J L 2005 *Energy Fuels* **19** 1212
[3] Wilhelms A and Larter S R 1994 *J. Marine Pet. Geol.* **11** 418
[4] Mullins O C 2011 *Annu. Rev. Anal. Chem.* **4** 393
[5] Goual L, Sedghi M, Zeng H, Mostowfi F, McFarlane R and Mullins O C 2011 *Fuel* **90** 2480
[6] Eyssautier J, Levitz P, Espinat D, Jestin, J, Gummel J, Grillo I, Barre L 2011 *J. Phys. Chem. B* **115** 6827
[7] Andreatta G, Bostrom N and Mullins O C 2005 *Langmuir* **21** 2728
[8] Alboudwarej H, Beck J, Svrcek W Y, Yarranton H W and Akbarzadeh K 2002 *Energy Fuels* **16** 462
[9] Povey M J W 1997 *Ultrasound Techniques for Fluids Characterization* (London: Academic Press)
[10] Challis R E, Povey M J W, Mather M L and Holmes A K 2005 *Rep. Prog. Phys.* **68** 1541

Acoustic attenuation spectroscopy for process control of dispersed systems

J G DosRamos

Matec Applied Sciences, 56 Hudson St., Northborough, MA 01532 USA

E-mail: dosramos@matec.com

Abstract. Acoustic Attenuation Spectroscopy (AAS) is a technique for the measurement of Particle Size Distribution (PSD) of concentrated colloids – in most cases without the need for sample dilution, i.e., at process conditions. While successful in the laboratory, AAS has not been widely employed for Process Control purposes. This paper describes a new flow-through (Process Online) measurement approach that can make AAS into a Process Control tool for industrial production settings, in particular for PSD measurement and control. This approach basically consists of splitting ultrasonic sound waves by using stepped reflectors and/or transducer delay rods which provide multiple measurement gaps without the use of moving parts.

1. Introduction

AAS is commonly used to measure particle size distributions (PSD) in the range 5 nm to 100 microns of aqueous, as well as, non-aqueous colloidal samples such as Chemical Mechanical Polishing (CMP) slurries, Ceramic Slips, Inks, Minerals, Catalysts, Metal Oxides, Pigments, O/W and W/O emulsions [1-5].

AAS consists of measuring the attenuation (zero to over 100 dB/cm) of sound waves (1-100 MHz) traveling through the colloid as a function of the Frequency of the sound waves. Measurements are typically performed at several accurate gaps for broader attenuation spectra. The colloids may be flowing and/or being stirred (thus, particle settling is prevented) at percent solids level of 0.1 to 60% volume although Greenwood et. al. have used long-path measurements on lower-concentration samples[6]. Sound speed spectra can be readily measured in addition to attenuation. Sound speed data can be employed to infer molecular-level phenomena in colloidal samples[7].

2. Current AAS laboratory systems

Despite the advantages AAS offers for PSD analysis in laboratory applications[4, 5, 8-13], it suffers from the following drawbacks for Process/Online application: (a) High-Resolution Motion Stage/Stepper motor required in order to perform measurements at multiple gaps, i.e., higher initial cost, and maintenance requirements which reduce uptime cycles; (b) Abrasion issues (Transducers/Reflectors, Seals) from flowing slurries. This factor is compounded by the fact that many slurries are designed to be highly abrasive, e.g., CMP Semiconductor slurries; (c) Relatively-Long Spectrum Acquisition (~couple of minutes) which prevents real-time data. One difficult scenario occurs when the flowing colloid's percent solids level fluctuates frequently which can reduce spectrum accuracy (attenuation level is directly proportional to particle volume fraction).

3. Proposed process online system

Desirable features of an AAS Process Online System are as follows: (i) No moving parts; (ii) Minimized abrasion; (ii) Attenuation measurements at multiple gaps; and (iv) Fast spectrum acquisition (< 30 seconds). These provide the following benefits: (i) Reduced initial and maintenance costs; (ii) Increased reliability; (iii) Maximum uptime; (iv) near Real-time data spectrum acquisition.

Figure 1 shows two different representations of proposed online sample cells for acoustic attenuation and sound speed spectrum measurements that meet these requirements. Figure 1(a)'s cell has a transmitter/receiver transducer plus a 3-step reflector (2 or more reflector surfaces, i.e., one or more steps, are desirable). This arrangement supplies three stationary mirror surfaces providing three different measurement gaps. More than one transducer element can be placed on the delay rod in order to widen the frequency range of the measurement. This arrangement will be referred to as AREPA (Acoustic Reflection Particle Analysis). It provides multiple acoustic paths without the need for a motion stage while minimizing abrasion effects. Thus, AREPA provides reduced initial and operational costs, increased reliability, and maximum uptime. Measurements are relatively fast (near real time) since there is no need to move reflectors or transducers.

Reed et. al. [3] used the delay rod's internal echo waveform as a reference to calculate attenuation data from a movable-reflector system. This approach enjoys increased accuracy through the use of cross correlation of the reference and liquid echo waveforms. AREPA measurements are more suitable for less concentrated colloid measurements as the sound travels twice in the gap.

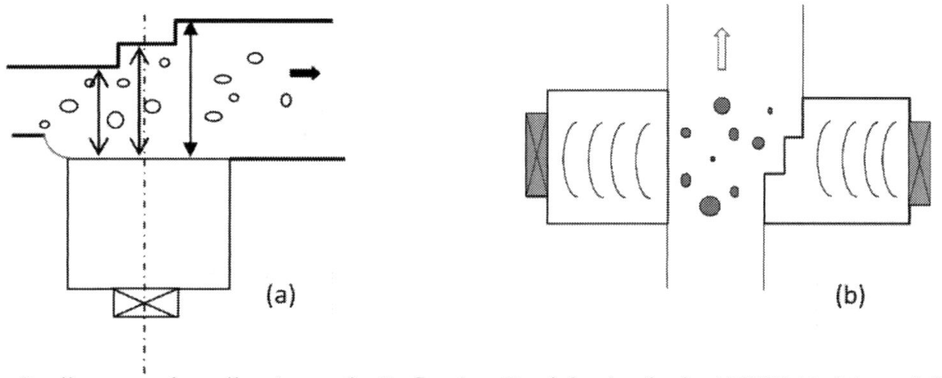

Figure 1. On line sample cells: Acoustic Reflection Particle Analysis (AREPA) (a), and Through-Transmission (TTAPAS) (b), show three different sound paths (2 or more desirable).

Alternatively, a Through-Transmission Acoustic Particle Analysis System (TTAPAS)-based sample cell would allow characterization of higher percent solids level samples as the sound waves travel only once in the gap as depicted in figure 1(b). The reflector and transducer steps can have various shapes, e.g., rectangular, circular, etc., including stacked/recessed disc-like surfaces; however, these would likely be more prone to abrasion issues. An alternative TAPPAS arrangement can be implemented as follows: instead of steps on a single transducer, multiple narrow-diameter transducers can be placed at different distances from the flat transducer (thus providing multiple measurement gaps).

These online systems could be either in process pipes (inline) or on a bypass. Three-way valves can be installed at either side of the flow-through AREPA/TTAPAS cell for periodic sensor calibration and/or cleaning by stopping colloid flow while allowing calibration solvent (without particles) to flow through the sensor. This calibration approach is similar to Reed et. al.'s calibration[3] except that it is extended to the use of non-aqueous solvent spectra (measured by lab units) as calibration fluids in order to avoid using water in non-aqueous applications.

4. Experimental data

Figure 2 shows echo waveforms from a 2-step (each 0.4 cm deep) reflector in DI water. The 0.64 cm wide middle reflector is nearly rectangular. Each reflector has the same surface area. Not shown is the reference echo within the 2.5 cm diameter quartz delay rod. Four echoes from the reflector surface closest to the transducer are 1, 2, 4, and 5. Echo 3 is the first echo from the middle reflector while the last echo is from the farthest reflector. These sound wave amplitudes/characteristics need to be incorporated into the sample-cell calibration for accurate acoustic attenuation spectrum data. Edge echoes are minimal for a 0.64 cm transducer element/center reflector setup (b).

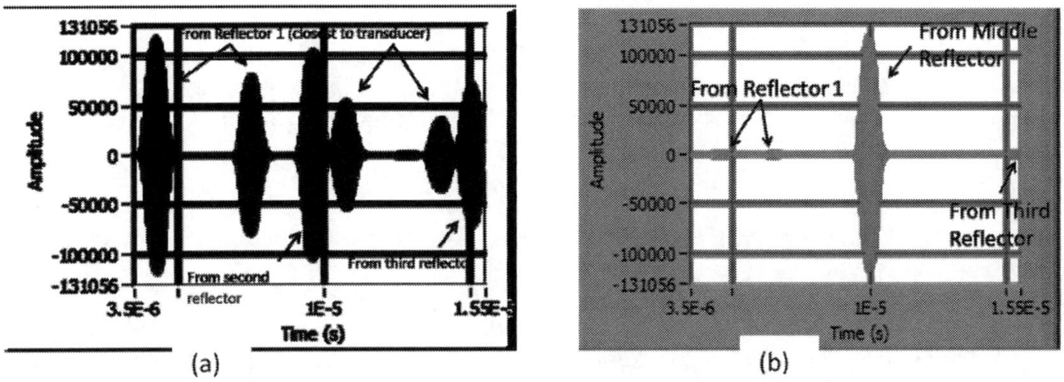

Figure 2. Waveforms (18 MHz, 21.1 °C.) from a two-step (three equal-area reflectors) AREPA cell on a DI-Water sample; step depths= 2 x 0.4 cm, total reflector diameter= 2.54 cm, center reflector width= 0.64 cm; transducer-element diameter= 1.9 cm (a), and 0.64 cm (b).

This arrangement divides the sound wave energy into as many reflector surfaces. Thus, each echo (or transmitted by TTAPAS) amplitude depends on each step's surface area. The farthest step can be made to have a larger area for increased echo amplitude. Radial location of each step is also important. The center of 3 steps (or TTAPAS delay rod) reflects more energy than the edge ones. Thus, the center reflector can have a smaller surface area.

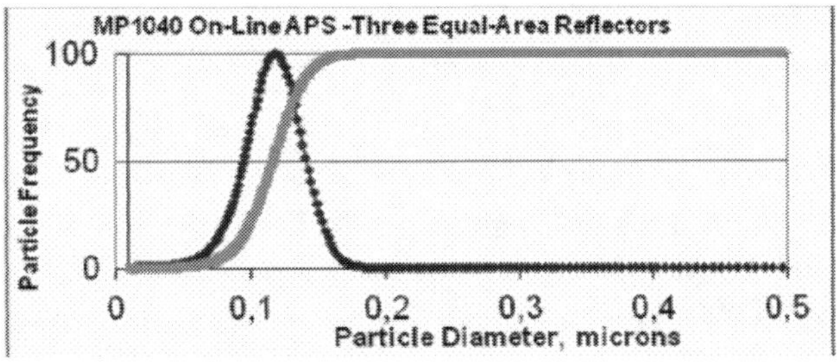

Figure 3. PSD data for a 4% wt silica sample using a three-reflector AREPA sample cell.

Figure 3 presents PSD data for a 4% wt MP1040 silica sample (Nissan Chemical) measured by a 3-reflector (2 x 0.4 cm steps) AREPA sample cell. PSD computations were done using the Carrasso et. al. [13] model.

5. Particle Size Distribution for Process Control

PSD curves are not practical as set points. However, mean particle size values (Volume/Weight, Area, or Number) can be misleading as any single mean particle size can be produced by an infinite number of PSD curves [14].

6. Conclusions

Acoustic Reflection Particle Analysis (AREPA), and Through-Transmission Acoustic Particle Analysis System (TTAPAS) show strong promise for PSD measurement in process control/Online scenarios. Both approaches eliminate the use of moving parts while minimizing abrasion. Two or more reflectors, i.e. one or more steps, would be preferable (for multiple gaps).

Area-average sizing data are less prone to sharp fluctuations from minor populations and thus would be more suitable as a set point for process control. Alternatively, the ratio of Volume/Area-weighted mean particle sizes would be more stable than the polydispersity index (PDI, ratio of volume to number average values). PSD Percentile figures, e.g. 90%, can also be useful.

Acknowledgements

The author would like to thank Prof. Robert W. Reed for his valuable feedback toward this work.

References

[1] Oja T, DosRamos J G and Reed R W 2004 Acoustic analysis of concentrated colloidal systems *Particle Sizing and Characterization* ed T Provder and J Texter (USA:Oxford University Press) chapter 15 pp 231-247
[2] Povey M J W 1997 *Ultrasonic Techniques for Fluid Characterization* (San Diego, California: Academic Press)
[3] Reed R W, DosRamos J G, and Oja T 2001 *Review of Progress in Quantitative. Nondestructive Evaluation* vol 21 ed D O Thompson and D E Chimenti (AIP Press) pp 1494-1501
[4] Alba F 1992 US Patent No. 5 121 629
[5] DosRamos J G, Reed R W, Oja T and Boulet G 2003 US Patent No. 6 604 408
[6] Greenwood M S, Adamson J D and Bamberger J A 2006 *Ultrasonics* **44** e461
[7] Andreatta G, Bostrom N and Mullins O C 2007 Ultrasonic Spectroscopy of Asphaltene Aggregation *Asphaltenes, Heavy Oils and Petroleomics* ed O C Mullins *et al* (New York: Springer) pp 231-257
[8] Epstein P S and Carhart R R J 1953 *J. Acoust. Soc. Am.* **25** 553
[9] Allegra J R and Hawley S A J 1972 *J. Acoust. Soc. Am.* **51** 1545
[10] Tempkin S J 1998 *J. Acoust. Soc. Am.* **103** 838
[11] Dukhin A S, Goetz P J and Hamlet C W 1996 *Langmuir* **12** 4998
[12] Tebbutt R E and Challis R E 1996 *Ultrasonics* **34** 369
[13] Carasso M, Patel S, Valdes J L and White C A 2000 US Patent No. 6 119 510
[14] DosRamos J G 2004 *Particle Sizing and Characterization* ed T Provder and J Texter (USA: Oxford University Press) pp 138-150

IOP Conf. Series: Materials Science and Engineering **42** (2012) 012024 doi:10.1088/1757-899X/42/1/012024

Acoustical Studies of L-leucine and L-asparagine in aqueous electrolyte through thermal expansion coefficient

S Jajodia[1], O P Chimankar[2,4], A Kalambe[2] and S G Goswami[3]

[1]Department of Physics, Chouksey Engineering College, Bilaspur (C.G.) -495 004
[2]Department of Physics, RTM Nagpur University, Nagpur-440033, INDIA
[3]Department of Physics, M.M.Science College, Nagpur

E-mail: opchimankar28@gmail.com
E-mail: drsangeetanj@yahoo.com

Abstract. Amino acids are the building blocks of the proteins; their study provides important information, about the behaviour of larger biomolecules such as proteins. The properties of proteins such as their structure, solubility, denaturation, etc. are greatly influenced by electrolytes. Ultrasonic velocity and density values have been used for evaluation of thermal expansion coefficient and adiabatic compressibility for ternary systems (amino acid/salt + water) namely L-leucine / L-asparagine each in 1.5 M aqueous solution of NaCl used as solvent for various concentrations and at different temperatures (298.15K – 323.15K). Present paper reports the variation of various thermoacoustical parameters such as Moelwyn-Hughes parameter (C_1), Beyer's non-linearity parameter (B/A), internal pressure (P_i), fractional free volume (f), available volume (V_a), repulsive exponent (n), molecular constant (r), van der Waals' constant (b), Debye temperatue (θ_D), etc. have been computed from the thermal expansion coefficient with the change of concentration and temperature. The variations of all these parameters have been interpreted in terms of various intermolecular interactions such as strong, weak, charge transfer, complex formation, hydrogen bonding interaction. The structure making and breaking properties of the interacting components existing in proposed ternary systems. It shows the associating and dissociating tendency of the molecules of solute in solvent.The hetromolecular interactions are present in both the ternary systems.

1. Introduction

Due to the complex molecular structure of proteins, direct study is somewhat difficult. Therefore, the useful approach is to study simpler model compounds, such as amino acids which are building blocks of proteins. The amino acids serve precursors of harmones, alkaloids, nerves, transition of all growth and many other biomolecules [1].

Ultrasonic study on the amino acids [2-3], proteins [4], peptides [5-6] in aqueous solutions and organic solvents have been carried out by number of researchers for investigation of solute-solute, solute-solvent intermolecular/interionic interactions. Also, the researchers have employed the ultrasonic technique to evaluate the thermodynamic properties of amino acids [7-8], peptides [9] and proteins [10].

[4] To whom any correspondence should be addressed

L-leucine is an essential amino acid, which means it can't be produced naturally by the body. So the body has to obtain it through diet or supplements such as beef, chicken, fish, dairy, eggs and dry fruits. It plays an important role in muscle gain and weight loss. L-asparagine is a non-essesntial amino acid which means it is manufactured from other amino acids in the liver. It participates in the functioning of brain and nervous system and helps to maintain equilibrium as well as transformation of amino acid in the liver. L-asparagine is commonly found in poultry, dairy, eggs, fish, meat, nuts, seafood and whole grains.

Thermodynamic properties of amino acids i.e. proteins in electrolyte solution provide information about solute- solvent interactions, those are important in understanding the stability of Proteins. Some of these interactions are found applicable in several biochemical and physiological processses in a living cell [11]. Such studies give useful information regarding protein unfolding and the extent of hydrophobic interactions of nonpolar side chains [12]. Salt- protein interaction induced electrostatic forces are used to play a very important role in modifying the protein structure by affecting properties like solubility, denaturation and activity of enzymes. In order to see influence of electrolytes on amino acids and their molecular interactions between salt & proteins functional group, more detailed investigations on thermodynamic behaviour are still needed.

The ultrasonic velocity data as such do not provide significant information about the nature and relative strength of various types of interactions (viz., ion-solvent, solute-solvent, solute-solute, ion-ion, etc). Hence their derived parameters such as thermal expansion coefficient (α), Moelwyn-Hughes parameter (C_1), Bayer's non-linearity parameter(B/A), internal pressure (P_i), repulsive exponent of intermolecular potential (n), Debye temperature (θ_D), cohesive energy density (ε) and intermolecular binding energy (δE) have been evaluated using standard relations[13] to detect possible inter-ionic interactions occurring between the amino acids and aqueous sodium chloride solution at different concentration and temperature.

2. Theoretical Method

The experimentally measured values of ultrasonic velocities (u) and densities (ρ) used from the literature [14] for evaluation of α for ternary systems L-leucine/L-asparagine each in 1.5M aqueous NaCl solutions for various concentrations and at different temperatures (298.15K–323.15K).The variation of molar volume or density with different temperature provide the thermal expansion coefficient. The various formulae are listed below, where β_a is the adiabatic compressibility, α is the thermal expansion coefficient, C_1 is the Moelwyn–Hughes parameter and n is the repulsive exponent.

$$\beta_a = 1/u^2\rho \tag{1}$$

$$\alpha = 1/\rho.(d\rho/dt) \tag{2}$$

$$C_1 = (13/3) + (\alpha T)^{-1} + (4/3)\alpha T \tag{3}$$

$$n = 3[2/f + 5] \tag{4}$$

3. Results and discussion

The theoretically evaluated values of all thermoacoustic parameters in aqueous sodium chloride solution of amino acid : L-leucine and L-asparagine as functions of molal concentration and temperature have been plotted in figure 1-8. The adiabatic compressibility values decreases with increase in concentration of solute as well as with temperature. This may be due to (i) an increase in the number of incompressible molecules/zwitterions in solutions and (ii) the formation of compact structure of zwitterions-water dipole and zwitterions-ions structure in solutions.

The similar abnormal behaviour of α (figure 1 & 2), P_i (figure 7 & 8), ε, δE and opposite nature of C_1 (figure 3 & 4), B/A, n in L-leucine and L-Asparagine with aqueous NaCl shows complex formation and strong intermolecular interaction in the mixures.The nonideality is more at higher concentration of L-Asparagine in aqueous NaCl due to the presence of hydrophobic nature of its side chain. This behaviour has been also confirmed by the variation of ultrasonic velocity and adiabatic compressibility with different concentration and temperature from previous literature [14].

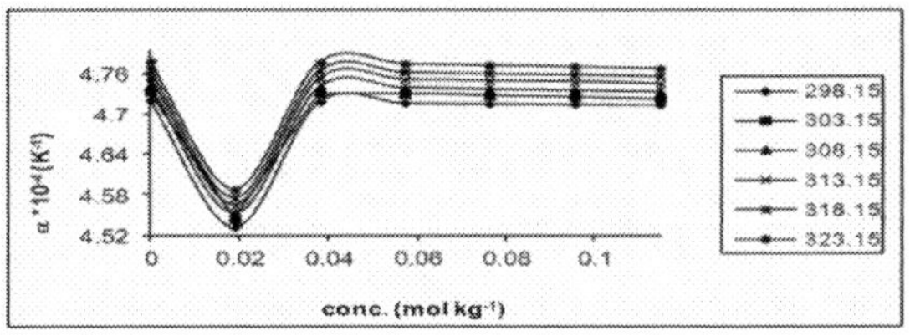

Figure 1. Variation of α with concentration for different temperature in L-leucine.

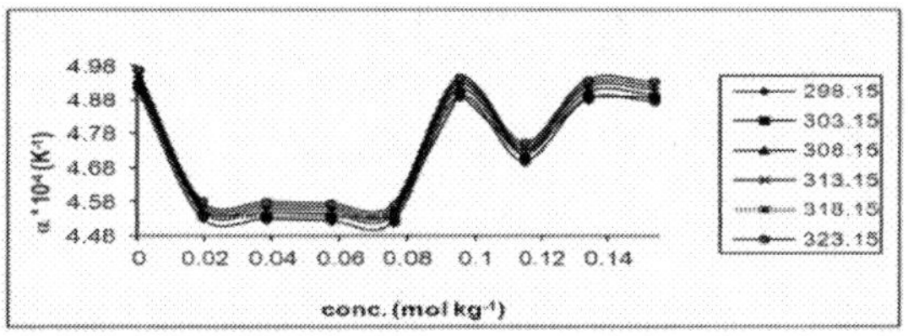

Figure 2. Variation of α with concentration for different temperature in L-asparagine.

In both the systems, the variation of C_l and α shows the opposite behaviour with the increase molal concentration and temperature i.e. when C_l increases, α decreases and vice-versa. This may be due to the associating tendency of the liquid mixture. In present systems, the variation of C_l & n found to be initial increasing trend at lower concentartion of L-Leucine and L-Asparagine, exhibit a peak at concentration 0.02 and then its variation is uniform at higher concentration of L-leucine in aqueous NaCl. The high non-ideal behaviour in L-Asparagine due to its hydrophobic nature of side chain at higher concentartion.

The initial increase of C_l, B/A and *n* shows the decrease in intramolecular modes of vibration and anharmonicity in both the systems under study. It also indicates the dissociating tendency and weak intermolecular forces in the amino acid solutions. The decrease in C_l and *n* from concentration 0.02 shows the increase in intramolecular modes of vibration and harmonicity in the solutions. This indicates associating nature in both ternary systems is attributed to strong intermolecular forces [15]. The values of *n* are found to be ~ 34, which is of the same order as polymer & fluorocarbon fluids**.** The high value of '*n*' shows the bulk nature of the molecules. The increase in *n* with molar concentration shows associating tendency of the molecules in the amino acid solutions. The low value of '*n*' shows repulsive forces at the slightly larger distances in the liquids than in solids.

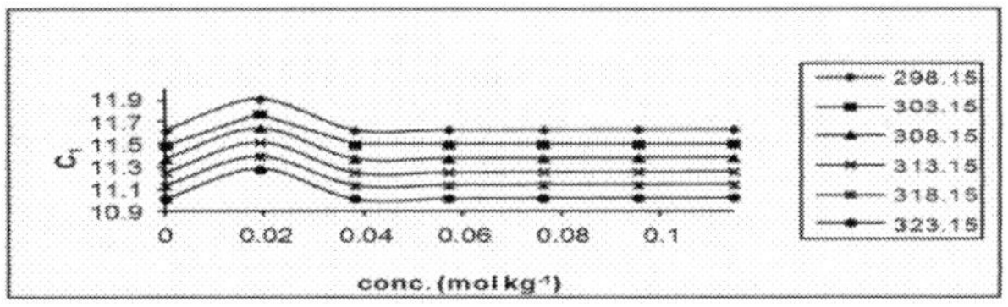

Figure 3. Variation of C_l with concentration for different temperature in L-leucine.

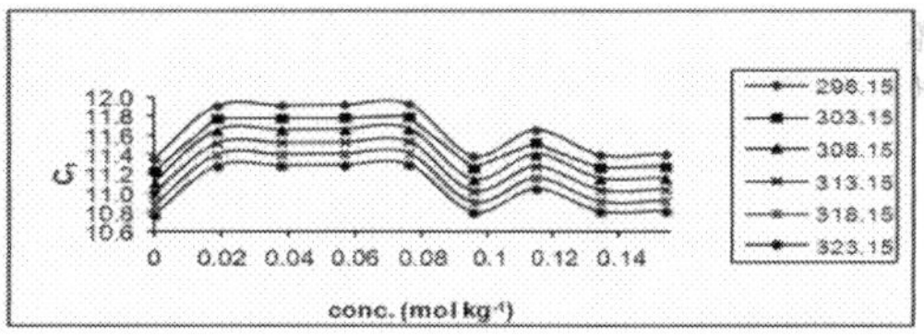

Figure 4. Variation of C_l with concentration for different temperature in L-asparagine.

The increase in θ_D (figure 5 & 6) with concentration of amino acid indicates the associating tendency among the amino acid (L-Leucine and L-Asparagine), NaCl and water molecules, owing to increase in the crystalline nature of liquid lattice.

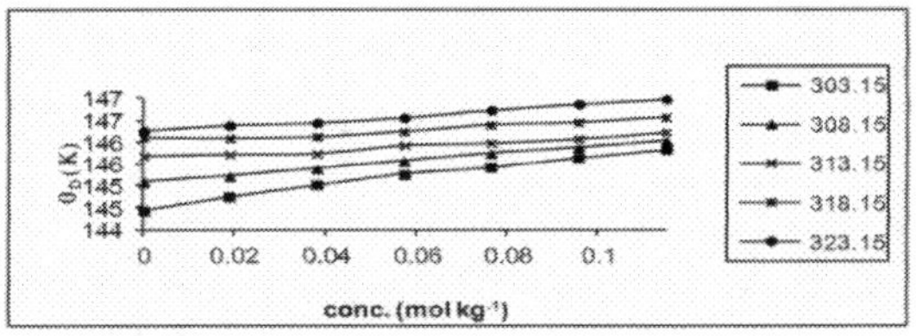

Figure 5. Variation of θ_D with concentration for different temperature in L-leucine.

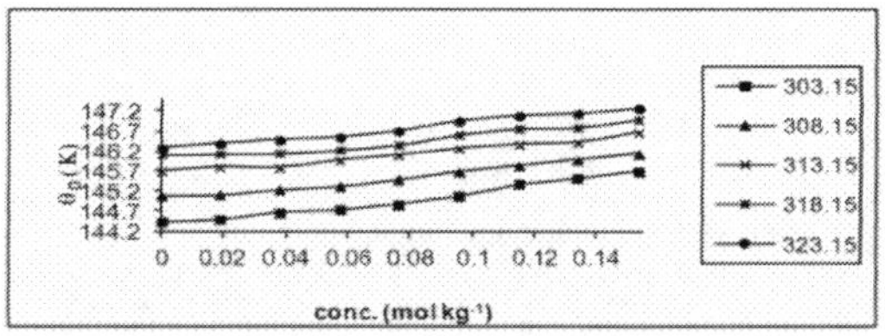

Figure 6. Variation of θ_D with concentration for different temperature in L-asparagine.

An increase in P_i and α from concentration 0.02 shows the orientation of the solvent (aqueous NaCl) molecules around the amino acid ions, which may be due to the influence of electrostatic field of ions. This means solution becomes less compressible. It indicates the associating tendency of the molecules in the amino acid solutions. Conversely, the initial reduction in P_i and α shows dissociating tendency of the molecules in the solutions.

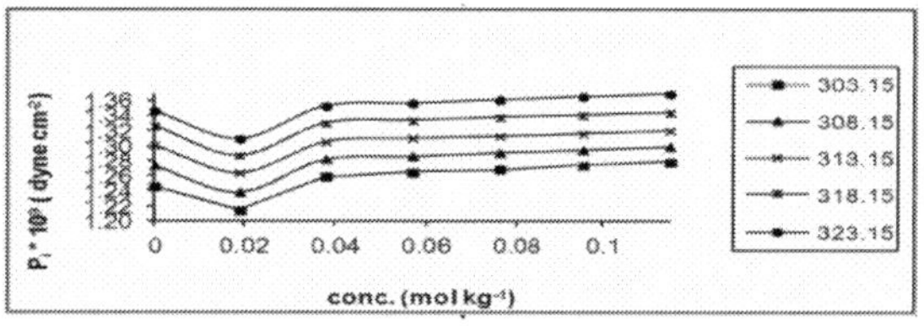

Figure 7. Variation of P_i with concentration for different temperature in L-leucine.

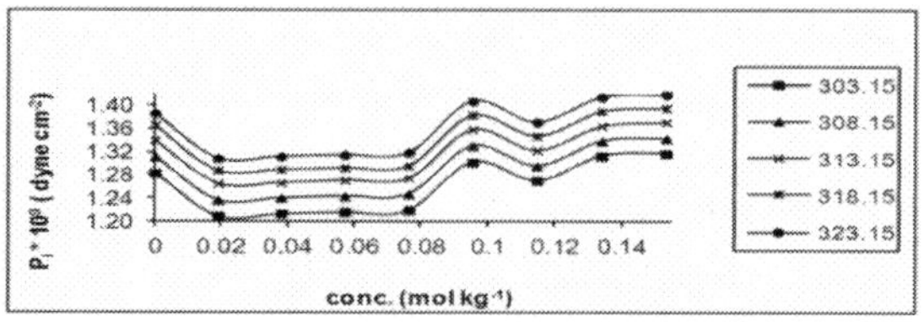

Figure 8. Variation of P_i with concentration for different temperature in L-asparagine.

The cohesive energy of a material is the energy required to disassemble it into its constituent parts, also known as binding energy. The increase in ε with increase in molal concentration shows associating tendency of molecules in the solutions. δE exhibits a similar behaviour as ϵ, Pi & α.

4. Conclusions

The observed nonlinearity confirms the presence of complex formation, solute- solvent interactions and weak association due to hydration. The existence of cohesive forces in the present systems may be the result of Zwitterions-ions and Zwitterions-water dipole interactions.The existence of ion-solvent or solvent – solvent interaction resulting in attractive forces promote the structure – making tendency (due to hydrophilic side chain) shows in L-Leucine, while ion-ion and solute – solute interaction resulting dipole-dipole, dipole induced dipole and electrostatic forces enhance the structure – breaking properties (due to hydrophobic side chain) in L-Asparagine at higher concentration in aqueous NaCl. Hence it is evident that the thermoacoustic measurement in the amino acid with aqueous electrolyte serves as a powerful probe in characterizing the physico-chemical properties of the medium.

Acknowledgement

The one of the author (OPC) are greatful to University Grant commission, New Delhi for providing financial assistance to this work through Major research project.

References

[1] Bhat R and Awuwalia J C 2003 *J. Phys. Chem.* **99** 1099
[2] Rodriguez H, Soto A, Arce A and Khoshkbarchi M K 2003 *J. Solution Chem.* **32** 53
[3] Banipal T S, Singh G 2004 *Indian J.Chem.***43**A 1156
[4] Waris B N, Hasan U and Srivastva N 2001 *Thermchim. Acta* **375** 1
[5] Banipal T S, Singh G 2004 *Thermochim. Acta* **412** 63
[6] Soto A, Arce A and Khoshkbarchi M K 2004 *J.Solution Chem.* **33**11
[7] Rohankar P G and Aswar A S 2002 *Indian J. Chem.* **41**A 312
[8] Rodriguez H, Soto A, Arce A and Khoshkbarchi M K 2003 *J.Solution Chem.* **32** 53
[9] Yasuda Y, Tochio N, Sakurai M and Nitta K 1998 *J. Chem. Eng.Data* **43** 205
[10] Taulier N and Chalikian T V 2002 *Biochim. Biophys. Acta* **48** 1595
[11] Millero F J, Surdo Lo A and Shin C 1978 *J.Phys. Chem.* **82** 784-792
[12] Enea O and Jolicoeur C 1982 *J. Phys. Chem.* **86** 3870
[13] Chimankar O P, Shriwas R S, Jajodia S and Tabhane V A 2011 *Advances in Applied Science Research* **2**(3) 500-508
[14] Riyazuddeen Bansal G K 2006 *Thermochimica Acta* **445** *40*
[15] Reddy, Reddy and Murthy 1987 *Acoustic Letters* **10** 128

IOP Conf. Series: Materials Science and Engineering **42** (2012) 012025 doi:10.1088/1757-899X/42/1/012025

Measurement of sound velocity profiles in fluids for process monitoring

M Wolf, E Kühnicke, M Lenz and M Bock

Technische Universität Dresden, Institut für Festkörperelektronik, Helmholtzstraße 18, D-01062 Dresden (Germany)

E-mail: mario.wolf@tu-dresden.de

Abstract. In ultrasonic measurements, the time of flight to the object interface is often the only information that is analysed. Conventionally it is only possible to determine distances or sound velocities if the other value is known. The current paper deals with a novel method to measure the sound propagation path length and the sound velocity in media with moving scattering particles simultaneously. Since the focal position also depends on sound velocity, it can be used as a second parameter. Via calibration curves it is possible to determine the focal position and sound velocity from the measured time of flight to the focus, which is correlated to the maximum of averaged echo signal amplitude. To move focal position along the acoustic axis, an annular array is used. This allows measuring sound velocity locally resolved without any previous knowledge of the acoustic media and without a reference reflector. In previous publications the functional efficiency of this method was shown for media with constant velocities. In this work the accuracy of these measurements is improved. Furthermore first measurements and simulations are introduced for non-homogeneous media. Therefore an experimental set-up was created to generate a linear temperature gradient, which also causes a gradient of sound velocity.

1. Introduction

The sound velocity depends on density and elastic properties of a medium. It can be used for an indirect measurement of many process parameters like temperature, concentration or density. Extensive research has being done with liquid metals to optimize mixing processes [1]. Since there is no conventional direct method to deduce the mixing state, the flow speed is measured. A time and local resolved measurement of sound velocity would yield this information directly.

2. Approach

In [2] a method to measure sound velocity without any reference reflectors at known position was introduced. It is based on the measurement of the echoes from moving scattering particles. Averaging over a sufficient number of amplitude echo signals allows considering a uniform distribution of scattering particles. The average amplitude is a measure of the extension of the focal area of the transducer, which could be confirmed by sound field calculations. The maximum position of this curve, the focus, depends on the sound velocity of the propagation medium and can be used as additional information besides the time of flight. Using a calibration curve allows determining sound velocity and the focal distance simultaneously from the time of flight. First examination with a focusing transducer reached an accuracy of 99,9 % for the measurement of sound velocity of water at different temperatures.

To measure the sound velocity non-invasively and locally resolved, an annular array is used. Driving the elements with different delay times, allows moving the focus position along the acoustic axis of the transducer. First measurements with a sparse annular array were presented in [3] for media with constant sound velocity.

3. Description of method

A half-analytical method has been developed to optimize transducers and to calculate the sound field in complex structures and layered media. It is based on time-harmonic GREEN's functions in a steepest descent approximation for a two-dimensional geometry [4]. For this reason, the source area is discretized and uniformly covered with point sources. The sound field that is irradiated from a finitely extended source is the superposition of the field of all point sources. This method is applied here to simulate the fields of the single elements of the annular-array.

For the experiments, a curved sparse annular array (see figure 1, center frequency of 9MHz) with six active ring elements of equal area (5.66 cm²) and five inactive rings was used. Measurements were done by focusing in emission and reception. Thereby the six elements are driven together where each element is delayed with a calculated time t_n. The received signals are time shifted with the same set of delay times and superposed. The time shifts are equal to the differences of the time of flight from each element to designated focal position *Fok* in a calibration medium. Here water at a temperature of 20 °C with a sound velocity of $c = 1485$ m/s was used.

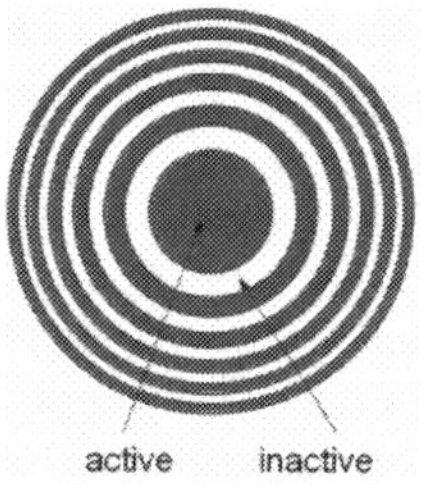

Figure 1. Scheme of a sparse annular array.

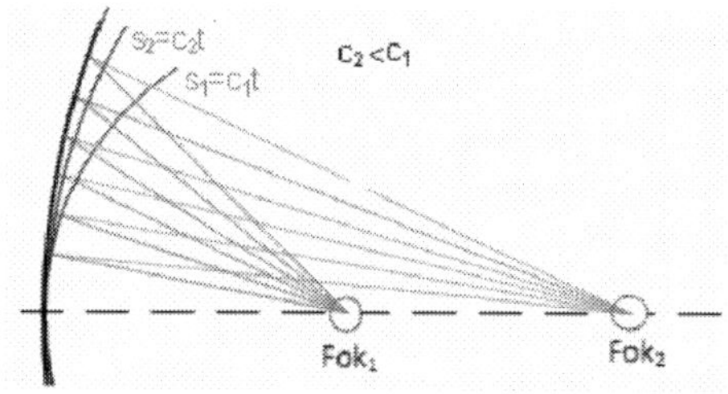

Figure 2. Focus position depending on sound velocity.

Figure 2 also shows the general effect of focus displacement when the sound velocity varies while driving the array with the same set of delay times. With increasing sound velocity the virtual curvature radius of the array increases and hence the focus moves back toward the transducer.

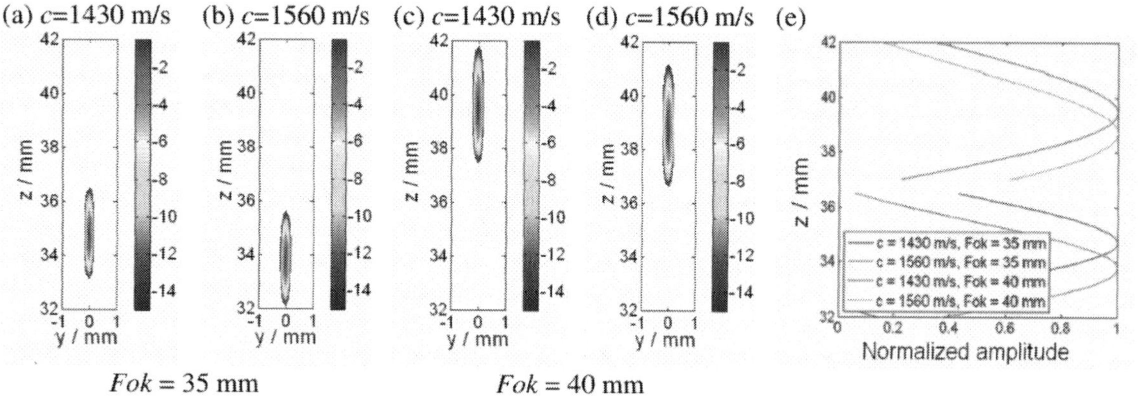

Figure 3. (a) – (d) Calculated sound fields for different sound velocities and time sets, (e) Amplitude at the rotation axis of the sparse array.

For one time set, the sound field of the transducer can be calculated for different sound velocities. This is done in Figure 3(a) and 3(b) for a focus of 35 mm and in figure 3(c) and 3(d) for a focus of 40 mm. Figure 3(e) shows the normalized absolute values of the sound fields at the acoustic axis of the transducer, which correspond to the measured average echo signal amplitudes, shown in figure 4(a). To compare simulations and measurements, the curves of figure 3(e) were converted to curves of amplitude as a function of time of flight and are also shown in figure 4(a). They are used to calculate calibration curves for constant sound velocity, which is illustrated in figure 4(b). The maximum of the amplitude curve left hand corresponds to one point in the diagram right hand. The calibration curves can be used, to determine the sound velocity from the used set of time lags and the time of flight to the maximum of the measured average echo signal amplitude.

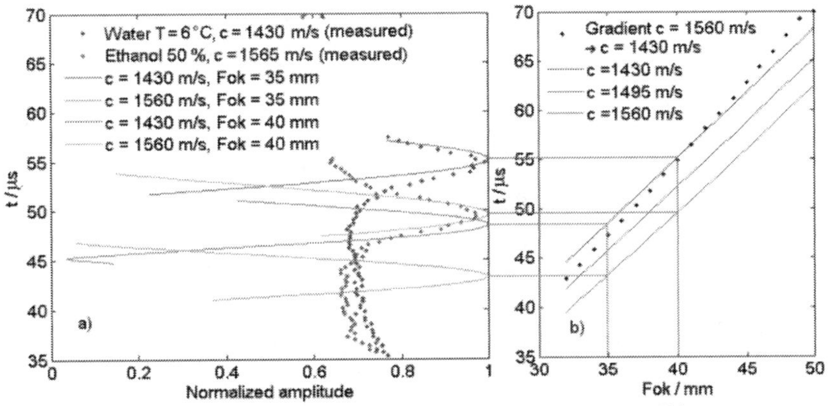

(a) (b)

Figure 4. (a) Normalized amplitude as a function of time of flight (b) Calibration curve for different sound velocities (continuous) and for a sound velocity gradient (dotted).

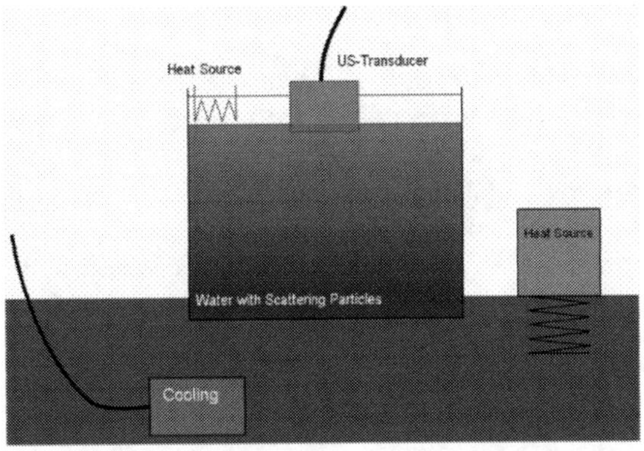

Figure 5. Experimental set-up to generate a temperature gradient.

Figure 4(b) also shows the calculated calibration curve for a fluid with a linear sound velocity gradient. According to Fermat's principle, a wave propagates along the way with the minimal time of flight. If sound velocity is not constant, this way may be curved. This causes changes in the sound field of each point source in comparison to the field of a source in media with constant sound velocity. The superposition of these fields produces a modification of the irradiated sound field. Therefore the calibration curve of a fluid with sound velocity gradient is not simply the linear slope between the calibration curves of minimal and maximal sound velocity.

4. Measurement results

An evident way to achieve a sound velocity gradient is to generate a temperature gradient with water, because the sound velocity as a function of temperature is well known for water. Figure 5 shows the general set-up. A basin located at the bottom is kept at a temperature of 6°C. A second smaller basin is fitted above. It has a metal plate at the bottom for good thermal conduction and a heat source at the top. This generates a vertical layered arrangement of warm water above cooler water, whereby a fluid flow is avoided. The temperature is measured with a single temperature detector.

Primarily sound velocity was measured at constant temperature, as described in section two, to compare measurements and simulations. It is striking that the measured curve for T=6°C fits very well to the simulated one, whereas the measured curve for T=60°C diverges. So additional measurements with water ethanol solutions were done. Different sound velocities can be adjusted with different concentrations of ethanol without a change in temperature [5]. It can be recognized that the measured curve of a water-ethanol solution with a sound velocity of 1565 m/s in figure 7 agrees to the simulated curve. This implicates a high temperature dependence of the transducer. It also explains the differences in the curves for calculated and measured sound velocity profiles. Furthermore it has to be pointed out, that the simulated profile does not model the temperature profile exactly, because the sound velocity is a nonlinear function of temperature. Although there is an approximately linear temperature gradient, a nonlinear sound velocity gradient appears, which had not been modeled yet.

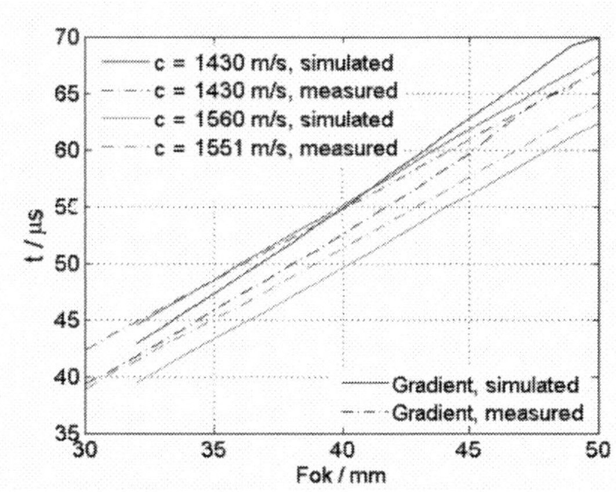

Figure 6. Measurements and simulations for different sound velocities, measurements are done at T= 6°C and T = 60 °C.

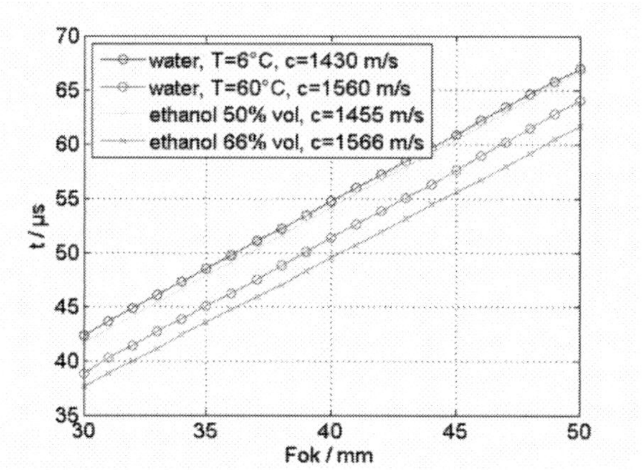

Figure 7. Temperature dependence of transducer: measurements with water of different temperatures and water ethanol solutions.

5. Conclusion

The accuracy of measurements in fluids could be improved significantly by focusing in emission and reception. The significant influence of the temperature of the transducers on the measurement results could be demonstrated. This necessitates to characterize transducers at different temperatures and to calculate adapted calibration curves at these temperatures. First simulations and measurements for media with a sound velocity gradient were presented. The occurred deviations necessitate further detailed examinations. The temperature in the small basin has to be observed locally resolved with a linear array of temperature sensors. The developed simulation algorithm for linear sound velocity gradients has to be advanced for nonlinear gradients. With the information of the temperature, a real sound velocity gradient can be simulated to obtain a better agreement.

References

[1] Galletti C, Paglianti A and Yianneskis M 2005 Observations on the significance of instabilities turbulence and intermittent motions on fluid mixing processes in stirred reactors *Chemical Engineering Science* **60** (8) (9) 2317-31

[2] Lenz M, Bock M, Kühnicke E, Pal J and Cramer A 2012 Measurement of the sound velocity in fluids using the echo signals from scattering particles *Ultrasonics* **52** (1) 117-124

[3] Kühnicke E, Lenz M and Bock M 2011 Non-invasive measurement of sound velocity profiles *IEEE Int.Ultrasonics Symp.* (Orlando, 18-21 October 2011) pp 2150-53

[4] Kuhnicke E 1996 Three-dimensional waves in layered media with nonparallel and curved interfaces: A theoretical approach *Journal J. Acoust. Soc. Am.* **100**(2) 709-716

[5] Burton C 1948 A Study of Ultrasonic Velocity and Absorption in liquid mixtures *J. Acoust. Soc. Am.* **20** (2) 186-199

Measurement of nonlinear viscoelastic properties of fluids using Dynamic Acoustoelastic Testing

C Trarieux[1], S Callé[1], A Poulin[2], J-F Tranchant[2], H Moreschi[1] and M Defontaine[1]

[1] INSERM U930 Equipe 5, Université François Rabelais, 10 bd Tonnellé, 37032 Tours, France
[2] LVMH Recherche, Département Innovation Matériaux et Technologies, 185 av de Verdun, 45800 Saint Jean de Braye, France

E-mail: trarieux.chloe@gmail.com

Abstract. A nonlinear ultrasound-based method called Dynamic Acoustoelastic Testing (DAET) is used to assess nonlinear viscoelastic properties of fluids. This method is based on the interaction between two elastic waves: a low-frequency (LF) sinusoidal wave (4 kHz) to successively compress and expand the liquid as a bulk stress, and ultrasound (US) pulses (1 MHz) to simultaneously probe the sample at different states of the quasi-hydrostatic pressure. The DAET method provides estimations of the elastic nonlinearities issued from the Time Of Flight Modulations (*TOFM*) of the US pulses. The *TOFM* is plotted as a function of the LF acoustic pressure, allowing an estimation of the nonlinear elastic parameter *B/A*. In this study, we first present the results obtained in Newtonian fluids such as water and silicone oils. Simple viscoelastic gels (Carbomers and Xanthan gums) have also been tested exhibiting the same behavior: *TOFM* linearly related to LF pressure amplitude corresponding to classical quadratic nonlinearity. Finally, preliminary DAET measurements have been performed in biphasic systems composed of hard glass beads in a gel-based matrix and in gelatin during a gelation process.

1. Introduction

Standard acoustical measurement techniques allow the investigation of fluids rheological properties, especially in a high frequency range. The use of shear and longitudinal waves leads to the determination of shear and bulk complex moduli, respectively [1, 2, 3, 4].

On the other hand, nonlinear acoustic and acousto-elastic techniques offer the opportunity to estimate nonlinear elastic parameters, particularly interesting for complex media characterization such as biphasic ones. The nonlinear elastic parameter *B/A*, describing “classical” nonlinearity is then derived from the use of longitudinal waves [5]. This parameter reaches values over 100-1000 in micro-inhomogeneous media such as: contrast agents [6], water-saturated glass beads [7], bones with micro-cracks [8, 9]. The *B/A* value is lower than 20 in homogeneous fluids [5, 10]. This parameter also presents a great sensitivity with the gelation process [11].

In this study the Dynamic Acoustoelastic Testing (DAET) method is employed to determine the nonlinear parameter *B/A*, and so quantify viscoelastic properties of different fluids.

2. DAET Method

2.1. *Experimental setup*

The DAET method has already been described in previous papers [8, 9, 12]. It is based on the interaction between two elastic waves propagating through the sample in perpendicular directions: a low-frequency (LF) wave and ultrasound (US) pulses as represented in figure 1.

This sample consists of the tested fluid held in a PMMA container (13mL) and placed in the interaction region of the two elastic waves. The LF wave consists of 100 sinusoidal periods inducing oscillatory compressions and expansions. A hydrophone measures over time the pressure amplitude applied to the sample. US pulses are emitted with a repetition rate 10 times higher than the LF wave frequency, in order to probe the sample at different states of the quasi-hydrostatic pressure. Time Of Flight Modulations (*TOFM*) of the US pulses, explicitly related to the elastic nonlinearities (see section 2.3.), are then calculated.

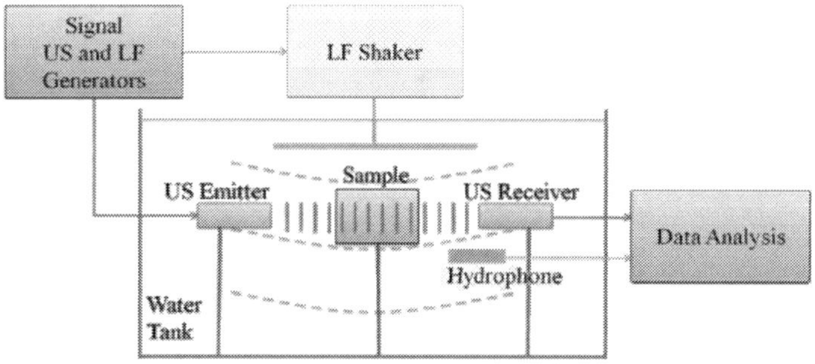

Figure1. Schematic view of the DAET system.

LF frequency:3700Hz
LF pressure: +/- 60kPa
US frequency: 1MHz
Sampling frequency:50MHz
Sequence duration: 100ms
Average over 10 sequences.

2.2. *Experimental protocol*

Firstly simple Newtonian fluids have been tested. Then viscoelastic gels of different concentrations and more complex biphasic systems have been investigated. In addition we also present results obtained during a gelation process. The products characteristics are summarized in table 1. A centrifugal process (1000 rpm) has been established in order to get rid of bubbles in the sample.

Table 1. Type of fluids investigated by the DAET method.

Fluid	Compositions investigated		
Water	-		
Castor oil	-		
Silicone oil	500mPa.s[a]	1000mPa.s[a]	30000mPa.s[a]
Carbopol gel	0.2%[b]	0.5%[b]	1%[b]
Rhodicare gel	0.5%[b]	1%[b]	
Gelatin gel	-		
2% Glass beads in Gelatin	110μm[c]	250μm[c]	

[a] dynamic viscosity [b] mass fraction of the polymer [c] mean diameter of beads

2.3. *Theory – nonlinear B/A parameter*

TOFM is determined by a cross-correlation method between each US pulse and the reference US pulse which propagates through the sample without LF wave perturbation [8, 12].

The propagation of the US pulse with index i is governed by the longitudinal elastic modulus M. With ρ_0 the density of the fluid and c_0 its velocity, *TOFM* can be written as follows: [12]

$$TOFM_{(i)} = \frac{-L}{2\rho_0 c_0^3} \Delta M_{(i)} \qquad (1)$$

The nonlinear elastic parameter *B/A* can be calculated from the relation between *TOFM* of each US pulse and the variations ΔP of LF pressure applied [8]:

$$\frac{B}{A} = \frac{-2\rho_0 c_0^3}{L} \frac{TOFM_{(i)}}{\Delta P_{(i)}} \tag{2}$$

From equations (2) and (3), the parameter *B/A* is related to the variations of *M*:

$$\frac{B}{A} = \frac{\Delta M_{(i)}}{\Delta P_{(i)}} \tag{3}$$

The complex longitudinal modulus M^* (with M' and M'' the loss and storage moduli, respectively), obtained from acoustic measurements, quantifies the volume viscoelastic properties of fluids [1, 2, 3]. The DAET method measures the variations of this longitudinal modulus, and so could quantify the variations of volume viscoelastic properties.

3. Results and discussion

The *TOFM* measurements are plotted as a function of the LF pressures on instantaneous diagrams.

3.1. Homogeneous fluids

The Newtonian fluids (Water, Silicone and Castor oils) and the viscoelastic gels (Carbopol, Rhodicare, Gelatin) have exhibited similar instantaneous diagram: *TOFM* measurements are linearly related to LF pressure amplitudes. Sometimes a hysteresis is observed (See figure 2).

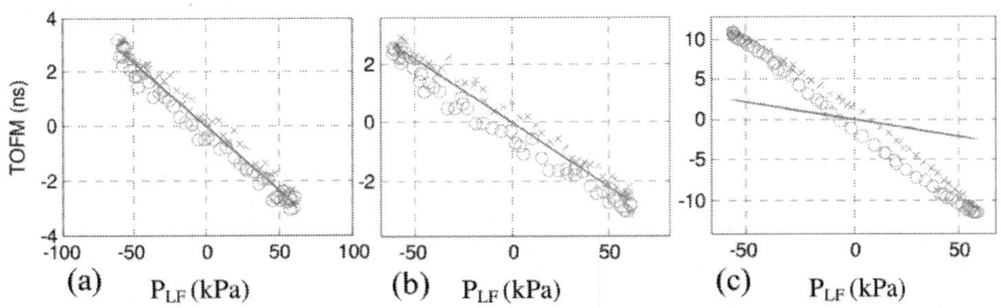

Figure2. Instantaneous diagrams of *TOFM* versus LF pressure for Water (a), Carbopol gel of 1% polymer concentration (b) and Silicone oil of 500mPa.s (c). The straight line represents the theoretical curve for water.

The experiments are carried out 10 times at least by container. Furthermore, 3 series of containers have been filled and tested for each product, except Gelatin. A good reproducibility of the curves is obtained. The temperature during experimentations is kept constant in a range between 20 and 25°C. Applying the equation (2), the nonlinear parameter *B/A* is extracted for each fluid (See table 2).

Gels and water present close *B/A* values, and no dependence with polymer concentration is observed in the investigated range. The *B/A* values of oils are higher and the Silicone oils exhibit a slight decrease as the dynamic viscosity increases. This trend is likely related to the molecular chain length increase. No correlation is found between *B/A* and dynamic viscosity values.

Table 2. *B/A* values of fluids investigated by the DAET method.

Fluid	*B/A*		
Water	5.0±0.1		
Castor oil	9.1±0.6		
Silicone oil	11.7±0.4 [500][a]	11.3±0.2 [1000][a]	11.0±0.4 [30000][a]
Carbopol gel	5.1±0.9 [0.2%][b]	4.9±0.7 [0.5%][b]	5.1±0.2 [1%][b]
Rhodicare gel	4.9±0.3 [0.5%][b]	4.9±0.5 [1%][b]	
Gelatin gel	5.2±0.1		
2% Glass beads in Gelatin	5.3±0.1 [110µm][c]	5.4±0.1 [250µm][c]	

[a] dynamic viscosity, [b] mass fraction of the polymer, [c] mean diameter of beads

3.2. Beads in Gelatin

Each of the two samples of 2% glass beads in Gelatin has been tested more than 10 times. The instantaneous *TOFM* diagrams show linear trends. The presence of dispersed beads (contactless) does not significantly affect the *B/A* value compared to Gelatin alone as shown in table 2. Increasing the beads concentration should exhibit higher level of nonlinearities thanks to an increased number of beads contacts [7]. In this study the beads-gelatin systems perform like homogeneous media.

3.3. Gelation process

After a 40°C heating the gelation process of Gelatin has been monitored by the DAET method. The variations of nonlinear parameter are shown in figure 3. Further studies need to be done in order to identify characteristic times. Moreover the contributions of temperature and effective microscopic network construction (transition sol-gel) have to be investigated. However the DAET method seems to be promising in order to study gelation process.

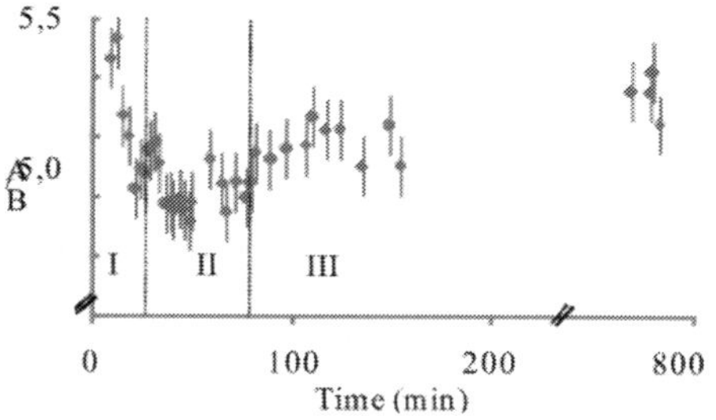

Figure 3. *B/A* values of Gelatin during a gelation process.Three steps can be distinguished visually: I-liquid phase, II-gelatinous phase, III-hard phase gel.

4. Conclusion

Based on *TOFM* measurements, the DAET method allows the determination of nonlinear elastic parameter *B/A* of different fluids. The *B/A* value of water is very close to the *B/A* values of viscoelastic gels (Carbopol, Rhodicare, Gelatin), but different from oils (Silicone, Castor). Furthermore the *B/A* values of oils seem to decrease when the length of the molecular chain increases. The *B/A* value of 2% beads in Gelatin may be slightly higher to the one without beads. Experimentations with higher beads concentrations in Gelatin will be carried out. Finally, the variations in *B/A* value observed during a gelation process should allow us to determine characteristic times of the transition sol-gel.

Acknowledgments
This work was supported by ANR (project ANR-2010-EMMA-019) and Region Centre FEDER

References

[1] Litovitz T A and Davis C M 1965 *Physical Acoustics* vol II A ed W P Mason (New York: Academic Press) Chapter 5 pp 281-349
[2] Dukhin A S and Goetz P J 2009 *J. Chem. Phys.* **130** 124519
[3] Marvin R S, Aldrich R andSack H S 1954 *J. of Appl. Phys.* **25**(10) 1213-1218
[4] Smith C, Kudryashov E D and Buckin V 2001 *Colloids and Surfaces A* **183-185** 517-526
[5] Beyer R T 1960 *The parameter B/A in Nonlinear Acoustics* ed M F Hamilton and D T Blackstock (Austin:Academic Press) pp25-39
[6] Moreschi H, Novell A, Calle S, Defontaine M and Bouakaz A 2009 *IEEE Int. Ultrasonics Symp.*(Roma, 20-23 september 2009) pp 251-254
[7] Renaud G, Calle S and Defontaine M 2010 *J. Acoust. Soc. Am.* **128**(6) 3344-54
[8] Renaud G, Calle S, Remenieras J P and Defontaine M 2008 *IEEE Trans. on UFFC* **53** 1497-1507
[9] Moreschi H, Callé S, Guerard S, Mitton D, Renaud G and Defontaine M 2011 *Proc. IME H J. Eng. Med.* **225** 282-295
[10] Banchet J, Cancian J and Cheeke J D N 2000 *Ultrasonics* **38** 301-304
[11] Robin G, Meulen F V, Haumesser L, Fortineau J, Feuillard G, Wilkie-Chancellier N, Martinez L and Griesmar P 2010 *IEEE Int.Ultrasonics Symp.*(San Diego,11-14 October 2010) pp 825-828
[12] Renaud G, Calle S and Defontaine M 2009 *Appl. Phys. Lett.* **94** 011905

IOP Conf. Series: Materials Science and Engineering **42** (2012) 012027 doi:10.1088/1757-899X/42/1/012027

Ultrasonic investigations of cermets elastic properties in dependence on steel concentration and temperature of sintering

A Abramovich

Physics Department of St. Petersburg State Technological University of Plant Polymers, I. Chernykh 4, St. Petersburg, 198095 Russia

E-mail: andrew@ns2740.spb.edu

Abstract. Cermets is a ceramic-metal composite usually produced by sintering a precompacted mixture of the initial powders. These composite materials were created for industrial applications to produce engineering structures possessing a high strength, thermal stability and resistance to aggressive media. In the present work elastic properties of cermets samples, obtained by sintering of corundum (α-Al_2O_3) and stainless steel powders were investigated in dependence on steel concentration 5 – 35% wt. and on temperature of sintering in vacuum 1400–1700°C. It was stated that values of elastic moduli are in complex dependence on concentration and temperature, reach maxima at steel concentration 15 – 20% wt. and increase with sintering temperature rise. In the work also the results of cermets microstructure researches and discussion of these results are presented. The results are discussed from stand view of ultrasound propagation through medium having grain boundaries which influence on the physical properties of composite.

1. Introduction

Composites based on ceramics and metals (cermets) have been being developed since the 1960s, and unique materials have been produced which combine the virtues of both components, such as heat resistance, temperature stability, wear resistance, and chemical stability. At the present time, cermets are already in service in the machine-building industry, the medical and food industries, etc. [1]. However, the development of cermets of new types and the prediction and investigation of their properties involve difficulties. Among these are the sophisticated high-temperature synthesis technology, not entirely known mechanisms for the grain formation and for the grain boundary influence on the strength and other properties of cermets, and the lack of methods providing reliable information about grain boundaries. In this work, we synthesized and studied by acoustic method cermets based on α-Al_2O_3 and industrial stainless steel (18 Cr, 9 Ni, 1 Ti, 72 Fe % wt.) in the larger concentration range in comparison with work [2] with the aim of (1) developing a comparatively simple and cheap technology for the fabrication of the cermets and (2) obtaining samples which have a high strength in combination with a low density. In addition we attempted to study influence of intergrain and interphase boundaries on elastic properties, because it is known that the mechanical properties of cermets depend heavily on their porosity and the quality of grain boundaries [1, 3]. We believe they are determined by the synthesis technique of the original material, as well as by the fabrication technology of the samples and their finishing thermal treatment. To investigate the influence of these factors, we applied ultrasonic waves (USW) method in this work.

2. Results and discussion

2.1 Experimental technique

To fabricate cermets, the initial fine grained mixture was prepared by milling α-Al_2O_3 powder and steel in a ball mill. Then, the mixture obtained was doped with a plasticizer and subjected to dry compaction under a pressure of 80–100 MPa and then sintered in a vacuum at 1400–1700°C. The cermets thus prepared ranged in volume porosity from 5 to 15% and had the steel content from 5 to 35% wt. The cermets samples were then cut into smaller parts of the required dimensions and polished, depending on the measuring method.

To investigate the elastic properties of the cermets, we measured their density by the hydrostatic method and the velocity of longitudinal (v_l) and transverse ultrasound waves (v_s) of a frequency of 2 MHz. With the data obtained, we calculated Young's modulus E and the shear modulus G from the well-known formulas of the elasticity theory for an isotropic medium [4]:

$$E = \frac{\rho v_s^2 \left(3v_l^2 - 4v_s^2\right)}{v_l^2 - v_s^2} \qquad G = \rho v_s^2 \tag{1}$$

We used the pulsed phase-interferometer method, which made it possible to determine the velocity of ultrasound within an accuracy of 0.1–0.2%. The measuring accuracy is: ≈ 5% for the elastic modules, ≈ 2% for density.

2.2 Results

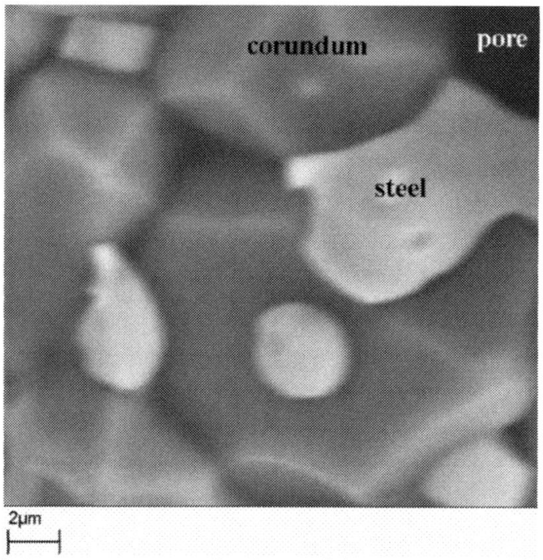

Figure 1. Microstructure of the cermets sample sintered in vacuum at 1600°C by SEM method (steel concentration – 20 % wt).

In figure 1 typical microstructure of studied cermets obtained by scanning electron microscopy (SEM) are presented. It is seen that, for the most part, corundum grains have isomeric hexagonal faces and the grain faces meet at an angle close to 120° at interface junctions; the average grain diameter is about 5-15 μm, the size of steel particle is 0.5-10 μm and pore size 2-5 μm. It should be noted that composition and structure grain and phase boundaries have been studied insufficiently because the general methods don't have necessary sensitivity. However, the volume share of the boundaries in high dispersion solid media is considerable [3]. It can suppose that in the sintering process some intermediate diffusion layer of definite compound is formed [1-3]. The thickness of the layer was estimated and averages several nm [3, 5]. The XRES method allowed finding spinel $FeAl_2O_4$ (at sintering temperature 1400°C) which can be presented in boundary medium.

Figure 2 shows concentration dependences of elastic moduli E and G for the samples of various sintering temperatures. As it is seen from the figure there are two steel concentration values at which E and G change noticeably: at 10% wt. curves 1 and 1* have maxima, curves 2, 2* и 3,3* - turning points; at 20% wt. maxima on curves 1,1*; 2,2*;3,3* have the greatest values E and G in the whole concentration range. The curves 4, 4* show unusual character of changes: continuous increase of elastic moduli in whole concentration range. Porosity of samples have an effect on elastic modules whereas it value reduce with of sintering temperature rise [1, 3]. Our estimation with the using of the calculations from [1] has shown that change of values of elastic modules due porosity is ≈ 10%. This value is noticeably less than concentration changes from figure 2.

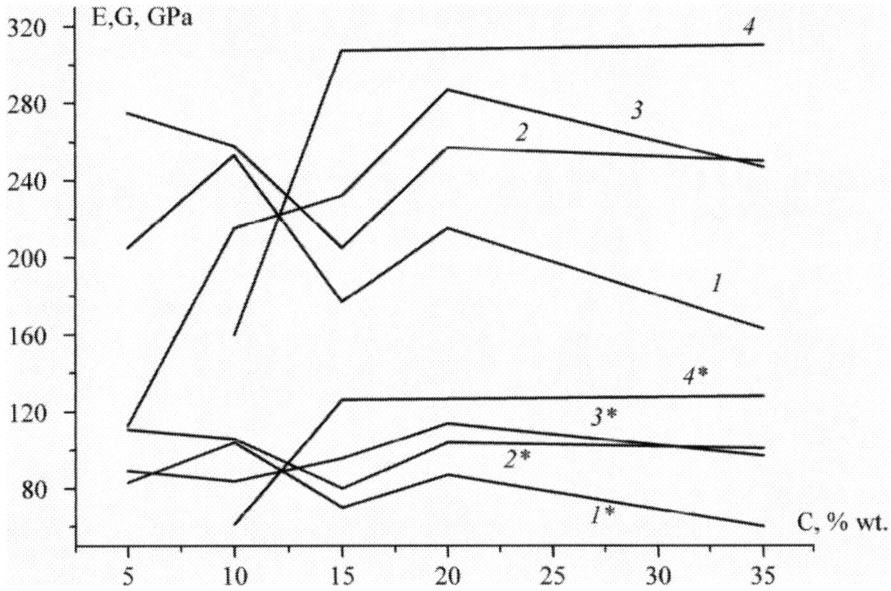

Figure 2. Concentration dependences of elastic modules E, G of cermets, sintered at various temperatures: 1400°C – curves 1, 1*; 1500°C – curves 2, 2*; 1600°C – curves 3, 3*; 1700°C – curves 4, 4* (1-4 correspond E, 1*- 4* correspond G).

2.3 Discussion

The interpretation of the obtained results is not unambiguous, first, due to a great number of factors affecting the elastic properties on the two-phase fine disperse composites in different ways (porosity, grain size and boundary, etc.); secondly, due to the large length of the USW if compared with the grain size. Therefore, only the integral elastic characteristics of the samples are observed in the experiment. Nevertheless, it is possible to draw preliminary conclusions.

The cermets which we have studied belong to class "ultra cermets" [1] in which metallic inclusions are either in volume of ceramic grains or in grain boundaries. This does not contradict the structure in figure 1. The properties of such cermets must dependence on metal and ceramic properties, relation of their volumes and adhesion between ones. In cermets it is necessary to realize strong interface coupling due to atoms of either first or second phase. The possible version of such coupling formation considered in [1] on example of elastic moduli dependences, observing for real composites. It is noted that for practical applications once are important only composites with noticeable increase of this moduli in comparison with moduli of green material. For cermets "corundum-metal" this effect can observe only at small concentrations of metal (5-10%wt.). We observe it at low synthesis temperatures (figure 2, curves 1, 1*). However at more high temperatures we stated considerably the greater effect for the largest concentrations 25-35% wt. (figure 2, curves 2, 2*) moreover values E and G become close to values "based" ceramics (E = 380 and G =170 GPa). This fact is not described in literature.

In [1] it is noticed that forming of strong coupling between ceramic and metal grains is accompanied by appearance of intermediate thin layer i.e. new phase which properties definite features of sintered composite however the layer thickness is estimated at several nanometers. Author of [5] have estimated elastic modules of grain boundary which can be used for calculations of phonon diffusion [3]. In our samples such layer is found with confidence by XRES method and its compound was definite as alumospinel $FeAl_2O_4$, sintered at low temperature 1400°C. Perhaps, such layer has the thickness of tens nm and possesses "crumbly" structure: these properties reduce the strength of composite in total and its modules (curves 1, 1* in figure 2). At high sintering temperatures (1500-1700°C) the grain boundary may be considerably thinner and therefore is not found by XRES method.

Decrease of elastic modules at growth of steel concentration can be explained by two reasons: 1) increase of volume steel share; 2) increase of layer thickness and change of grain boundary structure, therefore decrease of the layer elasticity. Thus the research of concentration dependences of moduli E, G allows choosing the optimal steel concentration for producing of composite with definite elastic properties.

4. Conclusions

Thus, in this work we synthesized and studied by the acoustic method a composite material with high elasticity, strength, and wear resistance. All these properties make this composite perspective for practice applications. One can state that acoustic studies of sintered composites have significant informational content, not only with the aim of obtaining material with optimal elastic and strength properties, but also to indirectly observe the formation of the intergrain and interphase boundaries.

References

[1] Kislykh P S *et al* 1985 *Cermets* (Kiev: Naukova Dumka)
[2] Abramovich A A *et al* 2010 *Acoustical Physics* **56** 790
[3] Ivanov S N 2001 *Physics of the Solid State* **43** 665
[4] Truell R *et al* 1969 *Ultrasonic methods in solid state physics* (NY-London: Academic Press)
[5] Iepnev A G 2007 *Tech. Phys. Lett.* **33** 936

IOP Conf. Series: Materials Science and Engineering **42** (2012) 012028 doi:10.1088/1757-899X/42/1/012028

Physico-chemical and excess thermodynamic properties of methanol & ethanol with 1, 4-dioxane at 308 K

G R Bedare[1], V D Bhandakkar[2] and B M Suryavanshi[3]

[1]Department of Physics, N. S. Science & Arts College, Bhadrawati Dist– Chandrapur (MS), India 442902
[2]Department of Electronics, Anand Niketan College, Warora Dist– Chandrapur (MS), India 442907
[3]Department of Physics, Institute of Science, Nagpur (MS), India 40001

Email: gr.bedare@gmail.com.

Abstract. The molecular interaction studies in the binary liquid mixtures of two aliphatic alcohols with 1, 4-dioxane has been carried out at 308 K using ultrasonic technique. Using measured values of ultrasonic velocity, density and viscosity, acoustical parameters such as adiabatic compressibility, free volume, free length and their excess values like V_f^E, β_a^E, L_f^E are evaluated. From the properties of these excess parameters, the nature and the strength of interactions in these binary systems are discussed. It has been observed that, weak dispersive type intermolecular interactions are confirmed in the systems investigated. Dipole inducement is found to be more predominant in methanol system.

1. Introduction

In the recent years, the ultrasonic investigation of liquids is very important in understanding the nature and strength of molecular interactions. In many industrial applications liquid mixtures, rather than single component liquid system, are used in processing and product formulations [1, 2]. Liquid mixtures consisting of polar and non-polar components are of considerable importance in industries. The formation and destruction of azeotropes in petrochemical industries, the biological activity of drug molecules and the activation energy of the metabolic process [3, 4] basically depend on the type and strength of the intermolecular interactions.

Ethanol is used in the manufacture of alcoholic drinks, solvent for paint, varnish and drugs, scents, flavorings, colorings, medicines, as a fuel, as the fluid in thermometers, and in preserving biological specimens. Methanol is an important industrial material and it is used in the manufacture of formaldehyde, as a solvent in the paint and varnish industry, and as an anti-Freeze in car radiators. Dioxane is primarily used as a stabilizer for 1, 1, 1-trichloroethane for storage and transport in aluminum containers.

Thermodynamic and transport properties of liquid mixtures have been extensively used to study the departure of a real liquid mixture behavior from ideality [5, 6]. The measurements of ultrasonic velocity and density have been adequately employed in understanding the molecular interactions in the liquid mixtures [7, 8]. The measurements of excess thermodynamic properties are found to be greatly significant in studying the structural changes associated with the liquids [9]. The 1, 4-dioxane is selected as a solvent in the present work since it finds a variety of applications. Alcohol plays an important role in many chemical reactions due to the ability to undergo self-association with internal structures [10]. In the present work, acoustical parameters of two non-ideal binary mixtures of 1, 4-dioxane with ethanol and methanol at 308K has been studied.

2. Materials and Methods

All the chemicals used were of Analytical Reagent (AR) grade with minimum assay of 99.9%. The ultrasonic velocity (*U*) in liquids have been measured using an ultrasonic interferometer (Mittal type, Model F-83) working at 2 MHz fixed frequency with an accuracy of $\pm$ 0.1 ms^{-1}. An electronically digital operated constant temperature water bath has been used to circulate water at the desire temperature. The density of liquids was determined using pycknometer by relative measurement method with an accuracy of $\pm$0.1 Kgm^{-3}. An Ostwald's viscometer was used for the viscosity measurement of liquids with an accuracy of $\pm$0.0001 NSm^{-2} . The temperature around the viscometer and pycknometer was maintained within $\pm$0.1 K.

The various acoustical parameters such as adiabatic compressibility, free length, free volume and their excess values have been calculated using the following standard expressions:

$$\beta_a = (U^2 \rho)^{-1} \quad (1)$$

$$L_f = K_T \beta_a^{1/2} \quad (2)$$

$$V_f = (M_{eff}\, U/\eta K)^{3/2} \qquad M_{eff} = \Sigma x_i m_i \quad (3)$$

Where, K_T *is* the temperature dependent constant having a value $209.5906{*}10^{-8}$ in MKS system at 308 K, K is temperature independent constant equal to $4.28{*}10^{9}$ in MKS system, b is a cubical packing fraction taken as 2 for all the liquids, R is the Universal gas constant, T is the experimental temperature. In equation (3) x_i is the mole fraction and m_i is the molecular weight of the component. The excess values like V_f^E, β_a^E, L_f^E calculated using the following standard relation given in equation (4) where A_i is any acoustical parameter and X_i is the mole fraction of liquid component.

$$A^E = A_{exp} - A_{id} \qquad A_{id} = \Sigma A_i X_i \quad (4)$$

3. Results and Discussion

The experimentally measured values of ultrasonic velocity, density and viscosity of pure liquids at 308 K are presented in table 1. From the table 2, it is observed that adiabatic compressibility (β_a) decreases with increase in concentration of 1, 4-dioxane as expected. This increase in structural order of ethanol and methanol may result in more cohesion, and leads to a decrease in β_a. It is observed that, L_f decreases with the concentration of 1,4-dioxane in system I and system II. Decrease in free length leads to positive deviation in sound velocity and negative deviation in compressibility. This indicates that the molecules are nearer in the system. The computed other parameter like free volume (V_f) of system-I and II are given in table 1 and 2. The variation in free volume (V_f) increases with increases in molality of 1, 4-dioxane in system-I and II. The free volume is the space available for the molecule to move in an imaginary unit cell. The observed decreases values of V_f in system-I and II are due to close association between solute and solvent. Thus, a progressive decrease in free volume and increase in internal pressure in methanol-1, 4-dioxane mixtures clearly indicates the existence of ion-solvent interaction, due to which the structural arrangement is considerably affected.

To ascertain the exact type of interaction existing between the components, respective excess parameters have been calculated and are presented in table 3. The trend of figure 1 and 2 shows that βa^E and L_f^E are increasingly positive as 1, 4-dioxane mole fraction is increased. Thus, strong dipolar interactions are restricted more and more whereas the weak dispersive interactions are predominating and are maximum around 0.8 mole fractions. As the molecular symmetry is retained to a greater extent at 0.9 mole fraction, βa^E and L_f^E are found to decrease. However V_f^E for both the systems are mostly negative, existing weak dispersive interactions. Such existence of weak interactions was confirmed by Palaniappan et.al [11]. Among the two alcohol systems considered, a more systematic variation in V_f^E

is observed in ethanol system whereas it is not so for methanol system. The presence of methyl group in methanol system produces more number of induced dipoles in 1, 4-dioxane, provided chair form of 1, 4-dioxane is available in the system as it is much more stable than boat form. Such restrictions are reflected in the observed trend of V_f^E in methanol system.

Table 1. The values of Velocity (*U*), Density (ρ) & Viscosity (η) of pure liquids at 308 K.

Liquids	Velocity(*U*) (m/s)	Density (*ρ*) (kg/m³)	Viscosity(η*10⁻³) (Ns/m²)
1.Ethanol	1229.6	769.46	0.715
2.Methanol	1063.2	770.35	0.439
3.1,4-Dioxane	1273.6	1014.97	0.401

Table 2. The experimental values of Velocity (U), Density (ρ), Viscosity (η) and calculated values of adiabatic compressibility (β_a), free volume(L_f) and free length(V_f) at 308 K for the System I & II.

X_1	X_2	U (m/s)	ρ (Kg/m³)	η *10⁻³(N)	β_a*10⁻¹⁰(Pa⁻¹)	L_f*10⁻¹⁰ (m)	V_f*10⁻⁷(m³mol⁻¹)
			System I : 1,4-Dioxane + Methanol				
0.1	0.9	1019.00	808.38	0.438	11.9	0.7105	0.9257
0.2	0.8	1029.00	823.35	0.434	11.5	0.6971	1.1729
0.3	0.7	1038.00	844.31	0.431	11.0	0.6824	1.4416
0.4	0.6	1046.00	883.23	0.428	10.3	0.6621	1.7345
0.5	0.5	1077.33	919.16	0.423	9.37	0.6302	2.1374
0.6	0.4	1082.66	928.14	0.418	9.19	0.6240	2.5060
0.7	0.3	1109.33	958.08	0.413	8.48	0.5995	2.9925
0.8	0.2	1164.80	973.05	0.408	7.57	0.5665	3.6735
0.9	0.1	1225.60	1002.99	0.405	6.64	0.5303	4.4553
			System II : 1,4-Dioxane + Ethanol				
0.1	0.9	1228.80	781.43	0.702	8.48	0.5992	0.9320
0.2	0.8	1233.66	790.40	0.670	8.31	0.5935	1.1343
0.3	0.7	1237.66	796.40	0.636	8.20	0.5893	1.3778
0.4	0.6	1241.66	805.38	0.594	8.05	0.5842	1.7018
0.5	0.5	1245.00	814.37	0.562	7.92	0.5793	2.0461
0.6	0.4	1249.33	832.33	0.527	7.70	0.5711	2.4813
0.7	0.3	1253.00	853.29	0.491	7.46	0.5624	3.0201
0.8	0.2	1256.00	889.22	0.457	7.13	0.5496	3.6613
0.9	0.1	1261.60	958.08	0.433	6.56	0.5271	4.3168

Table 3. The excess values of adiabatic compressibility (βa^E), free volume (V_f^E), free length (L_f^E) at 308 K for the System I and System II.

X_1	X_2	β_a^E*10⁻¹⁰ (Pa⁻¹)		L_f^E*10⁻¹⁰ (m)		V_f^E*10⁻⁸ (m³mol⁻¹)	
		Methanol	Ethanol	Methanol	Ethanol	Methanol	Ethanol
0.1	0.9	0.9706	0.1315	0.0325	0.0054	-2.9774	-3.1317
0.2	0.8	1.0687	0.2224	0.0383	0.0094	-5.0206	-5.6004
0.3	0.7	1.1318	0.3587	0.0427	0.0149	-6.8481	-7.6556
0.4	0.6	1.0283	0.4673	0.0414	0.0195	-8.4344	-8.9073
0.5	0.5	0.5948	0.5871	0.0282	0.0243	-8.9204	-9.9548
0.6	0.4	0.9539	0.6147	0.0414	0.0257	-9.7497	-1.0093
0.7	0.3	0.7846	0.6339	0.0357	0.0266	-9.3995	-9.1965
0.8	0.2	0.4186	0.5502	0.0215	0.0234	-7.1048	-7.2752
0.9	0.1	0.0224	0.2314	0.004	0.0103	-3.8012	-5.2102

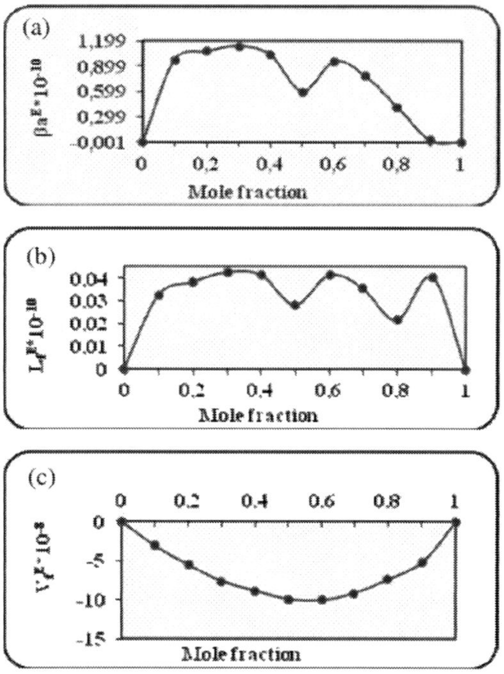

Figure 1. Graph for System I: 1,4-Dioxane + Methanol.(a)$\beta_a^E*10^{-10}$ vs. mole fraction (b) $L_f^E*10^{-10}$ vs. mole fraction (c) $V_f^E*10^{-8}$vs. mole fraction.

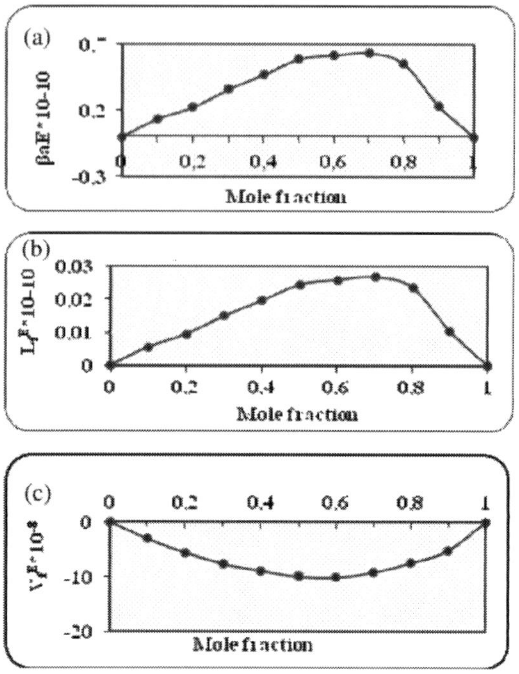

Figure 2. Graph for System II: 1,4-Dioxane + Ethanol. (a) $\beta_a^E*10^{-10}$ vs. mole fraction (b) $L_f^E*10^{-10}$ vs. mole fraction (c) $V_f^E*10^{-8}$vs. mole fraction.

4. Conclusions

Weak dispersive type intermolecular interactions are confirmed in the systems investigated. Dipole inducement is found to be more predominant in methanol system. Components maintain their individuality in the mixtures of both the systems investigated. All the experimental determinations of acoustic parameters and their excess values are strongly correlated with each other.

Acknowledgement

Authors (GRB [File No.:47-1774/10(WRO)] and VDB [File No.: F-47-919/09(WRO)]) acknowledge the financial assistance from UGC, in the form of Minor Research Project grant.

References

[1] Kalidas M and Murthy R S 1997 *J. Pure & Applied Ultrasonic* **19** 9
[2] Sheshagiri Rao M C 1971 *Indian J. Pure & Appl. Physics* **9** 169
[3] Bhandakkar V D, Tabhane V A and Ghosh S 2003 *Ind. J. Pure & Appl. Phy.* **41** 849
[4] Edward I Peters 1982 *Introduction. to Chemical Principles* 3rd ed (Philadelphia :CBS College Publishing) chapter 14 pp 324 -326
[5] Bedare G R, Bhandakkar V D and Suryavanshi B M 2011 *Arch. of Appl. Sci. Res.* **3**(5) 323
[6] Bhandakkar V D, Bedare G R, Muley V D and Suryavanshi B M. 2011 *Adv. in Appl. Sci. Res.* **2**(4) 338
[7] Bedare G R, Bhandakkar V D and Suryavanshi B M 2012 *J. Of Chem. & Pharm. Res.* **2** 1028
[8] Bhandakkar V D et. al 2011 *Ind. J. Pure & Appl. Phy.* **49** 250
[9] Bhandakkar V D 2011 *Adv. in Appl. Sci. Res.* **2**(3) 198
[10] Joshi Deepa, Bhatnagar Deepa, and et. al 2009 *J. of Metrology Soc. Of India* **24**(4) 215
[11] Tiyagarajan R and Palaniappan L 2008 *Indian J. Pure & Applied Physics* **46** 852

IOP Conf. Series: Materials Science and Engineering **42** (2012) 012029 doi:10.1088/1757-899X/42/1/012029

Nonlinear elastic properties of materials with residual stresses

A Korobov, A Romanov and A Morozov

Department of Acoustic, Faculty of Physics, M.V. Lomonosov Moscow State University, Leninskie gory 1, GSP-1, 119991, Moscow, Russian Federation

E-mail: aikor42@mail.ru

Abstract. The non-linear elastic properties of rock samples and metal samples of microcrystalline aluminium alloy were studied using NRUS method. The residual shear strains were initially introduced in metal samples. In these samples the effect of slow dynamics was also investigated. Based on the analysis of experimental results it was concluded: the effect of slow dynamics, observed in the experiment, can not be explained solely by thermoelastic effects, but it is associated, in our opinion, with the slow relaxation of the internal structure of the samples.

1. Introduction

The study of nonlinear effects is one of the most interesting and promising directions in modern acoustics. Effects due to the presence of large scale inhomogeneities and defects are of the most interest. The presence of inhomogeneities in solids is the source for the new types of physical properties otherwise absent in homogeneous solids. The best examples are such quantum effects as negative magnetoresistance, quantum Hall effect, etc. Large scale inhomogeneities and defects carry a significant impact on elastic properties of solids. In such bodies the nonlinearity due to periodization inconsistency in crystal lattice coexists with structural (non classical) nonlinearity. The specifics of such nonlinearity is that it exceeds the physical nonlinearity by 2-4 orders of magnitude [1]. The huge nonlinearity is caused by several effects, namely, by Hertzian contact between grains [2] and micro_asperities of rough grain surfaces [3], by cracks inside specimen, or by weak bonds between rigid sections of medium [4]. The hysteresis of elastic properties and effect of "slow dynamics" is observed in samples with inhomogeneities and defects along with large "non classical" nonlinearity [5, 6]. "Slow" means that when thermodynamical equilibrium in sample is broken due to excitation with high amplitude sound waves the return of elastic properties of the sample to the prior excitation values may take time on the scale from seconds to hours or even days. The softening of the materials and decrease in sound velocity is typical right after the excitation is applied. Effect of slow dynamics is most noticeable in structurally inhomogeneous materials like polycrystallic metals with residual stresses and rock samples. The later poses a significant interest for application in geoscience and seismology [7]. To this day there is no complete theoretical model that explains all aspects of slow dynamics. Some of the most commonly used models for the interpretation of experimental results are thermoelastic approach [7] and Preisach-Mayergoyz [8] elastic model. Current work shows the experimental results for the study of nonlinear elastic properties of materials with residual stresses.

2. Setup and samples

The non-linear elastic properties of samples of rocks and metal samples of microcrystalline aluminum alloy were studied using NRUS (nonlinear resonance ultrasound spectroscopy) method. The experimental setup using NRUS method is shown on figure 1. Sample in a shape of a thin rod (1), with emitting and receiving transducers attached to corresponding ends is firmly connected to a heavy base (2). Such setup allows the sample to work as $\Lambda/4$ resonator (Λ-wavelength in the sample). The resonance frequencies f_n of such resonator depend on the sample length L and elastic wave velocity V and are described by the relation:

$$f_n = [V(2n+1)]/4L \tag{1}$$

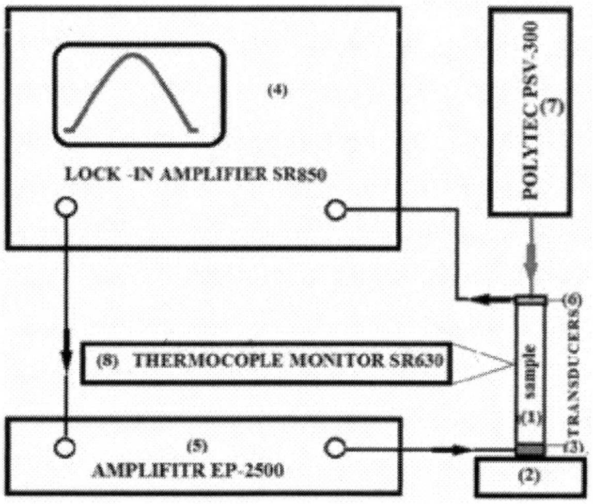

Figure 1. Setup for NRUS.

Emitting transducer (3) receives signal from Lock-In Amplifier *SR850* (4) that is amplified by amplifier (5). The result is frequency-modulated signal (FMS). The central frequency of FMC was chosen to be close to the main resonance frequency of the sample f_n. Transducer (3) emits longitudinal elastic waves. The receiving transducer is located on the free end of the sample (6). Electrical signal from (6) proportional to the wave amplitude in the sample is sent to Lock-In Amplifier SR850 (4) where the signal was digitally stored for the future analysis on PC. Such measurement allowed the precise resonance curve of the sample to be constructed. Higher harmonics appearing during excitation of the sample at the resonance frequency were measured using Polytec PSV-300 laser vibrometer (7). Temperature of the sample during the experiments was monitored using thermocouple monitor SR630 (8) with precision of 0.1K.

Experiment was conducted using samples from two materials: granite rods 120mm in length and 15mm in diameter, polycrystallic rods from aluminum alloy D16 120mm in length, 10mm in diameter. Metal samples were of two types: the original unaltered factory cast rod and rods with manually introduced residual shear deformations. The residual shear deformations were made by firmly fixing one end of the rod and twisting the other end around rod's axis. The residual shear deformation could be accurately estimated by the twist of the line that was drawn on the side of the rod prior to the introduction of deformation. In our samples the residual shear deformations were equal ≈0,03.

3. Experimental Results and Discussion

Before conducting the further experiments resonance frequencies of all samples were measured using the lowest possible signal amplitudes. That allowed to measure the wave velocities V_{rod} corresponding to each rod geometry. Granite rods had $V_{rod} \approx 2.88*10^3$ m/sec. Metal samples of both types yielded $V_{rod} \approx 4.7*10^3$m/sec. Last result shows that manually introduced residual stresses have no effect on linear elastic properties of metal samples within the error of the experiment. The measurements of resonance curves at different excitation amplitudes were taken. For metal samples without residual stresses there was no change of resonance frequency due to change of excitation amplitude. For the samples of granite rods and metal rods with residual stresses resonance curves are shown at figure 2. Resonance curves for the granite samples show presence of two resonance frequencies. Analysis employing the laser vibrometer showed that higher frequency resonance peak is due to flexural mode of the rod and the lower frequency peak corresponds to longitudinal mode figure 2(b).

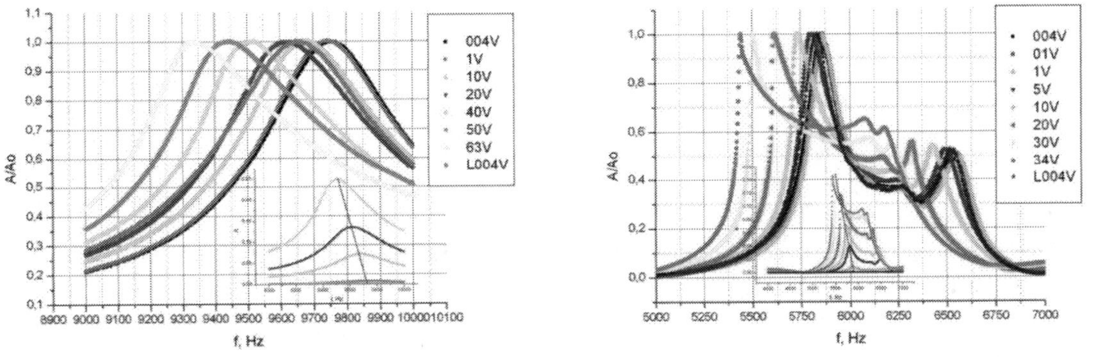

(a) (b)

Figure 2. Resonance curves for various excitation amplitudes: (a) in metal rod with residual shear deformations; (b) in granite rod.

As its shown in figure 2 the increase of excitation amplitude *A* for both samples causes their respective resonance frequencies f_n to decrease linearly:

$$f(A) = f_n - \alpha A \tag{2}$$

(α= -6.8 Hz/V-metal rod with defects, α = -12.6 Hz/V for granite rod). Linear decrease of resonance frequency can not be explained by classical square or cubic elastic nonlinearities due to non harmonical nature of crystal structure. The square elastic nonlinearity doesn't cause the change of resonance frequency due to increase of excitation amplitudes. Samples with cubic elastic nonlinearity display the dependence of the resonance frequency shift proportional to the square of excitation amplitude. Linear dependence in the frequency shift accounts for the significant presence of structural nonlinearity due to residual stresses and defects in the samples [6].

A few words on how slow dynamics experiments were conducted. Samples were subject to excitation amplitude of 60V for 10-15 minutes. That caused the resonance frequencies of the samples to decrease. Measuring the temperature of the samples using thermocoupled probe with precision $\pm 0.1^{0}$□ did not reveal any heating with temperatures remaining constant throughout the experiment. At the same time the amplitudes of main frequency and second harmonic in the samples were measured using laser vibrometer. Typical amplitude at main frequency for granite sample was 200nm and 1100nm for metal sample at 60V excitation amplitude. Amplitudes of the second harmonic for granite and metal samples with residual deformations were 220pm and 1200pm respectively. Maximum deformation value ε_m at main frequency with excitation amplitude 60V: $\varepsilon_m \approx$ 2.6E-06 for granite and $\varepsilon_m \approx$1.7E-05 for metal. Significant instability of amplitude of second harmonic exceeding 20dB was also observed. The amplitude of the main frequency remained constant. That was observed for both type of samples – granite and metal with residual deformations. Instability of the second harmonic measured in the metal sample without residual deformations did not exceed 10dB. Immediately after excitation at 60V the behavior of resonance frequency with time was measured at very low excitation amplitudes. Slow increase of resonance frequency in time was observed.

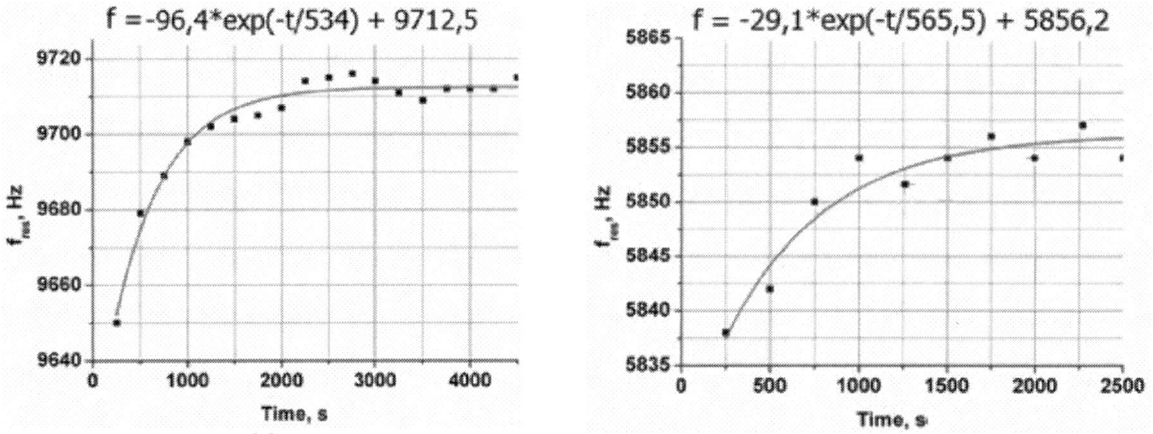

(a) (b)

Figure 3. Slow dynamics effect: (a) in metal sample with residual shear deformations; (b) in granite sample.

The increase of resonance frequencies in time for granite samples and metal samples with residual stresses followed the exponential law in figure 3:

$$f_n(t) = f_n(t \to \infty) - C \cdot \exp(-t/\tau) \quad (3)$$

Some of the most commonly used models for the interpretation of experimental results are thermoelastic approach [7] and Preisach-Mayergoyz [8] elastic model. The slow dynamics effects observed in our experiments can not be explain solely by thermoelastic effects. The main reason is the absence of temperature change in samples after continuous excitation with high amplitude signal. In our opinion the effect of slow dynamics is mainly due to the delayed relaxation of internal structure of the samples previously stressed with high amplitude signal. This conclusion goes along very well with observed prolonged return of resonance frequency to its former pre excitation value.

Acknowledgment
The present work was supported through research grants from Russian Foundation for Basic Research.

References

[1] Rudenko O V *Phys. Usp.* 2006 **49** 69
[2] Rudenko O V and Hedberg C M 2004 *Nonlinear Dynam.* **35** 187
[3] Rudenko O V and Wu C An 1994 *Acoust. Phys.* **40** 593
[4] Rudenko O V and Russ J 1993 *Nondestruct. Test.* **29** 583
[5] Guyer R A and Johnson P A April 1999 *Phys. Today* **30**
[6] Ostrovsky L A and Johnson P A 2001 *Rivista Del Nuovo Cimento* **24** (7)
[7] Zaitsev V, Castagnede B and Gusev V 2003 *Phys. Rev. Lett.* **90** 075501
[8] Bentahar M, El Agra H, El Guerjouma R, Giffa M and Scalerandi M 2006 *Phys. Rev.* B **73** 014116

IOP Conf. Series: Materials Science and Engineering **42** (2012) 012030 doi:10.1088/1757-899X/42/1/012030

Ultrasonic velocity and absorption study of binary mixtures of cyclohexane with acrylonitrile by interferometric method at different frequencies

N R Pawar[1], O P Chimankar[1,3], V D Bhandakkar[2] and N N Padole[1]

[1]Department of Physics, RTM Nagpur University, Nagpur-440033, INDIA
[2]Department of Electronics, Anand Niketan College, Warora, INDIA

E-mail: opchimankar28@gmail.com

Abstract. The ultrasonic velocity (u), absorption (α), density (ρ), and viscosity (η) has been measured at different frequencies (1MHz to 10MHz) in the binary mixtures of cyclohexane with acrylonitriile over the entire range of composition at temperature 303K. Vander Waal's constant (b), adiabatic compressibility (β_a), acoustic impedance (Z), molar volume (V), free length (L_f), free volume, internal pressure, intermolecular radius and relative association have been also calculated. A special application for acrylonitrile is in the manufacture of carbon fibers. These are produced by paralysis of oriented poly acrylonitrile fibers and are used to reinforce composites for high-performance applications in the aircraft, defense and aerospace industries. Other applications of acrylonitrile are in the production of fatty amines, ion exchange resins and fatty amine amides used in cosmetics, adhesives, corrosion inhibitors and water-treatment resins. Cyclohexane derivatives can be used for the synthesis of pharmaceuticals, dyes, herbicides, plant growth regulator, plasticizers, rubber chemicals, nylon, cyclamens and other organic compounds.

In the view of these extensive applications of acrylonitrile and cyclohexane in the engineering process, textile and pharmaceutical industries present study provides qualitative information regarding the nature and strength of interaction in the liquid mixtures through derive parameters from ultrasonic velocity and absorption measurement.

1. Introduction

The ultrasonic studies in binary liquids are essential for utilizing them in medical technology[1-3].Therefore it was though worthwhile to understand ultrasonic study on some binary liquids and their mixtures.Cyclohexane and acrylonitrile has large number of applications in medical and industries. Cyclohexane is used as a nonpolar solvent for the chemical industry, and also as a raw material for the industrial production of adipic acid and caprolactam, both of which are intermediates used in the production of nylon. It is a raw material in the manufacture of acrylic fibers, styrene plastics, and adhesives. Such fibers and plastics are components of clothing, furniture, appliances, construction materials, motor vehicles, and food packaging. Due to large number of applications in all areas cyclohexane and acrylonitrile were taken for experimental ultrasonic studies in the present paper.

Ultrasonic parameters are extensively being used to study molecular interactions in pure liquids binary liquid mixtures[4-5] and ionic interactions in single and mixed salt solutions of bio-liquids [6-

[3] To whom any correspondence should be addressed.

10]. The experimental investigations have shown that derived parameters such as the adiabatic compressibility (βa), ultrasonic absorption and their deviation from the additive rule provide a better insight into molecular processes[11-15].In the present work, the authors have been measured ultrasonic velocity, absorption, density and viscosity of pure binary liquids cyclohexane, acrylonitrile and their mixtures in the molar concentration range 0.1 to 0.9 and in the frequency range 1MHz to 10 MHz at 303 K. From these data other ultrasonic parameters have been calculated using standard formulae [13, 15].

Adiabatic compressibility is a measure of intermolecular association or dissociation or repulsion. It also determines the orientation of the solvent molecules around the liquid molecules.

$$\beta_a = 1/u^2\rho \quad kg^{-1}ms^2 \qquad (1)$$

The free length is the distance between the surfaces of the neighbouring molecules. The intermolecular free length has been calculated by the formula given by Jacobson (equation (2)). K is Jacobson's constant, a temperature (T) dependent parameter given by equation (3).

$$L_f = K\,(\beta_a)^{1/2} \quad m \qquad (2)$$

$$K=(93.875 + 0.345T)\ x10^{-8} \qquad (3)$$

Acoustic impedance is important in the determination of acoustic transmission and reflection at the boundary of two materials having different acoustic impedance

$$Z = U\rho \quad kg\ m^{-2}s^{-1} \qquad (4)$$

Vander Waal's constant (b) also called co-volume in the Vander Waal's equation is expressed in equation (5), where R is the gas constant.

$$b = V\,[1-(RT / Mu^2)]\,[1+Mu^2/3RT)^{1/2}] \quad m^3mol^{-1} \qquad (5)$$

Free volume is defined as the average volume in which the centre of the molecules can move inside the hypothetical cell due to the repulsion of surrounding molecules, where M is the effective molecular weight and K is temperature independent constant and its value is 4.28 x 10^9.

$$V_f = [M\,u/K\eta]^{3/2} \quad m^3mol^{-1} \qquad M= (x_1m_1+ x_2m_2) \qquad (6)$$

Internal pressure is a fundamental property of a liquid, which measure the change in internal energy of liquid or liquid mixtures, as it undergoes a very small isothermal change. It is a measure of cohesive or binding forces between the solute and solvent molecules. K is the temperature independent constant.

$$P_i = bRT\,(K\eta/u)^{1/2}(\rho^{2/3}/M)^{7/6}\ Nm^{-2} \qquad (7)$$

Intermolecular radius can be determined by using relation:

$$r_0 = \left(\frac{3M}{16\pi\rho N}\right)^{\frac{1}{3}}\left[1-\left(\frac{\gamma RT}{Mu^2}\right)\left\{\left(\frac{Mu^2}{3\gamma RT}+1\right)^{\frac{1}{2}}-1\right\}\right]^{\frac{1}{3}} \qquad (8)$$

The property, relative association (R_A) is influenced by breaking up of the associated solvent molecules on addition of solute in it

$$R_A = \left(\frac{\rho}{\rho_0}\right)\left(\frac{U_0}{U}\right)^{\frac{1}{3}} \qquad (9)$$

2. Materials and Methods

The liquids used were of AR grade and were redistilled in the laboratory. In this study the measurements have been made at a temperature 303K. The temperature of the liquid mixture was kept constant by the use of thermostat U-10 with ± 0.1^{0}C accuracy. Density measurement was carried out using specific gravity bottles ±1x10^{-5} gm/cm^3 and digital mono pan-balance with an accuracy of 0.001mg. Ultrasonic velocity measurements were made with an ultrasonic multi frequency interferometer (Mittal enterprises, New Delhi) at a frequency range 1MHz to 10MHz with an accuracy of ±0.01m/s The time of descent of the liquid between the viscometer marks was measured using electronic timer with accuracy of ±0.01sec for digital clock.

3. Results and discussion

From the plot of acoustic parameters verses molar concentration ultrasonic velocity (U), acoustic impedance (Z) and free volume (V_f) increases, however adiabatic compressibility (β_a), intermolecular free length (L_f), relative association (R_A), intermolecular radius (r_o), internal pressure (P_i) and Van der Waal's constant decreases as the molar concentration of acrylonitrile increases in cyclohexane.

In this system molecules of acrylonitrile are polar with dipole moment 1.68D and cyclohexane molecules are non-polar dipole moment is 0D. The nature and the strength of heteromolecular cyclohexane-acrylonitrile or acrylonitrile- cyclohexane interaction is determined by the interacting molecules. The permanent dipoles are present in the molecules of the acrylonitrile. In cyclohexane there is no chemical active group but its polar geometry has vital role for its association. The nature of the forces responsible due to stronger heteromolecular dipole-induced dipole interaction between the unlike molecules are responsible for the association in this binary system. Interstitial accommodation, induction and London dispersion forces in cyclohexane molecules may leads association in this binary liquid system.

It is also observed that ultrasonic velocity increases and free length (L_f) decreases with increase in concentrations . The decrease in free length (L_f) with molar concentrations may be due to the gain of dipolar association. This may be the reason for the observed increase in ultrasonic velocity (U) with decrease in adiabatic compressibility (β_a), in the liquid mixtures.The observed decreased in adiabatic compressibility (β_a) and Van der Waal's constant (b) with molar concentration shows increase in the degree of association among the molecules in the liquid mixture. Hence the intermolecular distance decreases with increase in molar concentration.

The decreasing values of adiabatic compressibility (β_a), internal pressure (P_i), relative association (R_A), and increase in the value of acoustic impedance (Z), free volume (V_f) decreases the intermolecular distance, making relatively less gap between the molecules.

The ultrasonic absorption coefficient (α/f^2) for liquid mixture of acrylonitrile in cyclohexane at frequency 1 to 10 MHz and at concentration 0 to 1 was studied for temperature range 10 to 50^{0}C. As the concentration of acrylonitrile increases, absorption coefficient gradually decreases and there is sudden decreased in absorption coefficient from 0.3 to 0.4 molar concentration and becomes nearly constant from molar concentration 0.4 to 1 This may be due to change in intermolecular arrangement. The decreased in observed ultrasonic absorption with increase in molar concentration which generally increases the inter proton distance between adjacent molecules. This is also due to the weakening of intermolecular forces.

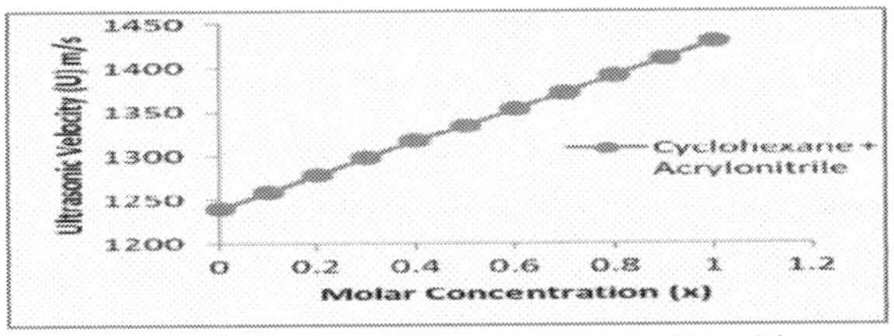

Figure 1. Variation of Ultrasonic Velocity with x.

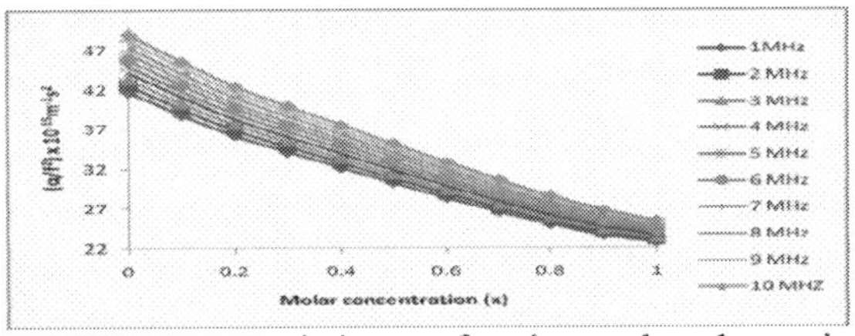

Figure 2. Variation of observed ultrasonic absorption with x.

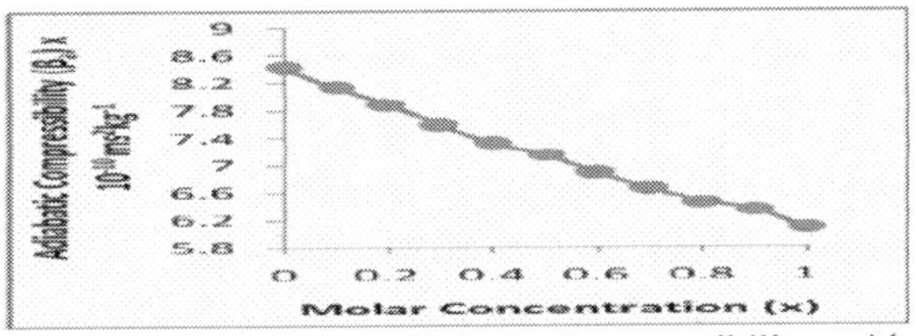

Figure 3. Variation of Adiabatic Compressibility with x.

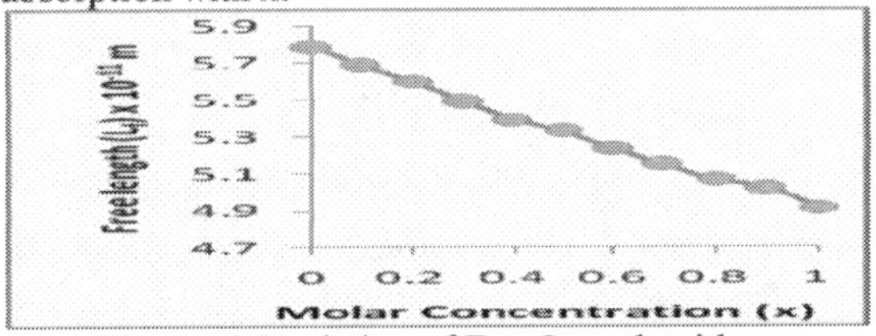

Figure 4. Variation of Free Length with x.

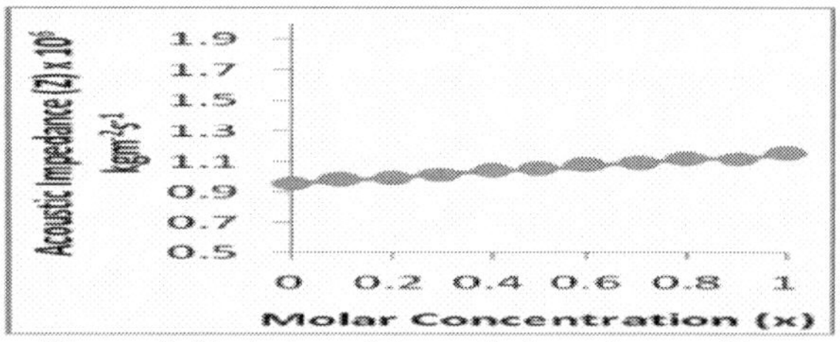

Figure 5. Variation of Acoustic Impedance with x.

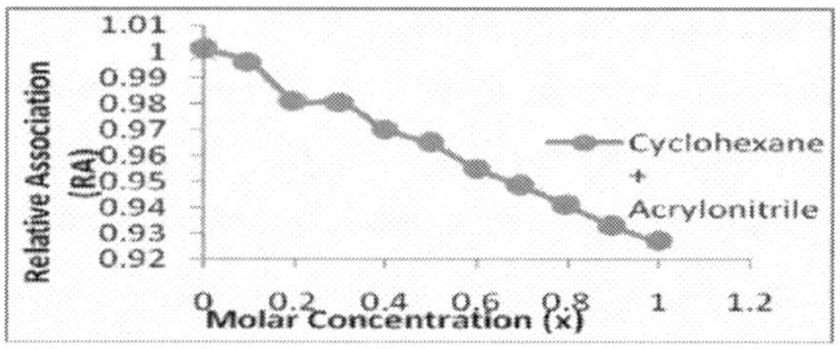

Figure 6. Variation of Relative Association with x.

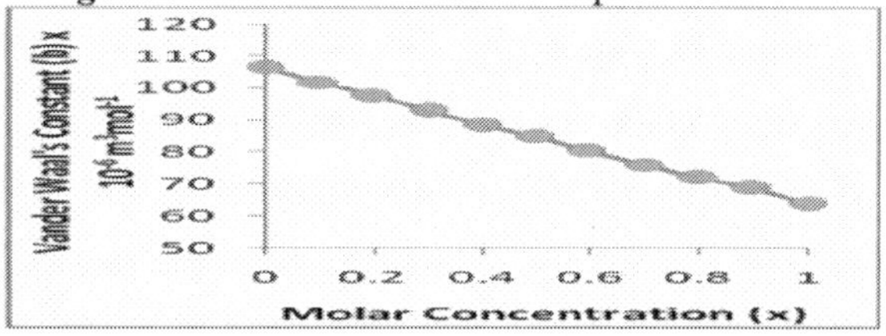

Figure 7. Variation of Vander Waal's constant with x.

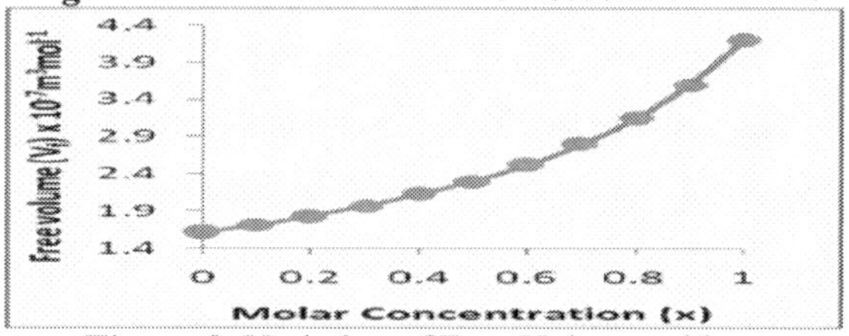

Figure 8. Variation of Free Volume with x.

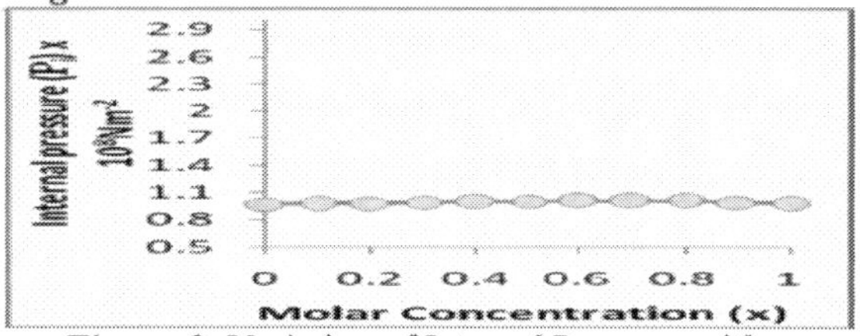

Figure 9. Variation of Internal Pressure with x.

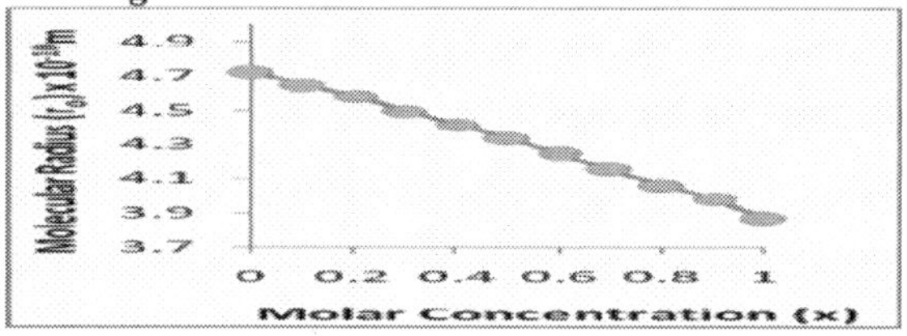

Figure 10. Variation of Molecular Radius with x.

4. Conclusions

Decrease in adiabatic compressibility and free length with increase in molar concentration is due to association. The observed decrease in ultrasonic absorption in the mixtures is indicated by the existence of weakening of intermolecular forces. Ultrasonic absorption study of binary liquid mixtures determines the energy losses in the compressions and decompressions in ultrasonic waves. The present investigation was undertaken in order to study the dependence of the relative strength of intermolecular interaction and their nature in pure liquids and their mixtures.

Acknowledgment

The authors (OPC) & (VDB) are greatful to University grant commission, New Delhi for providing financial support to this work through Major and Minor research project.

References

[1] Tabhane V A 1983 *Acoustic Letter* **8**
[2] Agnihotri P K and Adgaonkar C S 1989 *Acoustics Letters* **12** 7
[3] Tabhane VA, Muley V D and Khasare S B 1995 *Acustica* **81**
[4] Ravichandran S & Ramanajchan K 2006 J Pure & Appl Ultrason **28** 40.
[5] Rath D C and Samal K 1994 *J. Pure Appl. Ultrason* **16** 6
[6] Chimankar O P, Tabhane V A and Baghel G K 2007 *J. Acoust Soc. of India* **34**(4) 126
[7] Tabhane VA, Chimankar OP, Manja S and Nambinarayanan TK1999*J Pure Appl.Ultrason* **21**(3)57
[8] Chimankar O P, Rewatkar K G and Tabhane V A 2001 *Indian J. of Phys.* **75**B(2) 141
[9] Bhandakkar V D Tabhane V A and Ghosh Sharda 2003 *Indian J. Pure & Appl. Phys.* **41** 849
[10] Manikiam B and Narasimham A V 1984 *Indian J. Pure & Appl. Phys.* **22** 29
[11] Bhandakkar V D, Chimankar O P and N R Pawar 2010 *J. Chem. Pharm. Res.* **2**(4) 873
[12] Bhandakkar V D, Chimankar O P and N R Pawar 2011 *Indian J. Pure & appl. phys.* **49** 550
[13] Chimankar O P, Shriwas Ranjeeta S and Tabhane V A 2010 *Arch of appl. science res.* **2**(6) 285
[14] Adgaonkar C S, Deogaonkar V S and Kadu P D 1997 *Indian J. Pure & Appl. Ultrason* **15**
[15] Chimankar O P, Jajodia S, Shriwas R S and Tabhane V A 2010 *Arch Appl. Sci. Res.* **3**(3) 252

IOP Conf. Series: Materials Science and Engineering **42** (2012) 012031 doi:10.1088/1757-899X/42/1/012031

Identification of piezoelectric complex parameters in rings for power ultrasound applications

N Pérez [1,3], M A Andrade[2], R C Carbonari[2], F Buiochi[2], and J C Adamowski[2]

[1]Agro-Industrial and Agri-Food research group. Paysandú University Center, University of the Republic, Paysandu 60000, Uruguay.
[2]Department of Mechatronic and Mechanical Systems Engineering, Polytechnic School, University of Sao Paulo, Sao Paulo 05508-030 – SP, Brazil.

E-mail: nico@fisica.edu.uy

Abstract. Power ultrasonic devices frequently use Langevin type transducers. These types of transducers are essentially constructed using a sandwich of piezoelectric rings and two metal masses at the ends. The whole assembly is tuned to resonate in a desired main frequency and the total length corresponds to a half of the wavelength of that frequency. Finite element simulations (FEM) are used in the design of such complex structures; however the accuracy of the results is limited by the knowledge of the constitutive properties for the materials used in the transducer construction. Metals like aluminum or steel are well characterized, but the complete set of piezoelectric parameters for piezoceramics are difficult to find in the literature. In the few cases where the manufacturer gives the complete set of parameters, strong differences are observed between simulated and experimental data. In this work a novel methodology proposed by our research group is applied in the case of piezoelectric rings made with a hard piezoelectric material. The results are evaluated in rings with internal diameter 8 mm, external diameter 27 mm and thickness 5 mm. Finally the results are validated using an optical interferometer showing a good agreement.

1. Introduction

A wide range of power ultrasound applications are based on Langevin transducers [1]. Theses transducers consist in a stack of piezoelectric ceramics sandwiched by two metal masses. The whole assembly is preestresed by a central bolt in order to avoid cracks in the ceramics when high deformations occur. The transducers are operated in the continuous wave mode, due to the low losses in all components the resonances of the structure are very narrow band. The frequency of operation is selected at the main thickness mode resonance. Depending on the application, other resonant structures are coupled to the Langevin transducer acting as mechanical amplifier [2]. To design these transducers several techniques were used, first simple one-dimensional models determinate the size of the elements and the main frequency, after some models to include 3D characteristics are proposed [3]. Actually the use of finite elements FEM is a common tool in the project of the transducers [4]. However the accuracy in the modeling is limited by knowledge of the material properties. Recently

[3] Present address: Polo Agroalimentario Agroindustrial, Centro Universitario de Paysandú, Universidad de la República, Ruta 3 km 363 - Paysandú – Uruguay. E-mail: nico@fisica.edu.uy

our research group presents a new methodology to determinate de real part of the full set of piezoelectric parameters [5]. In the case of a piezoceramic disk polarized in the z axis, the symmetry reduces the number of parameters to five elastic, three piezoelectric and two dielectric constants. In this work the complete set of constants are obtained in the case of PZT8 piezoelectric rings. PZT8 is a very low losses material frequently used in power transducers; this material presents a lot of spurious modes. Another difficulty is the ring geometry, in the reference [5] the solution is implemented in piezoelectric disks, here the sensitivity analysis and the detailed algorithm to determinate the properties is implemented in the case of rings. To obtain the properties a set of five "identical" samples is selected, the details of the geometry are showed in table 1.

Table 1. Geometry density of the samples.

Sample	Internal diameter Mm	External diameter Mm	Thickness mm	Mass gr.
S1	7.73	27.3	5.08	20.804
S2	7.76	27.4	5.10	20.950
S3	7.71	27.3	5.08	20.755
S4	7.71	27.4	5.10	20.898
S5	7,70	27,3	5,10	20,745

In this work only the real part of the parameters is optimized, however the parameters in the model are complex numbers. At this moment our research group is working in the development of a methodology to optimize the imaginary part of the model, at the moment an approximate value is used, for this reason differences in amplitude of an individual mode can exists. Table 2 shows the values used for the five samples.

Table 2. Imaginary part of the piezoelectric model constants.

Sample	c^E_{11} 10^6 N/m^2	c^E_{12} 10^6 N/m^2	c^E_{13} 10^6 N/m^2	c^E_{33} 10^6 N/m^2	c^E_{44} 10^6 N/m^2	e_{31} C/m^2	e_{15} C/m^2	e_{33} C/m^2	$\varepsilon^S_{11}/\varepsilon_0$	$\varepsilon^S_{33}/\varepsilon_0$
Mean	100	1	1	10	20	0.002	0.001	0.001	0	0.1

2. Sensitivity analysis

The methodology to determinate the piezoelectric properties is based on the minimization of the error between the numerical FEM analysis and the experimental data. This FEM analysis can is implemented using axisymmetric elements; this makes the iterative simulations possible. In this case of PZT8 the numerical solution converges slowly due to the low attenuation, here 30 elements are used in the thickness of the sample (60 elements in the thickness frequency wavelength). The first step in the methodology is a sensitivity analysis of the real part, in this analysis the properties of the model are changed (± 25% from a star value) and the evolution of the maximum of resonance and; divide the problem to found an approximate solution using only the most sensitive parameters. When this point is reached a traditional optimization technique based in the Neadler-Mead algorithm antiresonance are plotted vs. the frequency. Using this information a first algorithm is constructed in order to obtain a solution close to the "real solution" avoiding local minimums [5]. This is the main part of the methodologyis used [6]. Figure 1 shows the sensitivity analysis for three selected properties, from here is can be noted that c_{33} is the most sensitive near the thickness frequency (around 420 KHz), and is used to adjust this frequency. Similarly, c11 is very sensitive in the radial modes and c44 is insensitive at these frequencies. However, c_{44} "controls" the separation between the between the fourth and fifth resonant modes.

IOP Conf. Series: Materials Science and Engineering **42** (2012) 012031 doi:10.1088/1757-899X/42/1/012031

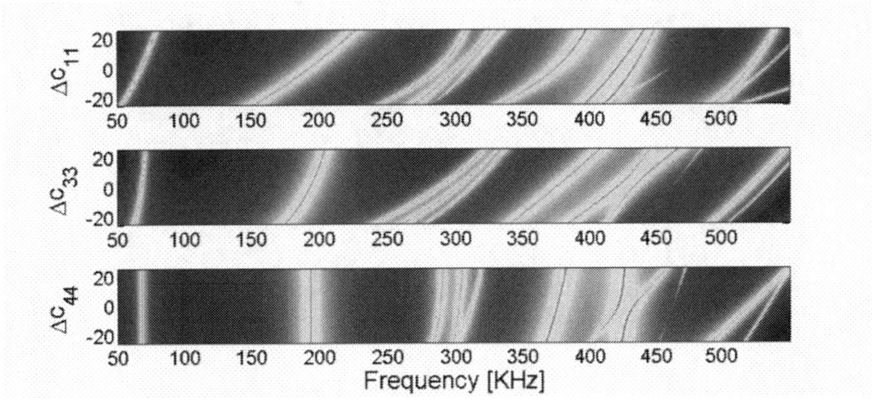

Figure 1. Sensitivity analysis for c_{11}, c_{33} and c_{44}.

3. Results

In this section the results of the optimization process are presented. The figure 2 shows the adjust of the electric impedance both in modulus and phase. The objective function used in the optimization is the mean error in the impedance modulus.

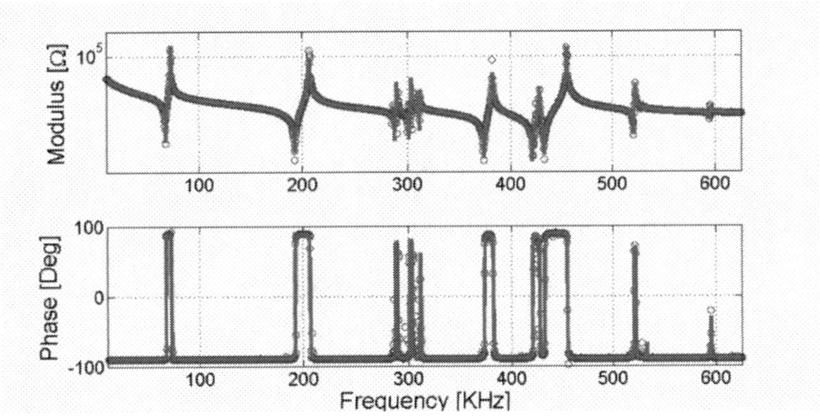

Figure 2. Results of the optimization in the sample number 1. Continuous line is the FEM simulation and the dotted data are experimental points.

Table 3. Real part of the piezoelectric model constants.

Sample	c^E_{11} 10^{10} N/m^2	c^E_{12} 10^{10} N/m^2	c^E_{13} 10^{10} N/m^2	c^E_{33} 10^{10} N/m^2	c^E_{44} 10^{10} N/m^2	e_{31} C/m^2	e_{15} C/m^2	e_{33} C/m^2	$\varepsilon^S_{11}/\varepsilon_0$	$\varepsilon^S_{33}/\varepsilon_0$
S1	14.02	7.32	7.50	12.63	3.05	-3.96	9.70	14.95	753	711
S2	14.03	7.33	7.50	12.63	3.05	-4.14	9.88	15.02	718	707
S3	14.00	7.31	7.48	12.63	3.05	-3.99	9.63	15.08	792	709
S4	14.03	7.36	7.49	12.59	3.02	-4.18	9.72	14.96	825	712
S5	14.06	7.36	7.44	12.65	3.04	-3.92	10.00	14.98	813	699
Mean	14.02	7.35	7.48	12.63	3.04	-4.06	9.78	14.97	802	701
Std %	0.2	0.03	0.25	0.15	0.4	3.5	1.5	1	6.5	2

To obtain the properties of the fabrication batch, the mean value is obtained over the set. Note that the elastic constants have a lower dispersion than the piezoelectric and dielectric constants. In order to verify the validity of the numerical solution, the displacement in a selected point (radial displacement in the mean plane of the ceramic) is compared with the experimental measure using an optical interferometer. Figure 3 shows the comparison in the selected frequency range.

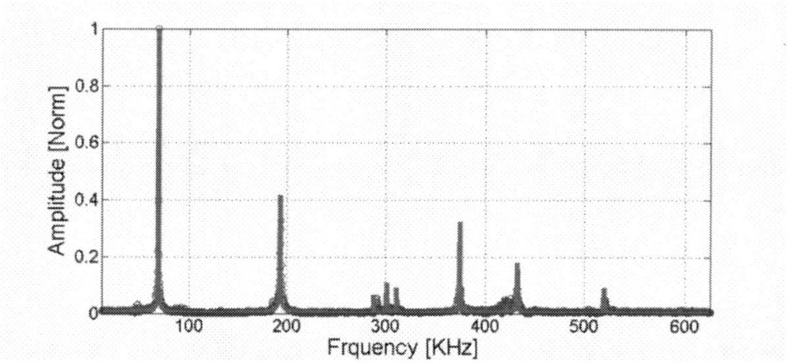

Figure 3. Comparison between experimental (dotted) and numerical (continuous line) results. The amplitude is normalized by the maximum.

4. Conclusions

First the results for the optimization in case of piezoelectric rings are presented and a good agreement is obtained. The material PZT8 presents a very low attenuation, that present and additional difficulty in the proposed methodology because the great amount of spurious modes ant the narrow resonances. This implies a great frequency discetization in order to reproduce all the modes. Also a great spatial discretization is needed in this case, for soft piezoceramics good results are obtained using 20 elements in the thickness, however in this case the results converges for 30 elements.

The results are verified in an independent way; the experimental data to make the optimization process are only electrical, modulus and phase of the electric impedance. At the end the simulated displacement is compared with the measured displacement showing a good agreement.

This is a first step in order to obtain the material properties for the PZT8, to obtain more reliable results different geometries must be used. Also, this work is the first practical application of the methodology [5] for samples with axisymmetric geometry, but the geometry is not a homogeneous disk. It is expected that these results can be used in the design of power ultrasonic transducers.

Acknowledgments

The authors would like to thank CSIC-UdelaR Uruguayan agency and the FAPESP Brazilian agency for the financial support that makes this work possible.

References

[1] Gallego-Juarez J A 1989 *J. Phys. E: Sci. Instrum.* **22** 804
[2] Gallego-Juarez J A, Rodriguez-Corral G, Riera-Franco E, Vazquez-Martinez F, Campos-Pozuelo C and Acosta-Aparicio V 2002 *Ultrasonics* **40** 889
[3] Iula A, Carotenuto R, Pappalardo M and Lamberti N 2002 *J. Acoust. Soc. Am.* **111** (6) 2675
[4] Iula A, Vasquez F, Pappalardo M and Gallego J A 2002 *Ultrasonics* **40** 513
[5] Pérez N, Andrade M A, Buiochi F and Adamowski J C 2010 *IEEE Trans. Ultrason. Ferroelectr. Freq. Control* **57** 2772
[6] Nelder J A and Mead R 1965 *Comput. J.* **7** 308

IOP Conf. Series: Materials Science and Engineering **42** (2012) 012032 doi:10.1088/1757-899X/42/1/012032

Magnetorheological fluid characterization using ultrasound measurements

J Rodríguez-López, L Elvira and F Montero de Espinosa

Centro de Acústica Aplicada y Evaluación No Destructiva (CAEND), CSIC-UPM, c/Serrano 144, 28006, Madrid, Spain.

E-mail: jaimerl@caend.upm.csic.es

Abstract. In this work the variations of velocity of sound and attenuation in magnetorheological (MR) suspensions have been studied when the temperature and the intensity of magnetic field have been varied and, also, when the suspension is observed for a long period of time. It has been shown that the behaviour of the MR fluids depends strongly on the fluid used as solvent when temperature is varied. Regarding the sedimentation process, it has been proved that the application of an external magnetic field enhances the stabilization process. Analyzing the hysteretic behaviour it is seen that the system does not recover its initial state when the magnetic field is removed, because the ordered microstructure does not disappear completely. As ultrasound parameters are sensitive to changes in the temperature, in the structure and also in the volume fraction, they are a promising tool to characterize MR fluids in order to improve its performances.

1. Introduction

Smart materials are materials which have one or more properties which can be changed in a controlled manner applying an external stimulus. Magnetorheological fluids (MRFs) are an example because their properties –rheological [1], optical [2], microstructural [3] or elastic [4]- change when an external magnetic field is applied. In the last decades, the MRFs, which are suspensions of magnetic particles in a fluid that can be rearranged into a more ordered structure due to the application of the external magnetic field, have been used in different kinds of applications, being the automotive applications, such as clutches, brakes and dampers the most developed [5].

These suspensions can work during long periods of time, under different temperatures and under different magnetic fields. Depending on the working conditions, different handicaps like the appearance of a sedimentation process or a hysteretic effect can take place varying the performance of the system. The sedimentation process in suspensions is usually studied by taking and then treating optical images. Concerning the rheological properties of these fluids when the temperature is varied, different studies have been made concluding that the changes are quite small. Finally there are several works studying the hysteretic behavior of these fluids highlighting the importance of this effect in practical applications such as dampers and clusters [6-8].

In this work it is shown that, because ultrasounds are sensitive to changes in volume fraction, temperature and the rearrangement of the microstructure, suitable techniques based on ultrasonic propagation can be developed to study the sedimentation process, changes in temperature and the presence of hysteresis. Furthermore, ultrasounds have the advantages that they can be used on line and they are inexpensive.

2. Materials and Methods

2.1. Materials

Three different fluids have been used:

1.-*MRaraldite25.* It is a noncommercial suspension of carbonyl iron powder (CIP) at 25% of volume fraction with a diameter between 1 to 10 microns, using an epoxy polymer –Araldite DBF (3M)- without hardener as a solvent (MRaraldite0%). At 30ºC, the density (1150 kg/m^3) and viscosity (1Pa.s) is high enough to make the suspension stable for several hours.

2.-*MRaceite9.* This noncommercial fluid is made of the same kind of particles -CIP- than the previous fluid but with a 9% of volume fraction; the solvent has a 91% of volume fraction of olive oil, an 8.6% of lithium grease and a 0.4% of fumed silica (MRaceite 0%). The viscosity of the host fluid is around 1 Pa.s and it has a density of 909 kg/m^3 making this suspension of particles stable.

3.-*MRC* is a commercial MR fluid (MRHCCS4-A) with a 20% volume fraction of particles provided by Liquid Research Co. This fluid is a stable suspension of magnetically polarisable micron sized particles suspended in synthetic hydrocarbon oil with some additives.

2.2. Methods

A transmission technique is used to measure the time of flight and attenuation variations of ultrasonic pulses with a central frequency of 960 KHz. For this purpose, two piezoceramics are placed stuck to a methacrylate cell with dimensions of 30x30x6 mm^3, being the smallest dimension the one which corresponds to the ultrasonic path. The experimental set up and the calculus to obtain the changes in the velocity of sound and attenuation are explained in reference [4].

The measurements described in sections 3.1 and 3.3 took place at a controlled temperature of 30 ºC into an air chamber. However, the experiments made to test the variations with temperature -section 3.2- were performed using a thermostatic controlled bath. The studied temperature range was from 25 to 45 ºC. Concerning the magnetic field source, permanent magnets providing an intensity of 212 mT were used in the experiments of sections 3.1 and 3.2, and an electromagnet providing intensities from 0mT up to 105mT was used to perform the experiments of section 3.3.

3. Results

3.1. Sedimentation

Ultrasonic properties are sensitive to the change in the volume fraction of particles in a suspension. Therefore, in these fluids, as there is a big difference between the densities of the host fluid and the particles, the latter tends to settle down causing that volume fraction of particles along the ultrasonic path change with time. In figure 1(a) and 1(b) the changes of the velocity of sound and attenuation with time are shown.

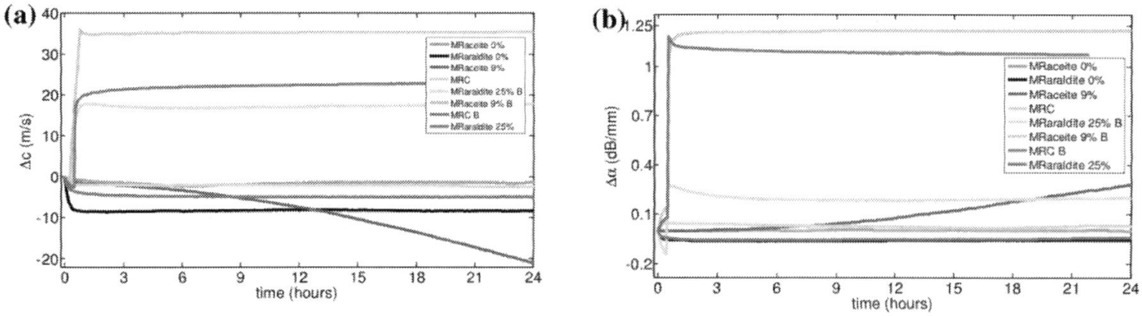

Figure 1. Study of the sedimentation process. (a)Velocity of sound and (b) attenuation variations as function of time for MRaraldite, MRaceite and MRC. *B* means that a magnetic field has been applied.

The first remarkable feature is that both the velocity and the attenuation are sensitive to the sedimentation process. The variations observed during the first hour correspond to the temperature stabilization. It can be seen that when there is not magnetic field, MRC and MRaraldite25 are very stable while MRaceite9 is not. When the magnetic field is applied, although there is a necessary time to get the structures formed, the stability is highly enhanced because the magnetic field makes the particles to interact with the magnetic field and between them avoiding the settle process.

3.2. Temperature

In this section, the variations of the velocity of sound and the attenuation with temperature are analyzed.

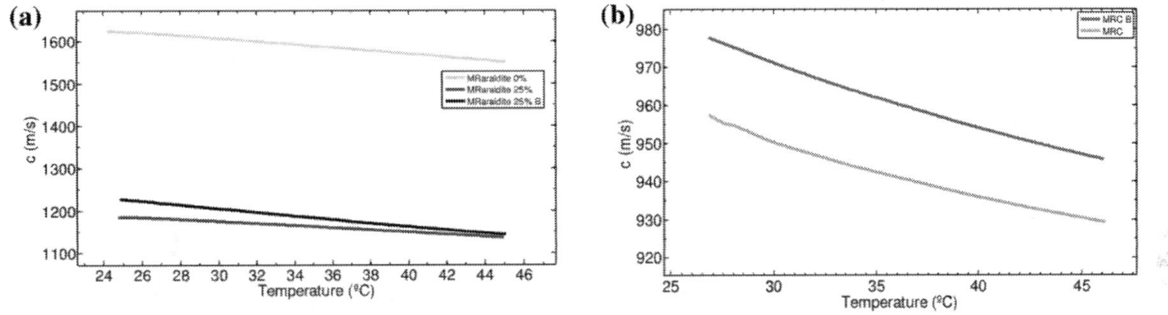

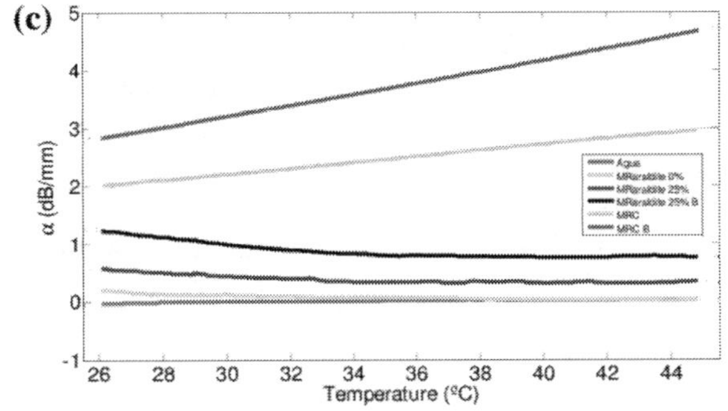

Figure 2. Effect of the temperature on MR fluids. Variation of the velocity of sound in MRaraldite (a) and MRC (b). Change of the attenuation with temperature (c) of both fluids and water. B means that a magnetic field has been applied. In (a) and (c) the properties of MRaraldite 0% (host fluid) is shown like a reference fluid.

From figure 2(a) and 2(c), where MRaraldite25 behaviour is shown, it can be seen that the dependence of the properties of the MR fluid on the temperature is basically influenced by the properties of the host fluid (MRaraldite0%) more than the particles suspended or the structures formed. When MRC and MRaraldite25 are compared, a decrease of the velocity of sound is noticed when the temperature is increased (figure 2(a) and 2(b)). Nevertheless, the behaviour in attenuation is different (figure 2(c)). The attenuation of MRaraldite25 decreases when the temperature is increased, as it does the host fluid (green line); however, the attenuation of MRC increases as the temperature does.

As it is well known, when a magnetic field is applied, a more ordered structure is induced. Taking into account that the properties are highly influenced by the host fluid rather than the suspended particles or structures formed, the properties with and without magnetic field vary in the same manner with temperature. Regarding the velocity of sound (figure 2(a) and 2(b)), it can be seen that when there is a magnetic field applied, the decrease is higher, being more evident in the case of the MRaraldite25 than in MRC. As the density is the same, even if a magnetic field is applied or not, and the properties of the host fluid will changed in the same way, the only explanation to this different behaviour is that when the structure is formed, the compressibility of the structure is more sensitive to the temperature variations than when particles are suspended. The thermal energy could also destabilize the structure formed, but due to the variations are small, it will not play an important role. Regarding attenuation, (figure 2(c)) it can be seen that the behaviour of MRaraldite25 is very similar with and without magnetic field, while when the MRC is observed, a rougher variation between both cases is shown, having the structure formed a more important effect than in the case of MRaraldite25.

3.3. Hysteresis

It is well known that MRFs do not recover their initial state when the applied magnetic field is removed. Due to the ferromagnetic properties of the suspended particles, there is a remanent magnetization of the particles which causes a memory effect. In figure 3(a) and 3(b), it can be seen how the acoustical properties shows this hysteretic behaviour.

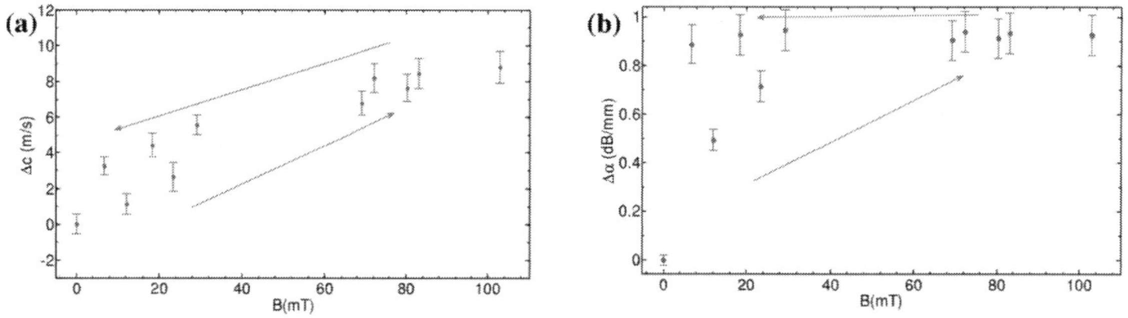

Figure 3. Hysteresis effect of MR fluids. Variations of velocity (a) and attenuation (b) of MRC when the intensity of magnetic field is varied from 0 mT to 100mT and back to 0 mT.

It is shown that the hysteretic behavior is much more evident when the attenuation is analyzed than when the velocity is studied. From these results, once the magnetic field is removed, it can be said that the remanent structure is not consolidated; therefore the velocity recovers the initial value because the compressibility is similar to the compressibility of the suspension. Concerning the attenuation, the fact that it does not recover the previous value highlights that the formed structures due to the magnetic field tends to stay, even non consolidated, at a similar position than when there were a magnetic field applied. The attenuation is much more sensitive to the shape of the structures than to its stiffness.

4. Conclusions

MRFs are systems widely used in mechanical devices. Its performance depends on different factors like "thickening in use" due to sedimentation, the effect of temperature or the appearance of hysteresis. It has been study applying ultrasonic techniques how the velocity and attenuation of sound are sensitive to changes of volume fraction, variations of temperature and changes in the microstructure of a system, resulting these techniques suitable to study operating conditions of MR fluids like stability, temperature effects, hysteresis effects or the application of a magnetic field.

Acknowledgements

This work was supported by a CSIC JAE fellowship, and by the Spanish Ministry of Science and Innovation (Plan Nacional de I+D DPI2010-17716 and PIB2010BZ-00570).

References

[1] de Vicente J, Klingenberg D J and Hidalgo-Alvarez R 2011 *Soft Matter* **7** 3701-10
[2] Melle S, Calderon O G, Rubio M A and Fuller G G 2003 *Physical Review E* **68** 11
[3] Mohebi M, Jamasbi N and Liu J 1996 *Physical Review E* **54** 5407-13
[4] Rodriguez J, Elvira L and Montero de Espinosa F 2012 *J. Magn. Magn. Mater.* **324** 222-30
[5] Carlson J D, Catanzarite D M and StClair K A 1996 *Int. J. Mod. Phys. B* **10** 2857-65
[6] Bramantya M A, Takuma H and Sawada T 2010 *Int. J. Appl. Electromagn. Mech.* **33** 1365-72
[7] Dominguez A, Sedaghati R and Stiharu I 2004 *Smart Mater. Struct.* **13** 1351-61
[8] Jedryczka C, Sujka P and Szelag W 2009 *Compel-Int. J. Comp. Math. Electr. Electron. Eng.* **28** 711-721

IOP Conf. Series: Materials Science and Engineering **42** (2012) 012033 doi:10.1088/1757-899X/42/1/012033

Study of miscibility and interaction in polyvinyl chloride-polyvinyl acetate blend using ultrasonic technique

P V Tabhane[1], O P Chimankar[1,4], R S Shriwas[1], C M Dudhe[2] and V A Tabhane[3]

[1]Department of Physics, RTM Nagpur University, Campus, Nagpur -440 033, INDIA
[2]Department of Physics, Institute of science, Nagpur -440 033, INDIA
[3]Department of Physics, University of Pune, Ganeshkhind, Pune -411 007, INDIA

E-mail: opchimankar28@gmail.com

Abstract. The propagation of ultrasonic waves and the measurement of their velocity in solutions form an important tool for the evaluation of various acoustical and thermo dynamical parameters which give an insight into the nature of miscibility and molecular interactions in polymer blend. In the present investigation, ultrasonic velocities, densities and viscosities have been measured in 0.1M solution of polyvinyl chloride in tetrahydrofuran using pulse echo overlap technique at 298K at different concentrations of polyvinyl acetate. Thermoacoustical parameters such as adiabatic compressibility, molar sound velocity, molar compressibility, expansion coefficient, acoustic impedance, van der Waals' constant and internal pressure have been computed from the experimental data. The variation of ultrasonic velocity and other thermo-acoustical parameters shows nonlinear increase or decrease with molar concentration which suggest immiscibility or semi compatibility among the component polymers. Two immiscible polymers are need to be compatibilized in order to be used in commercial applications. The nature of solvent-polymer-polymer interaction and the effect of concentration on the molecular interaction of polyvinyl chloride and polyvinyl acetate in tetrahydrofuran have been studied.

1. Introduction

There are various techniques for studying the compatibility of polymer blends, ultrasonic velocity and viscosity measurements reveal compatibility or miscibility of blends. Many workers [1,2] have carried out pioneering work on polymer-polymer compatibility using ultrasonic techniques. These provide powerful, effective and reliable tools to investigate properties of blends. Blending has been the most effective way to get better polymeric systems; they can offer better properties, cost and processing advantages in their applications. There is a strict correlation between the chemical structure, physical structure and the molecular mobility of polymers and the acoustical parameters. It is thus obvious that the ultrasonic parameters are useful for the study of molecular interactions.

[4] To whom any correspondence should be addressed.

The present paper deals with the ultrasonic studies of polymer solution of polyvinyl chloride (PVC) in tetrahydrofuran (THF) as a function of increasing molar concentration of polyvinyl acetate (PVAc). Polyvinyl chloride is a thermoplastic polymer. It is a vinyl polymer constructed of repeating vinyl groups (ethenyls) having one hydrogen replaced by chloride. Poly (vinyl acetate) is the most widely used polymer of the vinyl ester. It is not only used as a plastic in form of emulsion, but also as the precursor for the two polymers, poly (vinyl Alcohol) and poly (vinyl Acetal) which cannot be prepared by direct polymerization. Polyvinyl chloride and polyvinyl acetate are synthetic polymers with the formulae $(CH_2CHCl)_n$ and $(C_4H_6O_2)_n$ respectively. PVC is widely used in construction because it is cheap, durable, and easy to assemble. While polyvinyl acetate is a polymer of industrial importance which includes its application as strengthening agents for cements.

2. Materials and experimental method

The solution of PVC was prepared by dissolving in 250 ml of THF. This solution, having concentration 0.1M, was used as a stock solution. Various concentrations of the PVAc is added by weight percentage in the known and fixed concentration of stock solution to make poly-blend solution. Total 50 ml solution was prepared for the measurement of ultrasonic velocity, density and viscosity. The chemicals used were of excelar grade. The velocity was measured by using Automatic Ultrasonic Attenuation Recorder AUAR – 102 (Innovative Instrument, Hyderabad) and frequency counter APLAB-1116. The measurements were carried out by using pulse echo overlap technique at frequency 4MHz. The accuracy in the measurement of ultrasonic velocity is ±0.01 m/s. The density (±0.1mg) and viscosity (±0.01sec for digital clock) were measured by employing hydrostatic plunger method and Ostwald's viscometer respectively. The temperature was maintained constant (298K) during the period of measurement by a water circulation system from the thermostat U-10 with thermal stability of ± 0.1° C.

3. Results and discussion

The ultrasonic velocity (u), density (ρ), viscosity (η) and absorption (α/f^2) were calculated experimentally while adiabatic compressibility (β_a), relaxation time (τ), Rao's molar sound velocity (R), Wada constant (Molar compressibility) (W), acoustic impedance (Z), van der Waal's constant (b), internal pressure (P_i), expansion coefficient (α) have been estimated by using standard formulae as given below-

$$R = \frac{M}{\rho} u^{1/3} \quad \text{cm}^{10/3}/\text{s}^{1/3} \tag{1}$$

$$W = \frac{M}{\rho} \beta_a^{-1/7} \quad \text{cm}^{19/7}/\text{dyne}^{1/7} \tag{2}$$

$$Z = \rho u \quad \text{gm/cm}^2\text{s} \tag{3}$$

$$\alpha = \frac{P_i \beta_i}{T} \quad K^{-1} \tag{4}$$

Where isothermal compressibililty (β_i) can be calculated as:

$$\beta_i = \gamma \, \beta_a \quad \text{cm}^2/\text{dyne} \tag{5}$$

Where γ = ratio of isobaric and isochoric heat capacities

Figure 1-8 shows the plots of experimental velocity, density, viscosity, absorption and the calculated values of all the above acoustical parameters at different molar concentration for polymer-polymer solution of PVC/PVAc with tetrahydrofuran.

Figure 1 shows non-linear variation of u & β_a with increasing concentration. The variation of u shows two peaks at concentration 0.02M & 0.06M. From the shape of the u vs. composition of the second polymer in the mixture, one can account for the compatibility or miscibility of the polymer-polymer blend. Here variation of u is observed to be S shaped indicating non-compatibility or immiscibility between composite polymers. In this blend system initially u increases with the addition of second polymer in the solution, this suggest association. Then with the increase of concentration of second polymer u decreases showing dissociation but then again u increases followed by decrease and thus forming *S-shaped* structure. Observed non-linear variation in u with increase of polymer in solvent indicates polymer-polymer interactions [3]. While variation in β_a shows nature opposite to that of u, this indicates complex formation in the system [4]. Variation of Z (figure 2) with concentration shows same trend as that of u. The maximum z value show strong molecular interaction.

Variation of ρ & η with concentration (figure 3) of polyblend system is observed to be varying non-linearly with increase in concentration of PVAc in THF/PVC mixture. This is because with the addition of PVAc in the mixture increases the number of chains in the mixture. Decrease in ρ with increasing concentration of second polymer in blend also indicates non-compatibility or immiscibility of constituent polymers. Variation of τ and $\alpha/f^2_{\text{classical}}$ (figure 4) with increasing concentration of PVAc in THF/PVC solution show non-linear variation which also suggests complex formation in the mixture. Variation of $\alpha/f^2_{\text{observed}}$ and $\alpha/f^2_{\text{excess}}$ (figure 5) shows similar trend of variation with two peak structured polymer blend system, indicates the presence of two phases in polyblend system. The first peak which we get is in the PVC-rich region while second peak is in

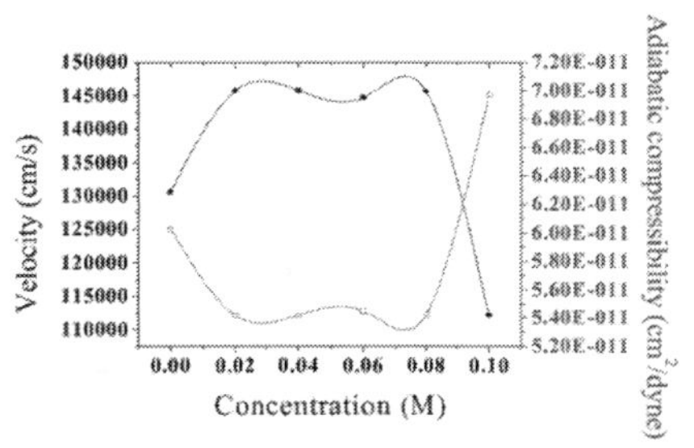

Figure 1. Variation of u and β_a with M.

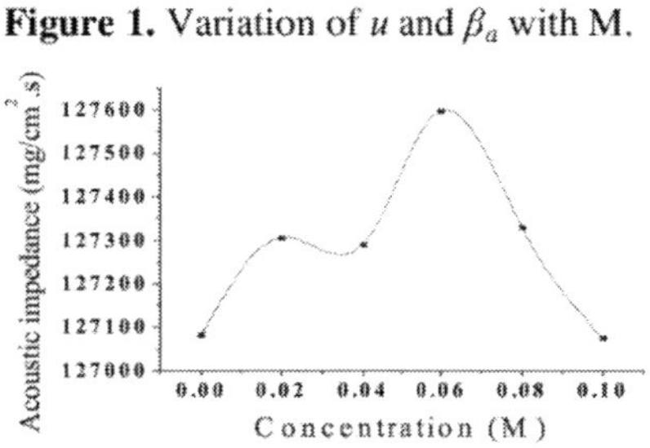

Figure 2. Variation of Z with M.

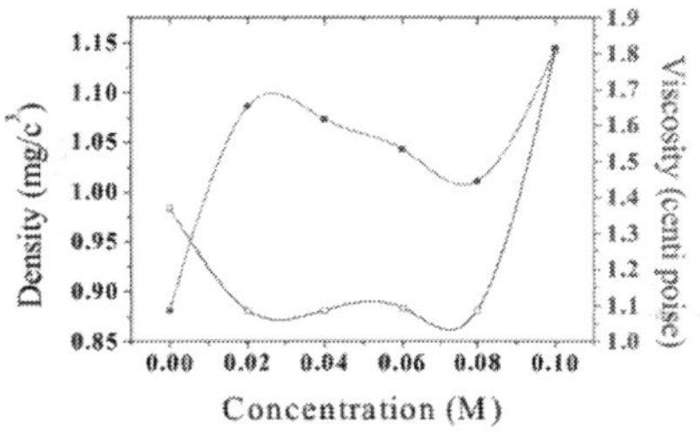

Figure 3. Variation of ρ & η with M.

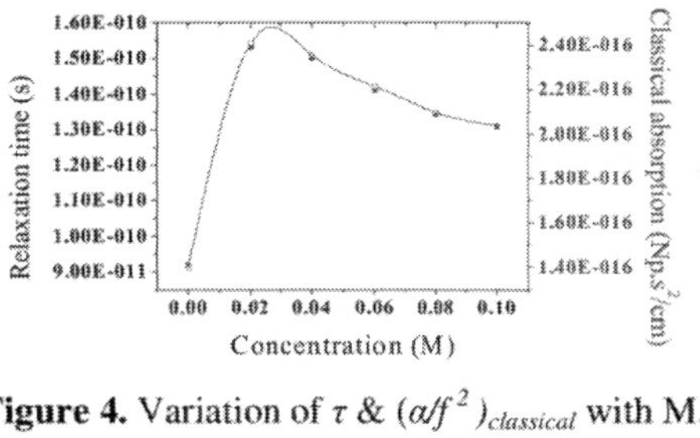

Figure 4. Variation of τ & $(\alpha/f^2)_{classical}$ with M.

the PVAc rich region. This suggests non homogeneity [5], indicated by double phase formation in the system. Also it suggests heterogeneity, indicated by double phase separation in the blend and thus it suggests non-compatibility of the polymers. The non-linear decrease in *b* (figure 6) & *R*, *W* (figure 7) may be attributed to dipole-dipole interaction, and complex formation between the molecules of solvent/polymer/polymer [6].

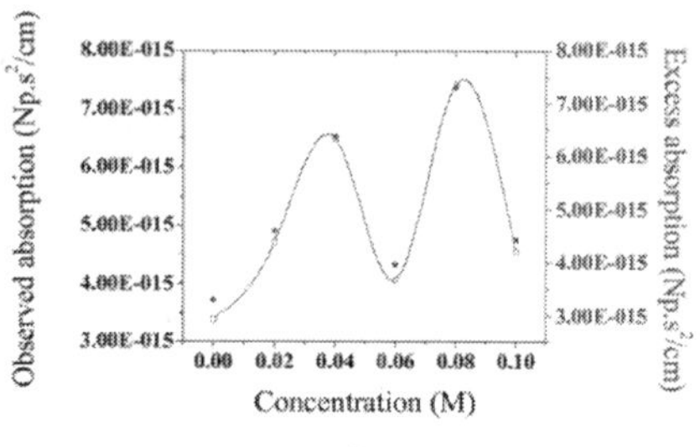

Figure 5. Variation of $\alpha/f^2_{observed}$ & α/f^2_{excess} with M.

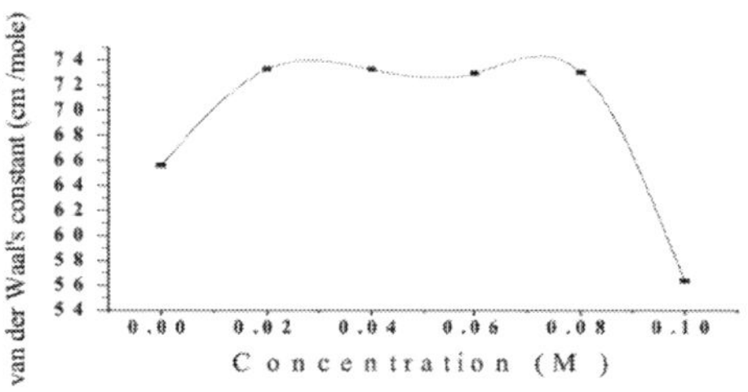

Figure 6. Variation of *b* with M.

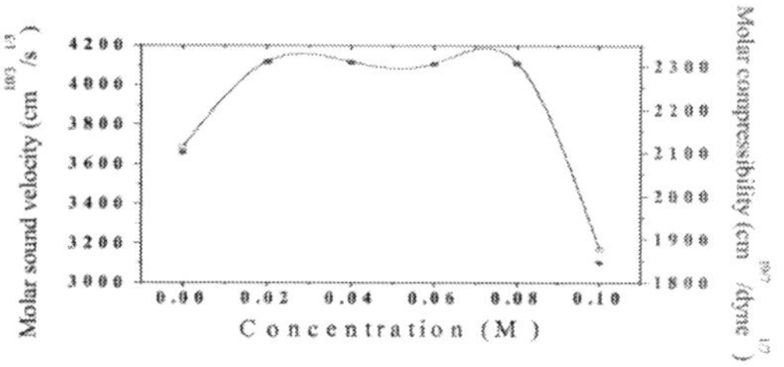

Figure 7. Variation of R & W with M.

The observed non-linear variation of P_i & α (figure 8) shows complex formation in the polyblend system, and semi-compatibility or immiscibility in the blend. The internal pressure parameter P_i of a liquid reflects the molecular interaction and thus is a sensitive parameter [7].

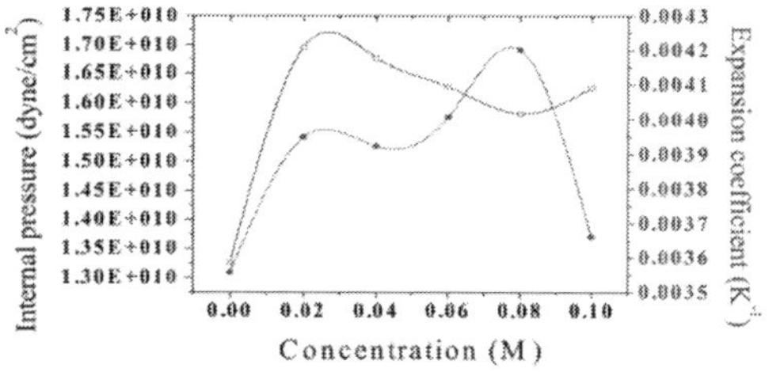

Figure 8. Variation of P_i & α with M.

4. Conclusions

The non-linear variation of ultrasonic velocity and other acoustical parameters suggest, heterogeneity, indicated by double phase separation in the blend, due to immiscibility or non-compatibility of constituent polymers.

Acknowledgments

The one of the author (OPC) is grateful to University grant commission, New Delhi for providing financial support to this work through Major research project letter F.No.39-456/2010 (SR).

References

[1] Singh R P 1993 *Acoust. Soc. Ind.* **21**(3) 159-170
[2] Rajagopalan S and Sharma S J 1981 *J. Pure & Appl. Ultrason* 3 1-3
[3] Pia K M E, Mallika S and Kalyansundaram S 2008 *J. Pure Appl. Ultrason.* **30** 54-56
[4] Tabhane V A, Agrawal S and Rewatkar K G 2000 *J. Acous. Soc. Of Indn.* **28** 369
[5] Tabhane PV, Chimankar OP, Dudhe CM and Tabhane VA 2012 *Der Chemica Sinica* **3**(4) 944
[6] Rao V, Ashokan PV and Shridhar MH 1999 *Polym.* **40**(25) 7167
[7] Tabhane PV, Chimankar OP and Tabhane VA 2012 *Modern Physics Letters B* **26** (27) 1250180

IOP Conf. Series: Materials Science and Engineering **42** (2012) 012034 doi:10.1088/1757-899X/42/1/012034

Ultrasonic Characterization of Superhard Material: Osmium Diboride

P K Yadawa
Department of Applied Physics, AMITY School of Engineering and Technology, Bijwasan, New Delhi-110 061, India

E-mail: pkyadawa@aset.amity.edu

Abstract. Higher order elastic constants have been calculated in hexagonal structured superhard material OsB_2 at room temperature following the interaction potential model. The temperature variation of the ultrasonic velocities is evaluated along different angles with unique axis of the crystal using the second order elastic constants. The ultrasonic velocity decreases with the temperature along particular orientation with the unique axis. Temperature variation of the thermal relaxation time and Debye average velocities are also calculated along the same orientation. The temperature dependency of the ultrasonic properties is discussed in correlation with elastic, thermal and electrical properties. It has been found that the thermal conductivity is the main contributor to the behaviour of ultrasonic attenuation as a function of temperature and the responsible cause of attenuation is phonon-phonon interaction. The mechanical properties of OsB_2 at low temperature are better than at high temperature, because at low temperature it has low ultrasonic velocity and ultrasonic attenuation. Superhard material OsB_2 has many industrial applications, such as abrasives, cutting tools and hard coatings.

1. Introduction

Microhardness measurements show that OsB_2 is extremely hard. So OsB_2 is called superhard material. *Ab initio* calculations show that this is due to formation of covalent bonds between boron atoms. OsB_2 is also a low compressibility material [1]. Superhard materials are useful in a variety of industrial applications, such as abrasives, cutting tools and hard coatings [2]. OsB_2 has been synthesized by Cumberland et.al. [3]. Gou et.al. [4] investigated the structural, elastic and electronic properties of OsB_2 and OsB using first-principle total energy calculations. Ultrasonic attenuation is a very important physical parameter to characterize the material, which is well related to several physical quantities like thermal conductivity, specific heat, thermal energy density and higher order elastic constants [5]. The elastic constants provide valuable information about the bonding characteristic between adjacent atomic planes and the anisotropic character of the bonding and structural stability [6, 7].

Therefore, in this work we predict the ultrasonic properties of hexagonal structured OsB_2 compound at different temperatures. The ultrasonic attenuation coefficient, acoustic coupling constants, higher order elastic constants, thermal relaxation time and ultrasonic wave velocities for OsB_2 for unique direction of propagation of wave are calculated as a function of temperature. The obtained results are analyzed in comparison to other hexagonal structured materials.

2. Theory

The second (C_{IJ}) order elastic constants of material are defined by following expressions.

$$C_{IJ} = \frac{\partial^2 U}{\partial e_I \partial e_J}; \qquad \text{I or J} = 1, \ldots\ldots 6 \tag{1}$$

where, U is elastic energy density , $e_I = e_{ij}$ (i or j = x, y, z, I=1, …6) is component of strain tensor. Equation (1) leads six second order elastic constants (SOEC) for the hexagonal structured materials [5, 8].

$$\left.\begin{array}{ll} C_{11} = 24.1\, p^4 C' & C_{12} = 5.918\, p^4 C' \\ C_{13} = 1.925\, p^6 C' & C_{33} = 3.464\, p^8 C' \\ C_{44} = 2.309\, p^4 C' & C_{66} = 9.851\, p^4 C' \end{array}\right\} \tag{2}$$

where p (axial ratio) = c/a: (a and c are lattice parameters)

$$C' = \chi a / p^5 \qquad B = \psi a^3 / p^3 \qquad \chi = (1/8)[\{n b_0 (n-m)\}/\{a^{n+4}\}] \qquad \psi = -\chi / \{6 a^2 (m+n+6)\} \tag{3}$$

m, n=integer quantity; b_0=Lennard Jones parameter.

The expressions for ultrasonic attenuation, ultrasonic velocities and allied parameters are given in our previous papers [8, 9].

3. Results and discussion

The unit cell parameters 'a' and 'p' for OsB_2, are 2.961Å and 1.09 [4] respectively. The value of m and n for OsB_2 are 6 and 7. The value of b_0 for OsB2 is 1.9×10^{-64} erg cm^7. The SOEC and bulk modulus (B) are presented in table 1.

Table1. SOEC and bulk modulus (B) in the unit of 10^{10}Nm^{-2} of OsB_2 at room temperature.

	C_{11}	C_{12}	C_{13}	C_{33}	C_{44}	C_{66}	B
OsB_2	45.76	11.24	10.44	53.65	12.52	17.95	23.18
[4]OsB_2	47.2	34.0	31.6	53.1			

It is obvious from table 1 that, there is good agreement between the present and reported theoretical SOEC of OsB_2 [4]. Also, the comparison can be made with the value of Debye temperature of OsB_2. The present value of Debye temperature for OsB_2 is 545 ^{0}K, calculated using SOEC. The Debye temperature of OsB_2 is 550^{0}K, which was determined by Singh et.al. [10]. Thus our theoretical approach for the calculation of SOEC for hexagonal structured compound at room temperature is well justified.

Table 2. Density (ρ: in 10^3 kg m^{-3}), specific heat per unit volume (C_V: in 10^6Jm^{-3}K^{-1}), thermal energy density (E_0: in 10^8Jm^{-3}), thermal conductivity (k: in Wm^{-1}K^{-1}) and acoustic coupling constant (D_L, D_S) of OsB_2.

Temp.	ρ	C_V	E_0	k	D_L	D_S
100	14.52	0.784	0.277	28.4	45.78	3.04
150	14.47	1.152	0.772	21.3	46.60	3.04
200	14.42	1.339	1.397	18.9	47.05	3.04
250	14.37	1.432	2.078	17.3	47.31	3.04
300	14.32	1.490	2.806	16.4	47.49	3.04

The temperature dependent density and electrical resistivity have been taken from the literature [4, 10]. The temperature dependent thermal conductivity is evaluated using temperature dependent electrical resistivity. The values of temperature dependent allied parameters are presented in table 2.

The computed orientation dependent ultrasonic wave velocities and Debye average velocities (V_D) at 300 K are shown in figures 1–2. Figure 1 shows that the velocities V_L (longitudinal velocity) and V_{S1} (quasi shear wave velocity) have minima and maxima respectively at 45° with the unique axis of the crystal while V_{S2} (pure shear wave velocity) increases with the angle from the unique axis. The combined effect of SOEC and density is reason for abnormal behaviour of angle dependent velocities.

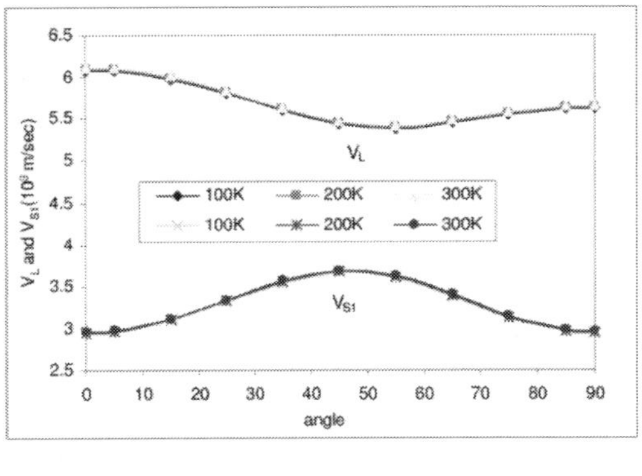

Figure 1. V_L and V_{S1} vs angle with unique axis of crystal.

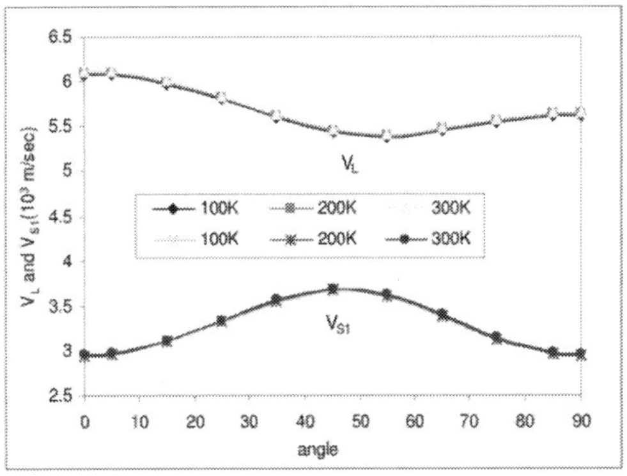

Figure 2. V_{S2} and V_D vs angle with unique axis of crystal.

The nature of the angle dependent velocity curves in the present work is found similar as that for third group nitrides, heavy rare-earth metals, laves-phase compounds and other hexagonal wurtzite structured materials [5, 8, 9, 11, 12]. Thus the angle dependencies of the velocities in OsB_2 are justified. V_D of these materials is increasing with the angle and has maxima at 55° at 300 K (figure 2). Since V_D is calculated using V_L, V_{S1} and V_{S2} [5, 9], therefore the angle variation of V_D is influenced by the constituent ultrasonic velocities. The maximum V_D at 55° is due to a significant increase in longitudinal and pure shear (V_{S2}) wave velocities and a decrease in quasi-shear (V_{S1}) wave velocity.

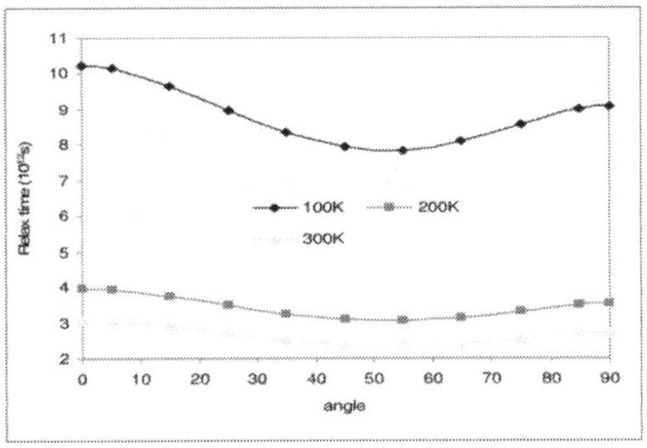

Figure 3. Relaxation time vs angle with unique axis of crystal.

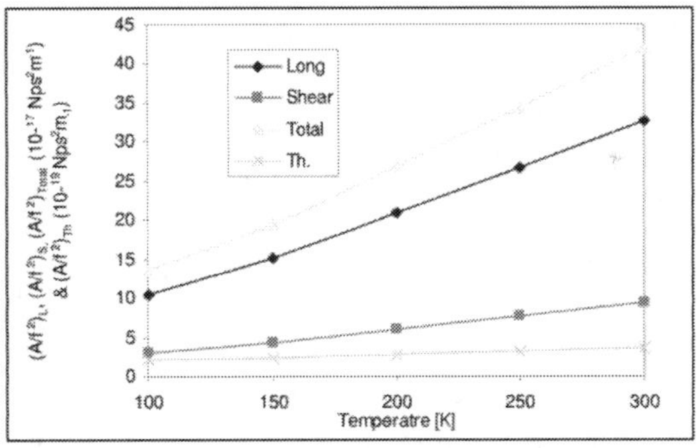

Figure 4. $(A/f^2)_L$, $(A/f^2)_S$, $(A/f^2)_{Total}$ and $(A/f^2)_{Th}$ vs temp. of OsB_2.

Thus it can be concluded that when a sound wave travels at 55° with the unique axis of these crystals then the average sound wave velocity is maximum. The calculated thermal relaxation time is visualised in figure 3. The angle dependent thermal relaxation time curves follow the reciprocal nature of V_D as $\tau \propto 3k / C_V V_D^2$. This implies that 'τ' for chosen compound is mainly affected by the thermal conductivity. The 'τ' for hexagonal structured materials is in the order of picosecond [12]. Hence the calculated 'τ' justifies the hexagonal structure of chosen compound. Thus the present average sound velocity directly correlates with the Debye temperature, C_V and E_0 of OsB_2. The attenuation coefficient over frequency square $(A/f^2)_{Akh}$ for longitudinal $(A/f^2)_L$, shear wave $(A/f^2)_S$ and thermoelastic loss$(A/f^2)_{Th}$ under the condition $\omega\tau<<1$ at different temperature are presented in figure 4. In the present investigation, the Akhieser type of loss of energy for longitudinal and shear wave and thermo elastic loss increases with the temperature of the material (figure 4). In the evaluation of ultrasonic attenuation, it is supposed that wave is propagating along the unique axis of OsB_2. $(A/f^2)_{Akh}$ is proportional to D (acoustic coupling constant), E_0, τ and V^{-3} [8, 9].

Hence, Akhieser loss in OsB_2 is predominantly affected by E_0 and k. Figure 4 indicates that the thermoelastic loss is very small in comparison to Akhieser loss and ultrasonic attenuation for longitudinal wave $(A/f^2)_L$ is greater than that of shear wave $(A/f^2)_S$. This reveals that ultrasonic attenuation due to phonon-phonon interaction along longitudinal wave is governing factor for total attenuation $((A/f^2)_{Total} =(A/f^2)_{Th}+(A/f^2)_L+(A/f^2)_S)$. The total attenuation is mainly affected by thermal energy density and k.

Thus it may predict that at 100K the OsB_2 behaves as its purest form and is more ductile as evinced by minimum attenuation while at 300K OsB_2 is least ductile. Therefore impurity will be least in the OsB_2 at 100K.

4. Conclusions

- Our theory of higher order elastic constants is justified for the hexagonal structured material.
- The order of thermal relaxation time for OsB_2 is found in picoseconds, which justifies their hexagonal structure.
- The acoustic coupling constant of OsB_2 for longitudinal wave are found five times larger than third group nitrides. Hence the conversion of acoustic energy into thermal energy will be large for OsB_2. This shows general suitability of chosen material.
- The ultrasonic attenuation due to phonon-phonon interaction mechanism is predominant over total attenuation, as a governing factor of thermal conductivity.
- The mechanical properties (yield strength, ductility, elastic properties etc.) of OsB_2 at low temperature (100K) are better than at high temperature (300K), because at low temperature it has low ultrasonic velocity and ultrasonic attenuation.

Acknowledgments

The author wish to express sincere thanks to Dr. Ashok K. Chauhan, President, AMITY School of Engineering and Technology (ASET), New Delhi, Prof. B. P. Singh, Senior Director, Prof. (Dr.) Rekha Agarwal, Director, ASET New Delhi, Prof. S. K. Kor and Prof. Raja Ram Yadav, Department of Physics, Allahabad University, for encouraging and providing necessary facilities in carrying out the work.

References

[1] Hebbache M, Stuparević L and Živković D 2006 *Solid State Communications* **139** (5) 227
[2] Kaner R B, Gilman J J and Tolbert S H 2005 *Science* **308** 1268
[3] Cumberland R W, Weinberger M B, Gilman J J, Clark S M, Tolbert S H and Kaner R B 2005 J. *Am. Chem. Soc.* **127** 7264
[4] Gou H, Hou L, Zhang J, Li H, Sun G and Gao F 2006 Appl. *Phys. Lett.* **88** 221904
[5] Yadawa P K, Singh D, Pandey D K and Yadav R R 2009 *The Open Acoustic Journal* **2** 80
[6] Ravindran P, Fast L, Korzhavyi P A, Johansson B, Wills J M and Eriksson O 1998 *J. Appl. Phys.* **84** 4891
[7] Louail L, Maouche D, Roumili A and Ali Sahraoui F 2004 *Mater. Lett.* **58** 2975
[8] Yadav A K, Yadav R R, Pandey D K, Singh D 2008 *Mat. Lett.* **62** 3258
[9] Pandey D K, Yadawa P K, Yadav R R 2007 *Mat. Lett.* **61** 5194
[10] Singh Y, Niazi A, Vannette M D, Prozorov R and Johnston D C 2007 *Phys. Rev.* B **76** 214510
[11] Pandey D K, Yadawa P K and Yadav R R 2007 *Mat. Lett.* **61** 4747
[12] Pandey D K, Singh D and Yadav R R 2007 *Appl. Acoust.* **68** 76

IOP Conf. Series: Materials Science and Engineering **42** (2012) 012035 doi:10.1088/1757-899X/42/1/012035

Ultrasonic monitoring of Iberian fat crystallization during cold storage

E Corona, J V García-Pérez, J V Santacatalina, R Peña and J Benedito[1]
Grupo de Análisis y Simulación de Procesos Agroalimentarios (ASPA).
Departamento Tecnología de Alimentos, Universitat Politècnica de València, Valencia, 46022, Spain. Tel: +34 96 387 91 47.

E-mail: jjbenedi@tal.upv.es

Abstract. The aim of this work was to evaluate the use of ultrasonic measurements to characterize the crystallization process and to assess the textural changes of Iberian fat and Iberian ham during cold storage. The ultrasonic velocity was measured in two types of Iberian fats (Montanera and Cebo) during cold storage (0, 2, 5, 7 and 10 °C) and in vacuum packaged Iberian ham stored at 6°C for 120 days. The fatty acid profile, thermal behaviour and textural properties of fat were determined. The ultrasonic velocity and textural measurements showed a two step increase during cold storage, which was related with the separate crystallization of two fractions of triglycerides. It was observed that the harder the fat, the higher the ultrasonic velocity. Likewise, Cebo fat resulted harder than Montanera due to a higher content of saturated triglycerides. The ultrasonic velocity in Iberian ham showed an average increase of 55 m/s after 120 days of cold storage due to fat crystallization. Thus, non-destructive ultrasonic technique could be a reliable method to follow the crystallization of fats and to monitor the changes in the textural properties of Iberian ham during cold storage.

1. Introduction

The overall quality of Iberian ham is highly dependent on the fat content, state and its fatty acid composition, which is greatly affected by the rearing system of pigs [1]. The most important traditional rearing systems are "Montanera" (diet based on acorns and grass) and "Cebo" (diet based on concentrate feeds). Cold storage is used during distribution and retail sale of Iberian ham and can exert a notable effect on textural properties [2]. Therefore, there is a great interest in seeking non-destructive techniques that allow monitoring the crystallization of fat and the effect of storage on the overall quality of Iberian ham. The aim of this work was the use of ultrasonic measurements to characterize the crystallization of Iberian fat and the textural changes of Iberian ham during cold storage.

2. Materials and methods

2.1. Raw material and sampling procedure

Two Iberian subcutaneous fats were used: Montanera and Cebo. Moreover, a total of 10 packages of vacuum packaged Iberian ham (10-20mm thickness) were analyzed. Different points were marked on

[1] To whom any correspondence should be addressed

the surface of each package and the ultrasonic measurements and fat content [3] were determined for each one. All samples were stored at 6±1 °C during 120 days.

2.2. Fatty acid composition
The fatty acid composition of Iberian fats was determined by gas chromatography (5890A, Hewlett Packard, USA) of the fatty acid methyl esters (FAME) following the method described by Niñoles et al [1].

2.3. Differential Scanning Calorimetry (DSC)
Iberian fat samples were monitored using differential scanning calorimetry (DSC) (DSC5200CU, Seiko Instruments, Inc., USA). The sample (20 mg) was subjected to cooling from 70 to -50 ° C using a wide range of cooling rates (0.2, 0.5, 1, 5 and 10 °C/min).

2.4. Instrumental textural analysis
The textural analysis was carried out (TA-XT2i, Stable Micro Systems, Survey England) using a conical probe in fat previously melted (70 °C for 30 min) and then placed in Petri dishes (60 mL). Three storage temperatures were tested: 0, 2 and 5 °C. Measurements were performed during 5 days.

2.5. Ultrasonic measurements

2.5.1. Iberian fats. The experimental set-up consisted of an aluminium cylindrical ultrasonic cell (diameter 17.8 mm, length 25 mm) with two piezoelectric transducers (1MHz, 0.5'' diameter, A303S, Panametrics, Walthman, MA, USA) inserted in each side of the cell, a ultrasonic pulser-receiver (5058PR, Panametrics, Walthman, MA, USA) and PC with a data acquisition card (PCI 5112, National Instruments, USA). The ultrasonic velocity, measured during the storage period, was obtained considering the time of flight and the distance between transducers (25 mm). The fat was melted (70 °C for 30 min) and placed into the ultrasonic cell set to the experimental temperature (0, 2, 5, 7 and 10 °C).

2.5.2. Iberian ham. The experimental set-up consisted of a couple of narrow-band ultrasonic transducers (1MHz, 0.5" diameter, A303S-SU, Panametrics, Walthman, MA, USA), a pulser-receiver (5058PR, Panametrics, Walthman, MA, USA) and a digital storage oscilloscope (TDS5034, Tektronix, Bearverton, Oregon, USA). A digital height gage was linked to the computer by a RS232 interface to measure the sample thickness. The ultrasonic velocity in the sample was computed from the time of flight obtained from the signal and the thickness provided by the height gage. The ultrasonic velocity was determined for all samples at 6 °C in a temperature-controlled chamber (±0.1 °C) (AEC330R, Infrico, Spain). Measurements were carried out at the beginning (1 day) and end of the storage (120 days).

3. Results and discussion
3.1. Fatty acid profile
The content of saturated fatty acids was significantly different ($p<0.05$) between both fats, being the content higher for Cebo (45.1 %) than for Montanera (41.8 %). For unsaturated fatty acids content (Monounsaturated and Polyunsaturated fatty acids), Montanera showed a significantly ($p<0.05$) higher content than Cebo (58.2 % versus 54.9 %) as a consequence of the rearing system. Montanera animals are mainly fed by acorns, which are a rich source of oleic acid, one of the most important unsaturated fatty acids [1].

3.2. Thermal behavior characterization of Iberian pork fat

Similar DSC curves were found for both fats for all the cooling rates (figure 1(a) and (b)), showing two main peaks (A and B). The peak A was related with the crystallization of the most saturated triglycerides (TGs), while the peak B with the crystallization of the most unsaturated TG fraction [2].

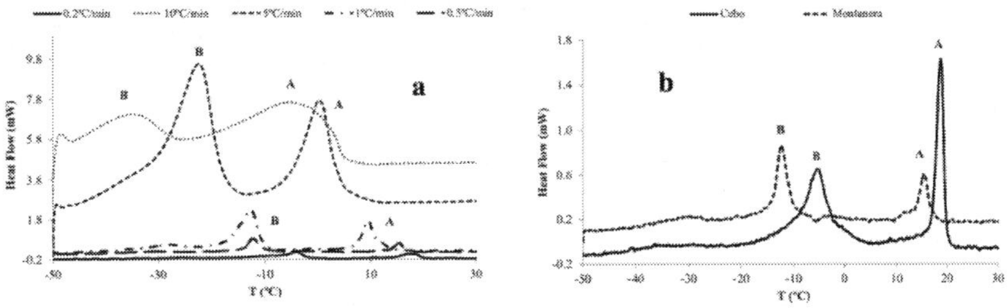

Figure 1. (a) DSC thermograms of Montanera crystallized at different cooling rates (10, 5, 1, 0.5 and 0.2 °C/min). (b) DSC curves of crystallization profiles of Cebo and Montanera at 1 °C/min.

As can be observed, both peaks (A and B; figure 1(a)) were broadened and displaced to lower temperatures when the cooling rate increased. On the other hand, Montanera showed significantly ($p<0.05$) lower peak (A and B; figure 1(b)) temperatures than Cebo. It was related to the higher content of unsatured TGs in Montanera.

3.3. Characterization of the textural measurements during cold storage

Figure 2 (a) and (b) highlights that for both fats, the penetration force increased during storage for all experimental temperatures.

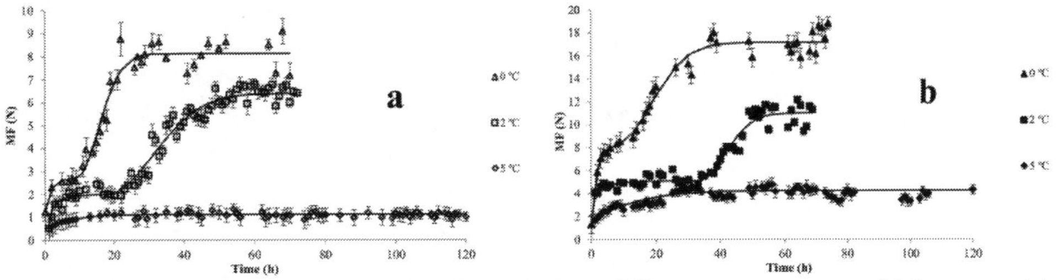

Figure 2. Changes of the maximum penetration force (MF) at different temperatures of Montanera (a) and Cebo (b) fats during cold storage. Average values ± Standard deviation are plotted.

For experiments at 0 and 2 °C, the curves showed two steep increases during the storage. The first increase was related with the crystallization of the most saturated TGs fraction found by DSC (peak A) and the second steep increase with the most unsaturated TGs fraction (peak B in DSC). At 5 °C, the most unsaturated TGs fraction did not crystallized during the storage. Figure 2 (a) and (b) shows that the lower the storage temperature, the higher the penetration force as a consequence of the formation of small fat crystals and a denser and harder structure [2]. For the same temperature, Cebo (figure 2(b)) was harder than Montanera (figure 2(a)) fat, which is explained by the fact that saturated fatty acids can easily align themselves to form a compact mass. The unsaturated fatty acids have kinks in their aliphatic chains, causing a zig-zag bending, which results in a less dense structure [4]. Therefore, the higher the saturated TGs content, the harder the texture during crystallization.

3.4. Monitoring of ultrasonic velocity during cold storage

The shape of the ultrasonic velocity curves during storage was similar to the textural ones (figure 3(a) and (b)). Thus, the crystallization of the two fractions of TGs was observed (0, 2, 5 and 7 10 °C) and related with the increase of the fat solid/liquid ratio, which results in a harder sample where ultrasound waves travel faster. For measurements at 10 °C, only the first increase of the ultrasonic velocity was observed. Moreover, the higher the storage temperature, the softer the crystalline structure and consequently the lower the velocity.

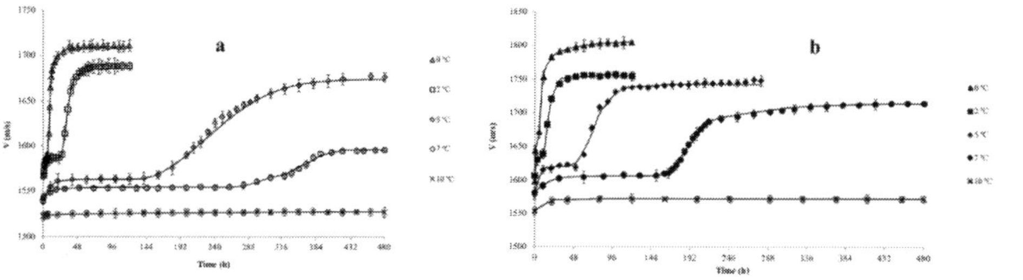

Figure 3. Changes of the ultrasonic velocity during cold storage at different temperatures of Montanera (a) and Cebo (b) fats. Average values ± Standard deviation are plotted.

Finally, when comparing between both fats, the highest values of ultrasonic velocity for the same storage temperature were always found in Cebo (figure 3(b)). As explained in section 3.3, the higher content of saturated TGs in Cebo results into a harder crystalline structure, where ultrasound propagates faster than in Montanera.

On the other hand, for Iberian ham (figure 4), the ultrasonic measurements were well related with the fat content (R^2=0.80) ($p<0.05$) in samples at 6 ºC, before and after cold storage. The ultrasonic velocity in Iberian ham showed an average increase of 55 m/s after 120 days of cold storage due to fat crystallization and therefore to an increase of hardness.

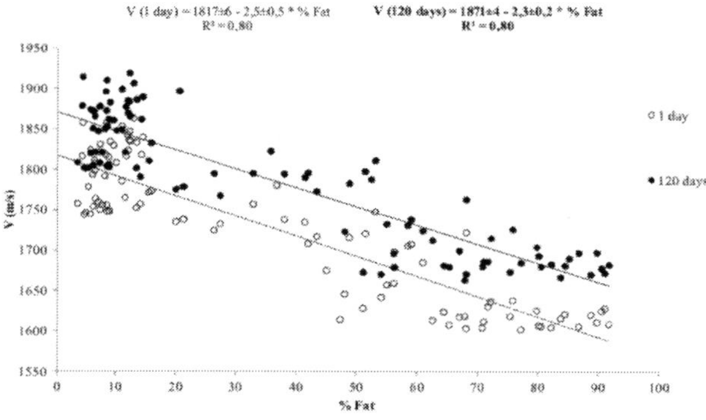

Figure 4. Influence of cold storage (6 ° C) on the ultrasonic velocity in samples of dry-cured Iberian ham.

4. Conclusions

Ultrasonic techniques could be used to identify overall quality in Iberian fats and ham and to monitor the textural changes during cold storage. This application is highly relevant considering ultrasound as a non destructive technology, being easily automated on-line.

Acknowledgements

The authors acknowledge the financial support from the ''Ministerio de Economia y Competitividad'' in Spain (Project RTA 2010-00029-C04-02).

References

[1] Niñoles L, Clemente G, Ventanas S and Benedito J 2007 *Meat Sci.* **76** 102-11
[2] Santacatalina J V, Garcia-Perez J V, Corona E and Benedito J 2010 *Food Res. Intl.* **44** 146-55
[3] AOAC International 1997 *Official Methods of Analysis of AOAC International* 16th ed (Washington, DC: Association of Official Analytical Chemist)
[4] Himawan C, Starov V M and Stapley A G F 2006 *Advn. in colloid and Interface Sci.* **122** 3-33

IOP Conf. Series: Materials Science and Engineering **42** (2012) 012036 doi:10.1088/1757-899X/42/1/012036

Beginning of fish defrosting by using non-destructive ultrasonic technique

M Malainine[1], B Faiz[1], A Moudden[2], D Izbaim[1], G Maze[3] and I Aboudaoud[1]

[1]Faculté des sciences d'Agadir - Université Ibn Zohr, B.P 8106 Agadir, Maroc, 80000 Agadir, Morocco
[2]Ecole Supérieure de Technologie d'Agadir, B.P 33/S Agadir, Maroc, 80000 Agadir, Morocco
[3]LOMC IUT Université du Havre, Place Robert Schuman, 76610 le Havre, France

E-mail: malaininemohamed@gmail.com

Abstract. During the experiments carried out on the monitoring and the study of fish defrosting by an ultrasonic technique, we have difficulties in detecting the beginning of the thawing which is an important criterion of fish quality control. To address this problem, we use the Singular Value Decomposition method (SVD) which is a mathematical tool that permits to separate the high and low energies of an histogram. The image representing low energy signals indicates the start of the thawing by showing an echo that was hidden in the original image for cod fish. Using transducers for central frequencies above 500 kHz the observed results are not very good. Therefore, this method is suitable for fish which fat content is medium or low.

1. Introduction

The state of freshness of the raw material is essential for the quality of the final fish product [1]. The use of low intensity ultrasonics (LIU) in food applications has been extended to, among others, meat, fish, dairy products, vegetables, fruits and fats and oils ([2], [3], [4],[5]). The retrodiffused signal by fish presents overlapping echoes. The aim of this work is to detect the beginning of the thawing, known to be an important criterion of fish quality control, by using the Singular Value Decomposition method (SVD) which is a mathematical tool that permits to separate the high and low energies of an histogram.

2. Materials and Methods

2.1. Materials

Cod fillets of 2 cm thickness are cut in cubes and are placed directly to freezer at temperature -18°C for a minimum of 24 hours. Then the fish sample is brought from the freezer and placed immediately into the vat, and the thawing is monitored during 150 min.

2.2. Methods

2.2.1. *Experimental set-up.* The experimental set-up is constituted by a panametrics broadband contact transducer of 500 KHz or 5 MHz central frequency, which is used successively as emitter and receiver, and placed under the vat containing the fish cube to study.

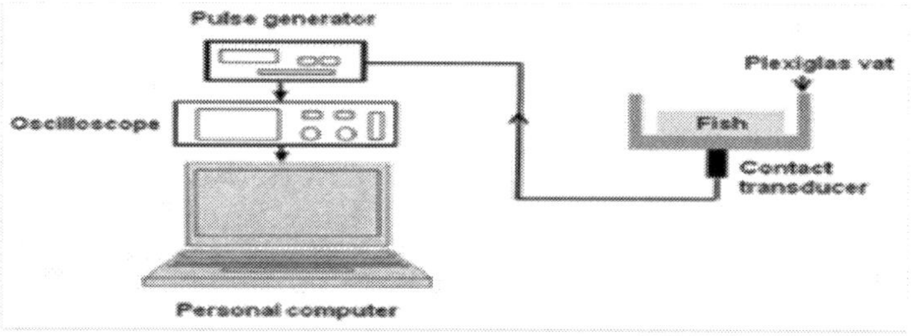

Figure 1. Experimental set-up.

2.2.2. *Example of a retrodiffused signal by cod fillet.*

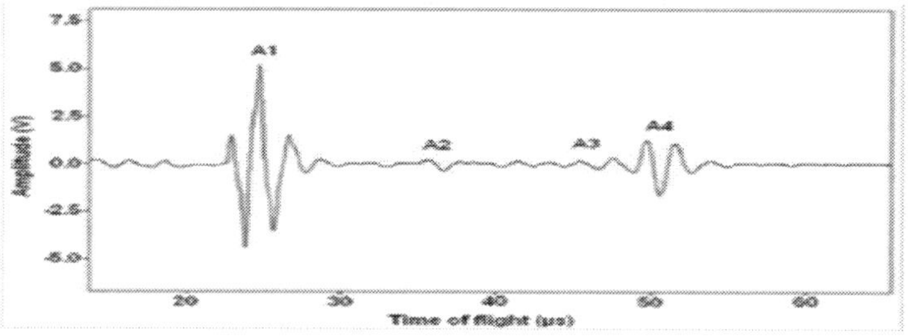

Figure 2. Type of the retrodiffused signal by fish during its defrosting.

- A_1 is the retrodiffused signal by the interface plexiglas/thawed part of fish.
- A_2 is the retrodiffused signal by the interface thawed part of fish / freezed part of fish.
- A_3 is the second back-and-forth in the plexiglas.
- A_4 is the retrodiffused signal by the interface thawed part of fish/air.

3. Results and Discussion

For the characterization of the cod first defrosting; the dimension of the matrix corresponding to the retrodiffused signals is 5000 x 150. The A_2 echo has a very low energy and is overlapped at the beginning of the thawing with the A_1 echo which has a high energy.

$$[USV]= svd(X) \tag{1}$$

The figure 3 shows the image of the result matrix of the cod first defrosting using the transducer of 500 KHz central frequency.

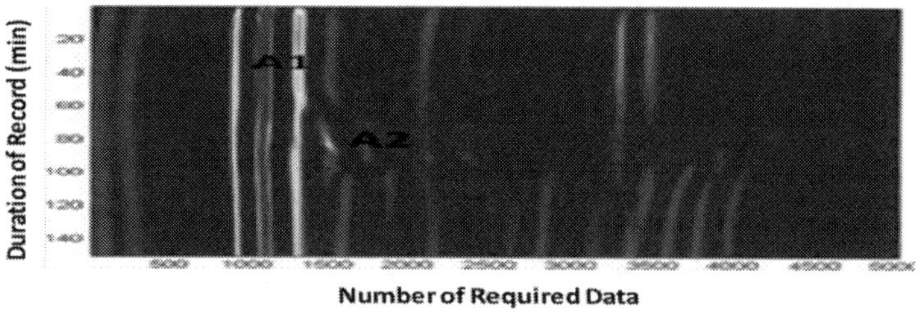

Figure 3. Superposition of the retrodiffused signals during cod first defrosting; transducer 500 kHz.

The SVD method is based on the calculation of singular values of the resulting matrix.

- Calculation of the singular values:
  The figure 5 shows the diagram of the singular values σ_i .

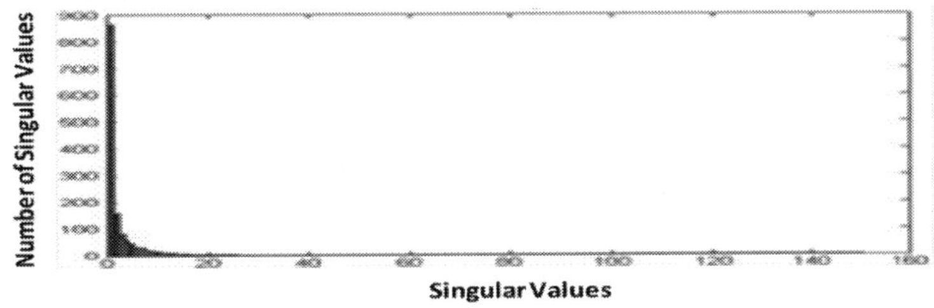

Figure 4. The singular values diagram of the Cod first defrosting.

X is equal to:

$$X = X_D + X_{LE} + N \tag{2}$$

Where D is for dominant, LE for low energy and N for noise.

- Extraction of the dominant energy contributions E1-4:
  The calculation of the dominant contribution is defined by the equation (3):

$$E_{1-4} = \sum_{i=1}^{4} \sigma_i * U_i * V_i^T \tag{3}$$

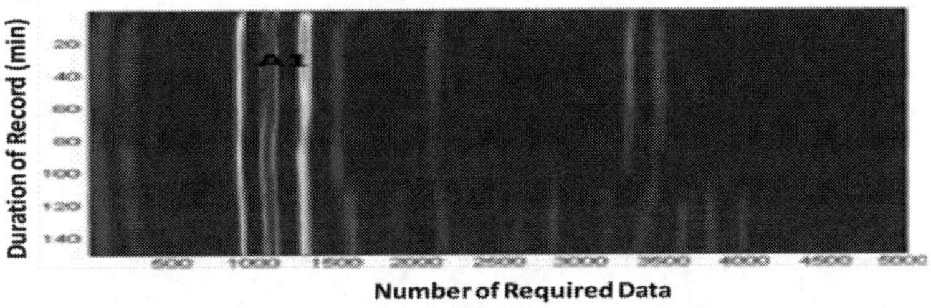

Figure 5. Image of the dominant energy contributions $E_{1\text{-}4}$.

We verify clearly the separation of the A_1 echo.

- Extraction of the low-energy contributions E5-20:

$$E_{5-20} = \sum_{i=5}^{20} \sigma_i * U_i * V_i^T \tag{4}$$

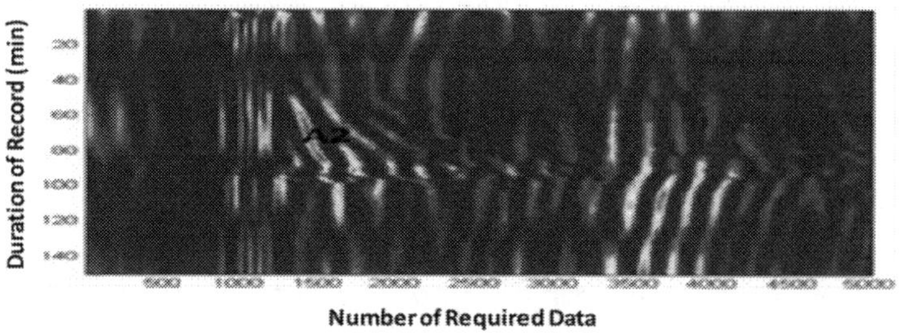

Figure 6. Image of the low-energy contributions $E_{5\text{-}20}$.

We note that the detection of the fish defrosting beginning is improved, it is equal to 33 min. While without treatment, it was equal to 78 min.

- Noise extraction:
  The calculation of the noise subspace is defined by the equation (5):

$$E_{21-150} = \sum_{i=21}^{150} \sigma_i * U_i * V_i^T \qquad (5)$$

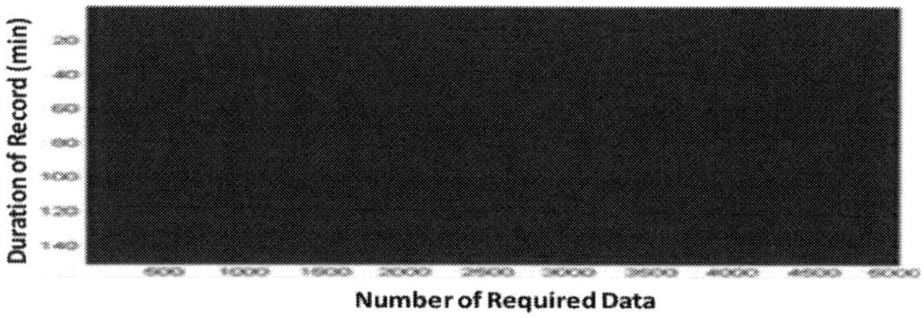

Figure 7. The noise subspace.

The noise contribution is very low.

4. Conclusion

The SVD method has allowed us to show the beginning of fish thawing for cod fish which has an average fat content. The results were good with a contact transducer of 500 KHz central frequency. Using transducers for central frequencies above 500 kHz the observed results are not very good.

References

[1] Olafsdottir G, Martinsdóttir E, Oehlenschläger J, Dalgaard P, Jensen B, Undeland I, Mackie I M, Henehan G, Nielsen J and Nilsen H 1997 Methods to evaluate fish freshness in research and industry *Trends in Food Science and Technology* **8** (8) 258–265

[2] McClements D J 1997 Ultrasonic Characterization of food and drinks: Principles, methods and applications *Critical Reviews in Food Science and Nutrition* **37** 1-46

[3] Mulet A, Benedito J, Bon J and Sanjuan N 1999 Low intensity ultrasonics in food technology. *Food Sci and Technology International* **5** 285-297

[4] Coupland J N 2004 Low-intensity ultrasound *Food Res. Int.* **37** 537-543

[5] Ninoles L, Sanjuan N, Ventanas S and Benedito J 2008 Ultrasonic and sensory characterization of dry-cured ham fat from Iberian pigs with different genetics and feeding backgrounds, *Meat Sci.* **80** 896-902

IOP Conf. Series: Materials Science and Engineering **42** (2012) 012037 doi:10.1088/1757-899X/42/1/012037

Use of an ultrasonic reflectance technique to examine bubble size changes in dough

A Strybulevych[1], V Leroy[1,2], AL Shum[3], HF Koksel[3], MG Scanlon[3] and JH Page[1]

[1] Department of Physics and Astronomy, University of Manitoba, Winnipeg, Manitoba, Canada R3T 2N2
[2] Laboratoire MSC, Université Paris-Diderot, CNRS (UMR 7057), Paris, France
[3] Department of Food Science, University of Manitoba, Winnipeg, Manitoba, Canada R3T 2N2

E-mail: anatoliy@physics.umanitoba.ca

Abstract. Bread quality largely depends on the manner in which bubbles are created and manipulated in the dough during processing. We have developed an ultrasonic reflectance technique to monitor bubbles in dough, even at high volume fractions, where near the bubble resonances it is difficult to make measurements using transmission techniques. A broadband transducer centred at 3.5 MHz in a normal incidence wave reflection set-up is used to measure longitudinal velocity and attenuation from acoustic impedance measurements. The technique is illustrated by examining changes in bubbles in dough due to two very different physical effects. In dough made without yeast, a peak in attenuation due to bubble resonance is observed at approximately 2 MHz. This peak diminishes rapidly and shifts to lower frequencies, indicative of Ostwald ripening of bubbles within the dough. The second effect involves the growth of bubble sizes due to gas generated by yeast during fermentation. This process is experimentally challenging to investigate with ultrasound because of very high attenuation. The reflectance technique allows the changes of the velocity and attenuation during fermentation to be measured as a function of frequency and time, indicating bubble growth effects that can be monitored even at high volume fractions of bubbles.

1. Introduction

The appearance and texture of many cereal products largely depends on the manner in which bubbles are created and manipulated in the dough during processing [1,2]. Understanding the effect of bubbles on the properties of dough is critical to controlling the texture of bread and hence its quality [1,2]. However, understanding how to control bubbles in a dough system is challenging because dough is an optically opaque bubbly system and the bubble distribution changes quite rapidly with time, a phenomenon that is true due to Ostwald ripening in dough made without yeast [3] or from inflation with carbon dioxide in dough made with yeast [4]. Because ultrasound is sensitive to bubbles in dough [5], the aim of this paper was to use ultrasound to monitor time-dependent changes in the bubble size distribution in two types of dough systems – those made with and without yeast.

2. Theoretical background

2.1. Acoustic resonances in dough as an effective medium

In a bubbly system such as dough, the phase velocity (v) and attenuation coefficient (α) exhibit a strong dependence on angular frequency (ω) due to the acoustic resonance of the bubbles. This resonant behavior can be described in terms of the complex effective medium wave vector $k = \omega/v + i\alpha/2$, which deviates from the wave vector of the matrix, k_0, by a contribution that depends on the number and sizes of the bubbles [6].

$$k^2 = k_0^2 + 4\pi \int N(r)dr \frac{r\omega^2}{\omega_M^2 - \omega^2 - i\omega\Gamma} \tag{1}$$

Comparing the predictions of this model with experimental data enables the bubble size distribution $N(r)$ to be determined, providing that we have information on the dough matrix properties, which influence the bubble resonant frequencies ω_M and the damping rate Γ (which depends on thermal, viscous, and radiative losses). We have shown in a variety of aerated soft materials that this approach is a promising line of attack [6-9].

2.2. Dough bubble size determination

It is assumed that the subdividing process occurring to air bubbles during mixing creates a log-normal distribution of bubbles in the dough [1], with median radius R, logarithmic standard deviation ε and gas volume fraction Φ. Using the above bubble resonance model [6], it can be shown that R and ε can be determined from measurements of the frequency at which the maximum in attenuation occurs f_{max} and the height of the attenuation peak α_{max} if the density of the dough (and hence Φ) has been determined independently, e.g., by using Archimedes principle [5].

$$\alpha_{max} = \frac{2\sqrt{3\Phi}}{R\exp(2\varepsilon^2)} \tag{2}$$

$$(2\pi f_{max})^2 = \frac{3\kappa P_0 + 4\mu'}{\rho_m R^2} \tag{3}$$

In these equations, κ is the polytropic index for the transformations undergone by the gas, P_0 is the pressure in the bubble, μ' is the real part of the complex shear modulus and ρ_m is the density of the dough matrix [6]. Full details of the analysis have been provided recently by Leroy et al [6].

3. Materials and methods

To investigate the dynamics of air bubbles introduced into the dough during mixing, doughs were made without yeast, from 100 g of hard red spring wheat flour, 57 g of water (for ease of sample preparation) and 2.4 g of salt. Dough was mechanically developed as previously described [5] by mixing for 4 minutes.

To investigate bubble expansion due to the addition of yeast, samples were prepared using a sponge and dough procedure [10] using 69.2 g of hard red spring wheat flour and 41 g of water for sponge and 29.7 g of flour and 4.8 g of water for dough (optimized water absorption).

Dough density was measured to ascertain the amount of air incorporated into the dough using 25 mL gravimetric bottles (Kimble Glass Inc., Vineland NJ) using 5 g sub-samples of dough.

Experiments were performed in a normal incidence wave reflection set-up using a transducer with a central frequency of 3.5 MHz, an acrylic delay line, to which the transducer was bonded, and a pulser. Reflected signals were detected by the same transducer, amplified and recorded on a digital oscilloscope.

A very sharp razor blade, lightly greased with mineral oil, was used to cut a disc approximately 9 mm thick from the centre of the dough sample. The cut surface was also lightly greased with mineral oil and placed greased side down on the acrylic delay line (so as to establish good acoustic contact). A metal weight was placed on top of the dough to improve contact.

The reflection set-up was used either at room temperature, but covered (to prevent moisture loss) for doughs made without yeast, or placed inside a proofing cabinet at normal proofing conditions (37°C, 83% relative humidity). In the latter case the set-up had been placed in the proofing cabinet for at least one hour prior to experiments to allow acclimatization to the proofing cabinet's temperature.

A reference signal (the reflected ultrasonic pulse with no dough on delay line) was measured at the beginning of all experiments. Acquisitions were then performed every minute after sample mounting. From the ratio of the Fourier transforms of dough and reference signals, the complex impedance of the dough was determined [6]. From the real and imaginary parts of the complex impedance, the longitudinal phase velocity and attenuation coefficient as a function of frequency were calculated over a broad frequency range.

4. Results and Discussion

4.1. Monitoring Ostwald ripening in unyeasted dough

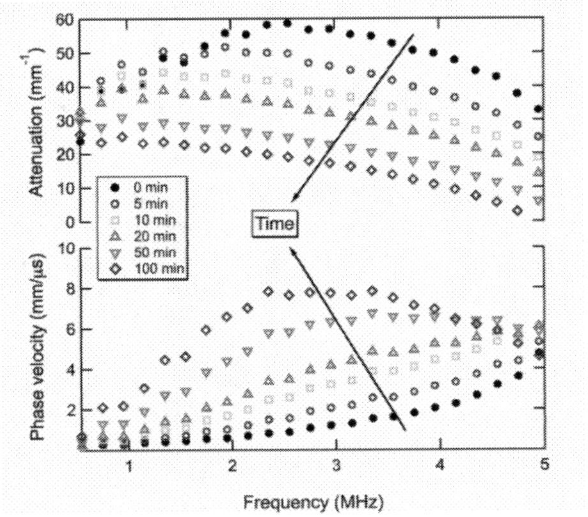

Figure 1. Changes in ultrasonic phase velocity and attenuation coefficient vs frequency for unyeasted dough.

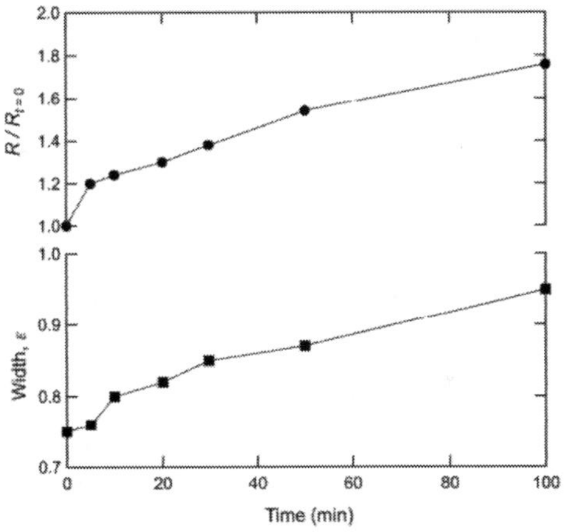

Figure 2. Increase in median bubble size and width of bubble distribution with time (determined from data in figure 1).

From figure 1, it can be seen that there are pronounced changes in phase velocity (bottom) and a peak in attenuation (top) due to the resonance of bubbles in the dough [11]. This peak in attenuation, which occurs at approximately 2 MHz, diminishes rapidly with time and shifts to lower frequencies.

By fitting the model described in section 2 [6-8] to the phase velocity and attenuation data in figure 1, we determined how the median bubble size and the width of the bubble distribution of the bubbles in the dough changes as a function of time (figure 2). It can be seen that both the median radius and the width of the distribution increase with time. Both these changes are characteristic features of Laplace-pressure driven growth of large bubbles at the expense of smaller ones, indicative of Ostwald ripening of bubbles within the dough [3].

4.2. Monitoring the effects of carbon dioxide on dough properties

The fermentation process is experimentally challenging to investigate with ultrasound because very high attenuation limits propagation of the ultrasonic signal, particularly at long proofing times [4]. With our reflection setup, signals with excellent signal to noise ratio can be detected throughout the entire proofing process, enabling the frequency dependence of the attenuation and velocity to be determined, and f_{max} and α_{max} to be measured. It can be seen from figure 3 that both f_{max} and α_{max} change in a non-linear manner as fermentation progresses.

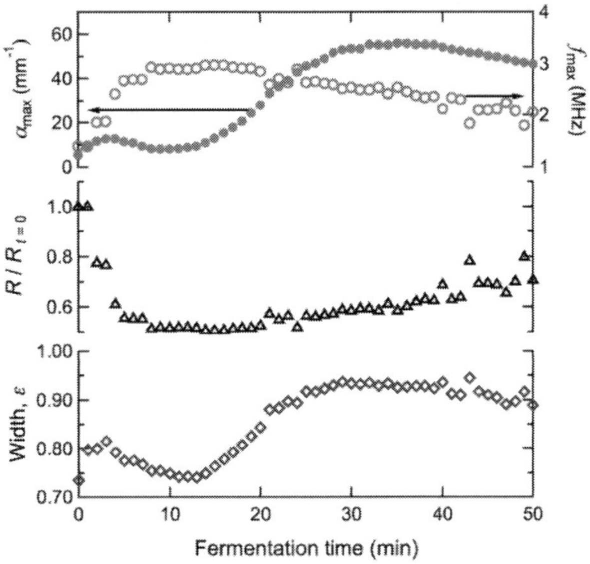

Figure 3. Changes in ultrasonic attenuation parameters that translate into changes in median bubble size and width of bubble distribution with proofing time.

An interesting phenomenon occurs for $5 < t < 10$ mins, where α_{max} decreases and the frequency at which the peak in attenuation occurs shifts to higher frequency (essentially a shorter wavelength). Lower attenuation at short fermentation times has previously been reported in low frequency experiments [12]. This is consistent with yeast being a facultative anaerobe, preferentially consuming oxygen when available. The air bubble volume will then shrink by ~20% (an effect also confirmed by initial increases in dough density).

At longer times, the yeast cells start to generate carbon dioxide as a result of fermentation activities. Attenuation is increased by the growth of bubble sizes due to inflation with CO_2, and the increase in bubble distribution width (observed up to 30 min fermentation) is monitored from ultrasonic reflection measurements.

5. Conclusions

Ultrasonic velocity and attenuation (as measured by a novel reflection technique) are sensitive to the size of bubbles in bread dough. It is shown that for unyeasted dough the median radius and the width of the bubble log-normal distribution increase with time due to Ostwald ripening. The reflectance technique allows changes due to bubble growth effects during fermentation to be measured as a function of time, even at high volume fractions of bubbles, where it is difficult to make ultrasonic measurements using transmission techniques.

References

[1] Campbell GM, Rielly CD, Fryer PJ and Sadd PA 1998 *Cereal Foods World* **43** 163
[2] Scanlon MG and Zghal MC 2001 *Food Research International* **34** 841
[3] van Vliet T 1999 *Bubbles Food* ed G M Campbell *et al* (St Paul, MN: AACC Press) p 121
[4] Elmehdi HM, Page JH and Scanlon MG 2003 *Trans. Inst. Chemical Engineers* C **81** 217
[5] Elmehdi HM, Page JH and Scanlon MG 2004 *Cereal Chemistry* **81** 504
[6] Leroy V, Strybulevych A, Page JH and Scanlon MG 2011 *Physical Review E* **83** 046605-1
[7] Leroy V, Strybulevych A, Scanlon MG and Page JH 2009 *European Physical Journal E* **29** 123
[8] Leroy V, Strybulevych A, Page JH and Scanlon MG 2008 *J. Acoustical Soc. America* **123** 1931
[9] Strybulevych A, Leroy V, Scanlon MG and Page JH 2007 *Soft Matter* **3** 1388
[10] Kilborn KH and Preston KR 1981 *Cereal Chemistry* **58** 198
[11] Leroy V *et al* 2008 *Bubbles in Food 2: Novelty, Health and Luxury* ed G M Campbell *et al* (St Paul, MN: AACC Press) p 51
[12] Elmehdi HM 2001 PhD Thesis, *University of Manitoba*, Winnipeg, MB

IOP Conf. Series: Materials Science and Engineering **42** (2012) 012038 doi:10.1088/1757-899X/42/1/012038

The maturity characterization of orange fruit by using high frequency ultrasonic echo pulse method

I Aboudaoud[1], B Faiz[1], E Aassif[1], A Moudden[2], D Izbaim[1], D Abassi[1], M Malainine[1] and M Azergui[1]

[1]Laboratoire de Métrologie & Traitement de l'Information - Département de physique, Faculté des Sciences – Université Ibnou Zohr ; B.P 8106 – Cité Dakhla Agadir, Morocco.
[2]Ecole Supérieur de Téchnologie d'Agadir – B.P 33/S Agadir – Morocco.

E-mail address : aboudaouddriss@gmail.com

Abstract. In this present work, we develop a new ultrasonic echo pulse method in order to study the feasibility of maturity assessment of orange fruit. This study concerns two varieties of orange (Navel and Mandarin) which are the most harvested in the region of Souss-Massa-Drãa in Morocco. We worked in the range of high frequencies by the means of a focusing transducer with 20MHz as a central frequency. By taking into account the strong attenuation of the ultrasounds in the texture of fruits and vegetables, we limited our study only to the external layer of orange peel. This control is based mainly on the measure of the ultrasonic parameters eventually velocity and attenuation in order to check the aptitude of this technique to detect the maturity degree of the fruit without passing by penetrometric and biochemical measurements which are generally destructives but the mostly correlated with human perception concerning the firmness of the fruit.

1. Introduction

Today the standards of quality are the important factors that must be respected in order to ensure the competitiveness of fruits and vegetables. Thus the quality control occupies an important place in the agricultural process. The producers and the agents in charge with the quality control must have a simple means which are applicable in the inspection point and requiring only material relatively inexpensive in order to measure the maturity. An objective index is more preferable than subjective index, and the ideal is to have recourse to non destructives methods. The penetration tests (Magness-Taylor, force/deformation, Kramer multiblade shear) applied directly to the surface of the fruit are the most acceptable methods to measure the firmness of the fruits and vegetables [1]. The penetrometrics measurements are largely used, because they are rather well correlated with the human perception of firmness, but there are problems involved in the use of probes of varied geometries. In addition, the devices used for measurements are expensive, and these techniques are slow, destructive and not very adaptable for the sorting on line of the fruits and vegetables.

In the last decades, the possibility of replacing the penetrometrics methods by non-destructive ultrasonic methods was documented in several studies [2, 3, 4, 5, 6, 7]. The propagation velocity and the attenuation are the basic ultrasonic properties which can be in correlation with the indices used for the evaluation of the quality of fruits and vegetables [1]. Accordingly, our major objective in this study is to develop the ultrasonic method by retrodiffusion with immersion in water like a new alternative technique for non-destructive maturity assessment of two varieties of orange most collected particularly in the region Souss-Massa-Drâa in Morocco. We worked in the range of high frequencies and by taking into account the strong attenuation of the ultrasonic waves in the texture of fruit and vegetable; we limited our study only to the external layer of orange peel. This layer is called the flavedo and it is distinguished by its small thickness. The use of the high frequencies allows us to perform a characterization of high resolution to the orange peel.

2. Experimental setup

2.1. The experimental device of ultrasonic measurement by immersion in water

The figure 1 shows the experimental setup of ultrasonic measurements used in this work. The transducer and the samples were immersed in a bath containing water. The focusing transducer of 20MHz as a central frequency (Panametrics PI20-2 20; 0.25 F=2.00 0690004) plays the role of transmitter and receiver at the same time. It transmits ultrasonic longitudinal waves to the orange peel through the water, after it receives the ultrasound waves backscattered from the sample. Thetransducer is excited with a pulse generator (Pulser-Receiver model 5052PR Sofranel) which also acts as an amplifier of the signal backscattered. The generator is connected to an oscilloscope (54600B HP, 100MHz) to visualize the backscattered signals. The signals displayed on the oscilloscope are recovered on a computer using a data card acquisition (National Instrument) through a GPIB connection (IEEE 488 Bus). Thus we developed a platform under the convivial environment of LabView software to analyze the ultrasonic signals acquired in order to perform the measurement of the ultrasonic parameters related to flavedo-layer.

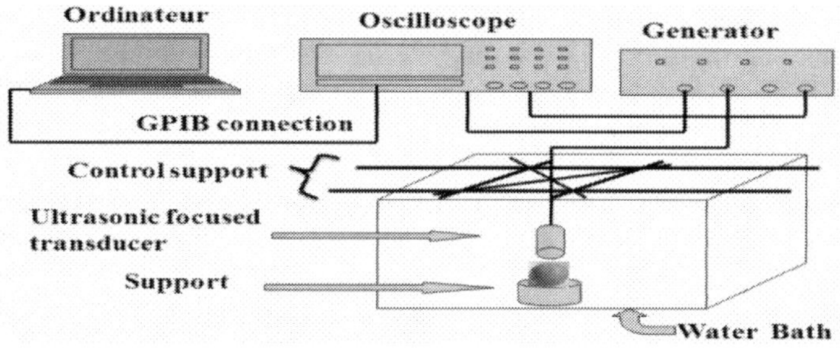

Figure 1. Diagram of the ultrasonic measuring device.

2.2. Ultrasonic measurements

2.2.1. Ultrasonic velocity measurement. To measure the speed of the ultrasonic waves in the flavedo-layer, we have used the following formula:

$$V = \frac{2d}{t_v} \tag{1}$$

Where: d is the thickness of the flavedo-layer and t_v is the flight-time.

To obtain a precise measurement of the flight-time, we used the cross correlation method. Being $R(t)$ the cross correlation function both the reference signal and the backscattered ultrasonic signal, and $H(t)$ its Hilbert transform, the cross correlation envelope is given by:

$$E(R(t)) = \sqrt{R(t)^2 + H(t)^2} \tag{2}$$

The envelope follows correctly the peaks each time an echo is present independently of its phase. In this way, the flight-time between the echos is exactly measured. After by applying the equation (1), we deduce the speed in the flavedo-layer.

2.2.2. Ultrasonic attenuation measurement. The ultrasonic attenuation coefficient α (dB/mm) is measured from the frequency domain. We deduce the attenuation coefficient by the following formula:

$$\alpha = \frac{20}{2d} * Log\left(\frac{A_2}{A_1}\right) \tag{3}$$

- A_1: Spectral amplitude of the first echo backscattered in the interface water / peel.
- A_2: Spectral amplitude of the second echo backscattered in inside the peel at the interface flavedo/albedo.
- d : Flavedo layer thickness

3. Results and discussion

The figures 1 and 2 show various ultrasonic signals backscattered by the studied samples and the measurements of the ultrasonic parameters corresponding to each sample are illustrated in the tables 1 and 2. The measurements collected in the two tables below make the average of three successive measurements in the equator of each sample.

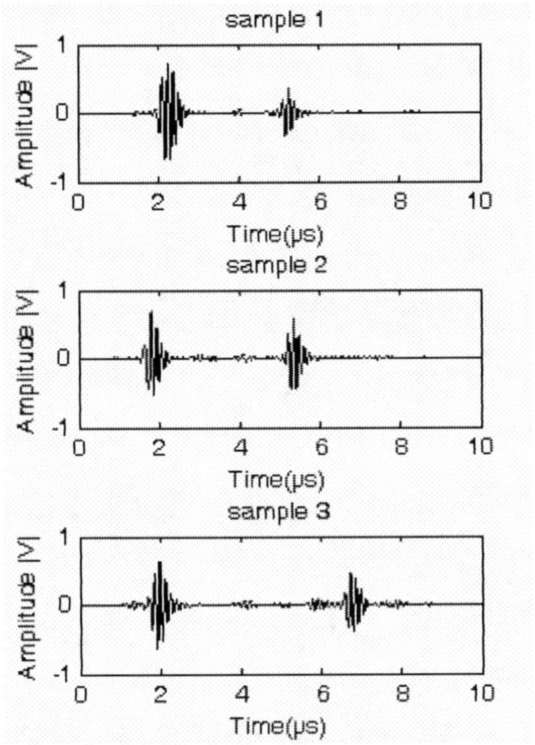

Figure 2. various ultrasonic signals backscattered by the samples 1, 2, and 3 of Navel.

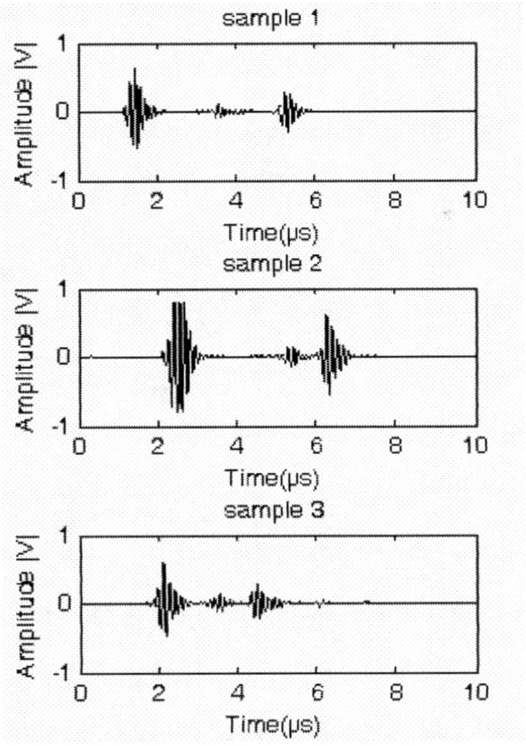

Figure 3. various ultrasonic signals backscattered by the samples 1, 2, and 3 of Mandarin.

Table 1. Ultrasonic parameters measurements of Navel.

	Flight Time(μs)	Thickness (mm)	Velocity (m/s)	Attenuation (dB/mm)
Sample1	1.51	1.8 ± 0.02	1325.08	4.15
Sample2	1.74	1.2 ± 0.02	1149.41	1.11
Sample3	2.41	1.9 ± 0.02	830.59	2.08

Table 2. Ultrasonic parameters measurements of Mandarin.

	Flight Time(μs)	Thickness (mm)	Velocity (m/s)	Attenuation (dB/mm)
Sample1	1.47	1.8 ± 0.02	1028.34	1.14
Sample2	1.63	1.9 ± 0.02	1030.06	0.99
Sample3	1.08	1.7 ± 0.02	1854.29	2.15

We notice that the values obtained for longitudinal speed differ largely with those found by Camarena and Martinez [4]. They found the longitudinal velocity included between 130 and 240 m/s in the peel of various types of orange fruit. We can explain this difference by two reasons: the first we worked in the high-frequency range contrary to Camarena and Martinez [4], they used the low-frequency range (200 KHz). The second reason is the difference about the detected thickness: the high-frequency range that we used allows us to carry out the ultrasonic measurements only on the flavedo-layer instead of the whole peel; Camarena and Martinez [4] have not this problem because the low-frequency range used allows the detection of the whole peel-thickness and they have also an advantage by using the transmission method, in this manner the problem of detected thickness is avoided . Therefore, it is obvious that the results of the velocity will be different because the detected thickness and the correspondent flight-time measured are also different.

Concerning the results about attenuation, we notice that the values obtained are in concordance with those obtained by Camarena and Martinez [4].They found the attenuation coefficient included between 1,8 and 3,7 dB/mm according to the orange type. It is practically in this same interval where are included the attenuation values which we measured: the attenuation varies between 2 and 5 dB/mm for the two types of orange fruit studied in this work.

4. Conclusion

In this work we try to develop a new ultrasonic technique in order to follow in real time the maturity degree of the orange fruit without passing by mechanical and biochemical tests generally destructives. The high frequency ultrasonic characterization provides a high spatial resolution for local measurements of speed and attenuation in the orange peel. The strong attenuation of ultrasounds in the peel has restricted our study only to the flavedo-layer instead of the whole peel-thickness. For this reason the obtained results differ to those previously published especially for the longitudinal speed. For the measurements of the attenuation coefficient, the results are in agreement compared to those already published. This method can be employed to study the differences of the water content in the peel with the age i.e. the analysis of the hydration state of the peel like a decisive criterion to detect the maturity degree of orange fruit.

References

[1] Abbott J A 1999 Quality measurement of fruits and vegetables *Postharvest Biol. Technol.* **15** 207–225

[2] Bechar A, Mizrach A, Barreiro P and Landahl S 2005 Determination of mealiness in apples using ultrasonic measurements *Biosyst. Eng.* **91** 329–334

[3] Camarena F and Martínez-Mora J A 2005 Potential of ultrasound to evaluate turgidity and hydration of the orange peel *J. Food Eng.* **75** 503–507

[4] Camarena F, Martinez-Mora J A and Ardid M 2006 Ultrasonic study of the complet dehydration process of orange peel *Postharvest Biology and Technology* **43** (2007) 115-120

[5] Mizrach A, Galili N, Gan-Mor S, Flitsanov U and Prigozin I 1996. Models of ultrasonic parameters to assess avocado properties and shelf life *J. Agric. Eng.* Res. **65** 261–267

[6] Mizrach A, Galili N and Rosenhouse G 1989. Determination of fruit and vegetable properties by ultrasonic excitation Trans. *Am. Soc. Agric. Eng.* **32** 2053–58

[7] Verlinden B E, De Smedt V and Nicolaï B M 2004 Evaluation of ultrasonic wave propagation to measure chilling injury in tomatoes *Postharvest Biol. Technol.* **32** 109–113

IOP Conf. Series: Materials Science and Engineering **42** (2012) 012039 doi:10.1088/1757-899X/42/1/012039

On-line characterization using ultrasound of pectin hydrolysis catalyzed by the enzyme pectinmethylesterase

C Aparicio, P Resa, C Sierra and L Elvira
Centro de Acústica Aplicada Y Evaluación No Destructiva (CAEND), UPM-CSIC, C/Serrano 144, 28 006 Madrid, Spain

E-mail: cristina.aparicio@caend.upm-csic.es

Abstract. The major problem in the fruit juice industry is associated with juice quality deterioration due to the cloud loss of juice concentrates by the enzymatic reaction of pectinmethylesterase enzyme (PME, EC 3.1.1.11). During pectin hydrolysis, pectin and water are transformed into polygalacturonic acid (pectate) and methanol by the action of PME. In this work, a low-intensity ultrasonic technique is used to monitor this enzymatic reaction, with PME both from orange peel and from *Aspergillus niger*. Changes in sound velocity during pectin hydrolysis (1% concentration of pectin, T=30°C and pH=4.5 and 7) with 0.25 ml of enzyme solution (PME) have been measured using a through-transmission technique. Sound velocity decreases as pectin is transformed into pectate and methanol and at the end of the process, the change in sound velocity reaches 0.3 m/s with PME from orange peel and 0.33 m/s with PME from *Aspergillus niger*.

1. Introduction

Pectinases are one of the most important groups of enzymes used in fruit and vegetable industry. One of the main problems in the fruit juice industry is the maintenance of juice turbidity. It is accepted that cloud loss is initiated by the enzymatic reaction of pectinesterase, which demethylates pectin [1]. Cloud stability is a critical orange juice quality parameter imparting characteristic flavor, color, texture and aroma. It is attributed to the suspension of particles composed of a complex mixture of protein, pectin, lipids, hemicellulose, cellulose and other minor components [2]. Pectinmethylesterase (PME, EC 3.1.1.11) is the first enzyme acting on pectin, a major component of plant cell wall. PME catalyzes the hydrolysis of the methyl esters of pectin to yield polygalacturonic acid (pectate) and methanol.

Pectinmethylesterase enzyme can be obtained from orange peel and also from *Aspergillus niger*. In several works, the pectin hydrolysis has been characterized by measuring the methanol concentration released by the action of PME activity, by spectrophotometric and electrophoretic methods [3], [4].

Ultrasonic techniques are finding increasing use in the food industry for the analysis of foods, providing information about physicochemical properties, such as composition, structure, physical state and flow rate. There are a number of reasons for the current interest on ultrasound: the food industry is becoming increasingly aware of the importance of developing new analytical techniques to study food materials and to monitor properties of food during processing for fully automating producing chains, rapid and precise measurements can be made by this technique which is non destructive, non invasive and can easily be adapted for on-line applications. In addition it can be used to analyze systems that are optically opaque and the rapid advances in microelectronics have led to the development of relatively inexpensive ultrasonic instrumentation that is now commercially available [5].

The objective of this work was to characterize on-line the pectin hydrolysis catalyzed by pectinmethylesterase enzyme from orange peel and from *Aspergillius niger*, using an ultrasonic measurement device based on the propagation of low-intensity pulsed ultrasonic waves developed to monitor biological process [6]. Although, there are lots of researches on pectin hydrolysis, no published data have been found in the literature on ultrasonic techniques to characterize on-line this process.

2. Material and Methods

2.1 Samples

Enzyme solution: pectinmethylesterase purified from orange peel (EC 3.1.1.11, P5400, Sigma-Aldrich Quimica SA) was purchased as a 7.81 mg of lyophilized power with 129 units /mg of solid (One unit was defined by Sigma as the release of 1 μequiv of acid from pectin within 1 min at a pH of 7.5 and a temperature of 30ºC). The enzyme solution was prepared with the mixture of 1 mg/ml (was weighed in a digital balance COBOS with a precision of ± 0.01mg) in 1.7 M Sodium Chloride solution (pH 7). The mixture was dissolved with a magnetic stirrer (COBOS Precision) during 10 hours. PME from *Aspergillus niger* (RAPIDASE FP SUPER, 20 kg drum) was a liquid preparation with ≥ 900 units/g for food use containing PME.

Pectin solution: pectin from citrus peel (galacturonic acid ≥ 74% P9135, Sigma-Aldrich Quimica SA) and Sodium Chloride was added to distilled and degassed water to prepare 1% (w/v) solution with different pH (4.5 and 7) and 10% (w/v) solution of pectin. A magnetic stirrer was used to dissolve the solutions during 2 hours. After that, the solutions were degassed.

Buffer pH=7: 1.2g of NaOH (Sigma-Aldrich, USA) and 6.81g of KH_2PO_4 (Sigma-Aldrich, USA) were added to 300mL of distilled water. 291 mL of this solution were added to distilled water up to 1L. Buffer pH=4.5: 13.6 g of KH_2PO_4 were added to 1 L of distilled water.

2.2 Experimental set up

Measurements of time of flight (TOF) along the pectin hydrolysis as a function of time were carried out by using the ultrasonic device developed and described in depth by [6]. The device consisted in a four-channel unit formed by four measurement cells; two in each climatic chamber (A and B, respectively). At the bottom of each cell there is an IKA mini MR Standard magnetic stirrer to homogenize the liquid samples. Thermal control is achieved by means of the temperature sensors and actuators driven by PID controllers. A through-transmission measurement set-up was incorporated to the system. Transducers consist of three PZT piezoceramic plates (PZ27, Ferroperm, Denmark) designed and constructed to emit and to receive ultrasonic waves. The software to control the measurement process was implemented using LabVIEW (National Instruments).

Commercial 125 ml glass bottles were used as sample containers. These bottles were placed inside the four measurement cells. The transducers were put on the opposite sides of the bottle and pressure-coupled through silicone layers.

The pH of the solution samples were measured with a pH-meter (HI-223, Hanna Instruments) equipped with a HI-1131P refillable electrode (resolution ±0.01pH).

2.3 Pectin hydrolysis

Two glass bottles were filled with 1% (w/v) pectin solution and pH =4.5 and placed in the cells of chamber A, and two bottles were filled with 1% pectin solution and pH=7 and placed in the cells of chamber B, both chambers at 30ºC. When temperature was stabilized 0.25 ml of PME from *Aspergillus niger* (225 U) were introduced in each bottle of the pectin solution placed inside chamber A and 0.25 ml of PME from orange peel (0.3 U) were introduced in each bottle of the pectin solution inside chamber B. The variation of TOF due to pectin hydrolysis by the action of the PME was registered and recorded continuously along the whole process.

3. Results and discussion

From variation of time of flight ($\Delta\tau$, ns), sound velocity (c, m/s) can be calculated from [7]:

$$\Delta c = -\frac{c_0^{\,2} \cdot \Delta\tau}{d} \tag{1}$$

where d (m) is the distance travelled by the ultrasonic pulses through the liquid and c_0 (m/s) is the sound velocity of water at 30ºC.

Figures 1 and 2 show the change in the sound velocity during hydrolysis due to the action of the PME from orange peel and from *Aspergillius niger*, respectively.

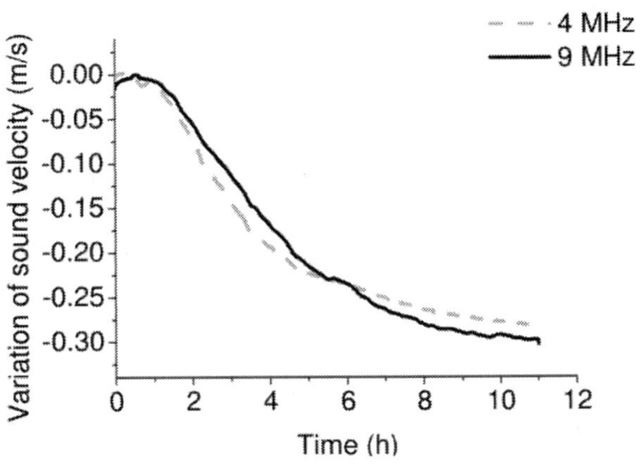

Figure 1. Pectin hydrolysis by the action of PME from orange peel in a buffer of pH=7 at 30ºC.

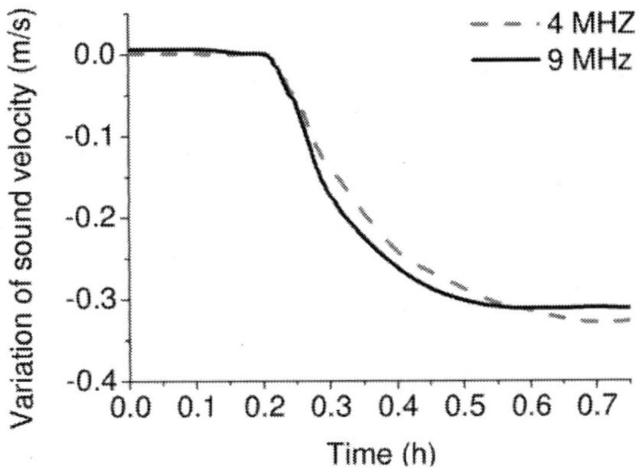

Figure 2. Pectin hydrolysis by the action of PME from *Aspergillus niger* in a buffer of pH=4.5 at 30ºC.

Figure 1 shows that after 8 hours the change of TOF in pectin hydrolysis by the action of PME from orange peel was stabilized. The change in sound velocity was 0.28 m/s for 4 MHz and 0.30 m/s for 9 MHz. In figure 2, pectin hydrolysis by the action of PME from *Aspergillus niger* was faster, around 30 minutes. This was expected, since there was more amount of enzyme (225 U). The change of sound velocity was 0.33 m/s for 4 MHz and 0.31 m/s for 9 MHz. It can be observed that the velocity decreases as pectin is divided into methanol and pectate.

There is no detectable sound speed dispersion for the frequencies used (4 MHz and 9 MHz) as the difference measured between them (±0.02 m/s) are below the device stability (±0.03 m/s). Whatever the origin of the PME enzyme is, a similar amount of pectin has been hydrolysed.

4. Conclusions

A device based on the analysis of ultrasonic waves propagating through a liquid was successfully used to characterize the pectin hydrolysis catalyzed by pectinmethylesterase enzyme. The process has been monitored on line, and a variation in time of flight has been obtained due to the hydrolysis reaction. This is an innovative method to characterize on line the pectin hydrolysis, which is relevant for the food industry due to the importance of the pectinmethylesterase enzyme in fruits and juices. It has been demonstrated that the change in sound velocity at the beginning and at the end of the pectin hydrolysis was 0.3 m/s.

The results obtained would be used in future works, to analyze on line the inactivation of the pectinmethylesterase enzyme by this continuous method.

Acknowledgments

This work has been supported by the project of National Plan of I+D DPI2010-17716. Cristina Aparicio and Pablo Resa have contracts from CSIC (JAEDOC2009-087 & JAE DOC2008-065).

References

[1] Van den Broeck I, Ludikhuyze L R, Van Loey . M and Hendrickx M E 2000 Inactivation of orange pectinesterase by combined high-pressure and –temperature treatments: a kinetics study *Journal of Agricultural Food Chemistry* **48** 1960-70

[2] Tiwari B K, Muthukumarappan K, O'Donnell C P and Cullen P J 2009 Inactivation kinetics of pectin methylesterase and cloud retention in sonicated orange juice *Innovative Food Science and Emerging Technologies* **10** 166-171

[3] Zocca F, Lomolino G, Curioni A, Spettoli P and Lante A 2007 Detection of pectinmethylesterase activity in presence of methanol during grape pomace storage *Food Chemistry* **102** 59-65

[4] Klavons J A and Bennet R D 1986 Determiantion of methanol using alcohol oxidase and its application to methyl ester content of pectins *Journal of Agricultural Food Chemistry* **34** 597-599

[5] McClements D J 1995 Advances in the application of ultrasound in food analysis and processing *Trends in Food Science & Technology* **6** 293-299

[6] Elvira L, Durán C, Sierra C, Resa P and Montero de Espinosa F 2007 Ultrasonic measurement device for the characterization of microbiological and biochemical processes in liquid media *Measurement Science and Technology* **18** 2189-96

[7] Resa P, Elvira L, Sierra C, Montero de Espinosa F 2009 Ultrasonic velocity assay of extracellular invertase in living yeasts *Analytical Biochemistry* **384** 68-73

Assessment of breadmaking performance of wheat flour dough by means of frequency dependent ultrasound

D Braunstein[1], JH Page[2], A Strybulevych[2], D Peressini[1] and MG Scanlon[3,4]

[1] Department of Food Science, University of Udine, via Sondrio 2/A-33100 Udine, Italy
[2] Department of Physics and Astronomy, University of Manitoba, Winnipeg, Manitoba, Canada R3T 2N2
[3] Department of Food Science, University of Manitoba, Winnipeg, Manitoba, Canada R3T 2N2

E-mail: scanlon@cc.umanitoba.ca

Abstract. Technological performance of wheat flour varies among different wheat varieties. Gluten plays a key role within the solid phase of dough in the formation and the retention of gas bubbles during breadmaking. Rheological tests are usually performed to predict breadmaking potential. The aim here was to investigate the ability of ultrasound to discriminate wheat doughs based on breadmaking qualities. The ultimate goal is the development of an on-line quality control system currently unavailable in the baked goods industry, rendering this work innovative. Samples were prepared from a strong wheat flour, with one control sample and one added with inulin and distilled monoglycerides, producing doughs of distinct breadmaking quality. Doughs were subjected to density determination, elongation tests, and ultrasound analysis. The ultrasound tests were performed in the frequency range of 300 kHz - 6 MHz. Ultrasonic phase velocity increased with increasing frequency to about 2 MHz, becoming constant and then decreasing from 3 MHz for the control sample. Distinct differences in attenuation coefficient between the fibre-enriched and control doughs were observed. Ultrasound can potentially add to a better understanding of dough quality and can discriminate between doughs of contrasting properties.

1. Introduction

Ultrasound is a useful technique for the non-destructive characterization of the physical properties and structure of materials. In recent years, there has been interest in using it to investigate the properties of wheat flour doughs [1-8]. These authors have shown that a number of useful technological parameters can be obtained from ultrasound measurements. For example, Bellido and Hatcher [6] showed that formulation changes in noodles that produced a change in ultrasonic properties were highly correlated with conventional measurements of mechanical properties, while Garcia-Alvarez et al [7] showed that dough consistency, an important criterion in the handling of breadmaking doughs, could be determined ultrasonically for a wide range of flour qualities.

Specific attributes of influential factors in breadmaking have also been the focus of ultrasonic studies of dough properties. As well as the highly influential effect of flour quality [9], ingredients

[4] To whom any correspondence should be addressed.

such as shortening have also been studied [10]. Bubbles, that are known to be critical influences on the properties of breadmaking doughs [11,12], have also been studied ultrasonically. Ultrasound is particularly sensitive to bubbles in dough because of large differences in the compressibility and density of the dough matrix compared to the gas in the bubbles [2,13].

Because of demonstrated health effects of dietary fibre, there has been considerable interest in enriching wheat flour doughs with dietary fibre [14,15]. However, issues with dough handling performance and the quality of the resulting loaf of bread have been reported [15,16]

The goal of this paper is evaluate whether an ultrasonic reflectance technique, that is able to interrogate the complex shear modulus at ultrasonic frequencies [17], can discriminate the properties of a fibre-enriched dough from a control dough using longitudinally polarized ultrasonic pulses.

2. Materials and Methods

2.1. Materials

Flour from common wheat with strong gluten properties (W= 325, P/L= 0.56) was used for this study. Farinograph water absorption and stability of flour were 52.4% (on 14% flour moisture basis) and 10.7 min, respectively. Distilled monoglycerides (MYVATEX MIGHTY SOFT LT) were kindly donated by Kerry Ingredients and Flavours (The Netherlands). Frutafit® TEX inulin (degree of polymerisation $\geq$ 22) was obtained from Sensus (The Netherlands).

2.2. Dough preparation

Control dough samples were mixed at 30°C to maximum development in a 50-g farinograph bowl, at optimum water absorption using flour (50 g on 14% m.b.) and tap water (26.2 mL). Dough was also prepared using flour (50 g on 14% m.b.), water (27 mL), inulin (4 g) and distilled monoglyceride (0.5 g). Ingredients were mixed until a consistency of 500 BU was reached. Mixing times were 4 and 14 min, and moisture contents 43.6 and 42.0% for control and fibre doughs, respectively.

2.3. Large deformation rheological properties

A texture analyser (TA.XT2) equipped with a 5 kgf load cell and Kieffer dough and gluten extensibility rig was used to perform a uniaxial extension test on the processed dough sample at 3.3 mms^{-1}. Extension was performed after a dough resting time of 75 min. Force-displacement curves were transformed into stress-strain data [18]. The fracture stress (σ_{max}), the Hencky strain (ε_H) at fracture stress, and the integrated area (A) under the stress-strain curve were taken as measures of resistance to extension, extensibility, and energy required to extension, respectively. An apparent strain hardening ($d \ln \sigma_{max} / d\varepsilon_H$) was computed in the strain interval of 20-95% fracture strain.

2.4. Density measurement

Dough density was measured to ascertain the amount of air incorporated into the dough using 25 mL gravimetric bottles (Kimble Glass Inc., Vineland NJ) using 5 g sub-samples of dough.

2.5. Ultrasonic experiments

A pulsed reflection technique [17], but employing longitudinally polarized ultrasonic pulses, measured the ultrasonic signal reflected from the dough. Dough samples were placed on a plexi-glass delay rod connected to a broadband transducer (0.3-6 MHz). A Fourier transform technique was used to extract phase velocity and attenuation coefficient data from the acquired signals as a function of frequency.

3. Results and Discussion

3.1. Repeatability of the reflectance technique

The ultrasonic phase velocity (UPV) and the attenuation coefficient (AC) for three replicates of the control dough (strong flour without additives) are shown in figures 1 and 2. The error bars are associated with the variability in the ultrasonic parameter between three sub-samples taken from a given replicate dough. It can be seen that variability depends on frequency, being more pronounced at the peak and at the higher frequencies.

The frequency dependent changes in both phase velocity and attenuation coefficient are very similar to those reported for strong breadmaking flours analysed by transmission experiments [8,13]. The peak in attenuation coefficient, that occurs at approximately 2 MHz, is indicative of a strong resonance arising from the bubbles entrained into the dough during mixing [13,19]. On the low frequency side of the peak, velocity and attenuation are lower, indicative of the long wavelength probing of an effective medium of matrix and bubbles that constitutes the dough [2,13]. Ultrasonic velocity at frequencies well above the resonance, where dough matrix properties are accessed [20], is somewhat higher than the velocity of sound in water (1500 m s^{-1}).

Thus, these results show that the reflectance technique is a repeatable means of measuring the ultrasonic properties of wheat flour doughs at frequencies (near resonance) where it is difficult to propagate ultrasound through samples in transmission experiments.

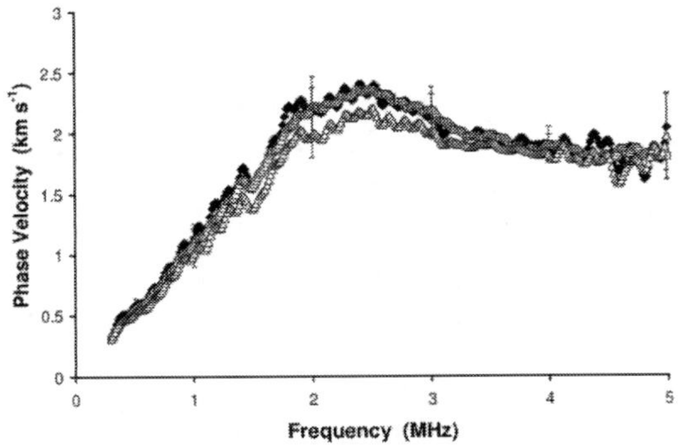

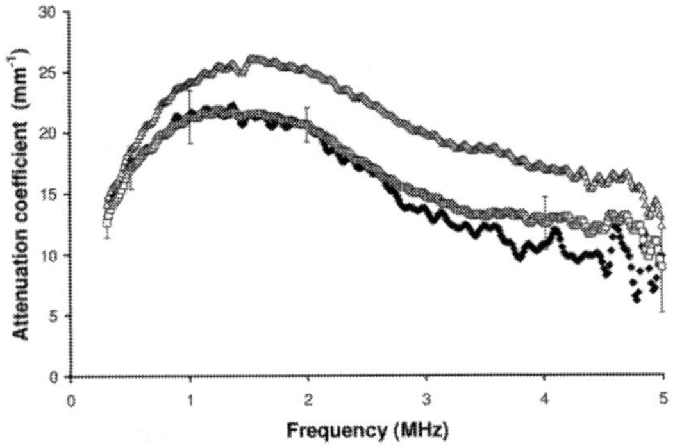

Figure 1. UPV vs frequency for control dough. **Figure 2.** AC vs frequency for control dough.

3.2. The effect of fibre on dough properties

The addition of inulin resulted in a decrease in optimum moisture content of the dough probably due to a lubricating effect of sugars and oligosaccharides [15,16]. Time required for dough development, or to reach a dough consistency equivalent to 500 BU, greatly increased as a consequence of fibre addition [15]. This effect can be mainly related to inulin's intrinsic water absorption ability, leading to a competition with the gluten proteins for available water to form a gel [21]. Fibre-enriched doughs were of slightly lower density (1197±0.7 kg/m^3) than the control (1200±0.6 kg/m^3).

Large deformation rheological properties of dough are reported in table 1. Extensibility and strain hardening, predictors of breadmaking performance, were substantially reduced by fibre addition, resulting in gas cells that are less stable and inflate to a lower volume during proving [15,22]. Changes in rheological behaviour could be interpreted as the result of a decrease in water availability for gluten in the presence of inulin.

Table 1. Fracture parameters calculated from the stress-strain curves of doughs.

Samples	σ_{max} (kPa)	ε_H (-)	A (kPa)	d ln σ_{max} /dε_H(-)
Control	39.0±4.0	2.08±0.14	30.3±2.0	1.32±0.11
Fibre	33.9±3.4	1.20±0.07	27.5±2.8	1.05±0.12

The differences in large strain properties of the doughs are also evident in small strain ultrasonic measurements. In figure 3, substantial differences in AC occur at bubble resonance and at high frequencies. Since dough densities are not very different, and thus gas content is similar, the increase in AC in the fibre-enriched doughs is likely due to fibre's effect on the properties of the dough matrix.

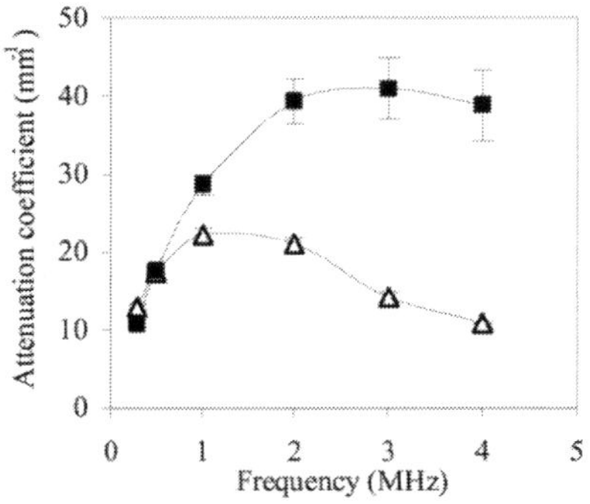

Figure 3. AC vs frequency for control (△) and fibre (■) doughs.

4. Conclusions

Ultrasonic reflectance measurements provide repeatable determinations of phase velocity and attenuation coefficient in complex industrial materials such as wheat flour doughs. Frequency-dependent evaluation of ultrasonic properties confirm that fibre has a pronounced effect on dough properties, effects that are manifest as large differences in the large strain material properties of the dough.

References

[1] Létang C, Piau M, Verdier C and Lefebvre L 2001 *Ultrasonics* **39** 133
[2] Elmehdi H M, Page J H and Scanlon M G 2004 *Cereal Chem.* **81** 504
[3] Ross K A, Pyrak-Nolte L J and Campanella O H 2004 *Food Res. Int.* **37** 567
[4] Alava J M, Sahi S S, Garcia-Alvares J, Turo A, Chavez JA, Garcia M J and Salazar J 2007 *Ultrasonics* **46** 270
[5] Skaf A, Nassar G, Lefebre F and Nongaillard B 2009 *J. Food Eng.* **93** 365
[6] Bellido G G and Hatcher D W 2010 *Food Res. Int.* **43** 701
[7] Garcia-Alvarez J, Salazar J and Rosell C M 2011 *Ultrasonics* **51** 223
[8] Scanlon M G et al 2011 *Proc. 4th International Wheat Quality Conference* ed R. N Chibbar and J E Dexter (Jodhpur, India: AGROBIOS (International)) p 271
[9] Garcia-Alvarez J, Rodriguez J M, Yanez Y *et al 2005 IEEE Ultrasonics Symposium* **1-4** 1480
[10] Mehta K L, Scanlon M G, Sapirstein H D and Page J H 2009 *J. Food Sci.* **74** E455
[11] Campbell G M, Rielly C D, Fryer P J and Sadd PA 1998 *Cereal Foods World* **43** 163
[12] Babin P, Della Valle G, Chiron H et al 2006 *J. Cereal Sci* **43** 393
[13] Leroy V et al 2008 *Bubbles in Food 2: Novelty, Health and Luxury* ed G M Campbell et al (St Paul, MN :AACC Press) p 51
[14] Angioloni A and Collar C 2009 *J. Food Eng.* **91** 526
[15] Peressini D and Sensidoni A 2009 *J. Cereal Sci* **49** 226
[16] Wang J, Rosell CM and de Barber CB 2002 *Food Chem.* **79** 221
[17] Leroy V, Pitura K M, Scanlon M G and Page J H 2010 *J. Non-Newtonian Fluid Mech.* **165** 475
[18] Dunnewind B, Sliwinski E L, Grolle K and van Vliet T 2004 J. *Texture Stud.* **34** 537
[19] Scanlon MG, Page J H, Leroy V et al 2008 *Bubbles in Food 2: Novelty, Health and Luxury* ed G M Campbell et al (St Paul, MN: AACC Press) p 217
[20] Povey M J W 1997 *Ultrasonic Techniques for Fluids Characterization* (SanDiego, CA: Academic Press)
[21] de Gennaro S, Birch G G, Parke S A and Stancher B 2000 *Food Chem.* **68** 179
[22] van Vliet T, Janssen AM, Bloksma A H and Walstra P 1992 *J. Texture Stud.* **23** 439

IOP Conf. Series: Materials Science and Engineering **42** (2012) 012041 doi:10.1088/1757-899X/42/1/012041

Analysis of ultrasonic techniques for monitoring milk coagulation during cheesemaking

E Budelli[1], N Pérez[2,4], P Lema[1], and C Negreira[3]

[1]Instituto de Ingeniería Química, Facultad de Ingeniería, Universidad de la República Julio Herrera y Reissig 565, CP 11300 Montevideo, Uruguay.
[2]Polo Agroalimentario Agroindustrial. Centro Universitario de Paysandú, Universidad de la República. Ruta 3 km 363, Paysandú, Uruguay.
[3]Laboratorio Acústica Ultrasonora, Instituto de Física, Facultad de Ingeniería, Universidad de la República. Iguá 4225 Montevideo 11400, Uruguay.

E-mail: nico@fisica.edu.uy

Abstract. Experimental determination of time of flight and attenuation has been proposed in the literature as alternatives to monitoring the evolution of milk coagulation during cheese manufacturing. However, only laboratory scale procedures have been described. In this work, the use of ultrasonic time of flight and attenuation to determine cutting time and its feasibility to be applied at industrial scale were analyzed. Limitations to implement these techniques at industrial scale are shown experimentally. The main limitation of the use of time of flight is its strong dependence with temperature. Attenuation monitoring is affected by a thin layer of milk skin covering the transducer, which modifies the signal in a non-repetitive way. The results of this work can be used to develop alternative ultrasonic systems suitable for application in the dairy industry.

1. Introduction

Development of objective methods for monitoring the evolution of industrial processes has become of great interest. These methods could improve repeatability of the process, allowing less dependence on operators. Ultrasonic techniques appear as an attractive alternative because they can be used on-line, they are non-destructive and have potential low cost of instrumentation [1].

Determination of a single ultrasonic property requires a controlled procedure and precise calibration since they are very sensitive to different characteristics of the medium such as temperature, elasticity and density [2]. In cheesemaking, precise determination of the cutting time is essential when optimizing the process and has attracted the interest of several authors [3-5]. The objective of this work was to study the suitability of ultrasonic techniques for on-line monitoring of milk coagulation at industrial scale. Experiments were run first in a laboratory scale cell. Then larger scale experiments were carried out using a submergible cell capable of being applied at industrial scale.

2. Materials and methods

2.1. Milk coagulation

[4] To whom any correspondence should be addressed.

Pasteurized whole milk, of known fat and protein composition, was purchased from a local dairy industry. In order to reproduce industrial cheesemaking practices, calcium chloride ($CaCl_2$) and rennet (*Escherichia coli* fermentation chymosin, 600 IMCU/ml) were added so that their concentration in milk was 0.02 g/L and 0.3 g/L respectively [6].

2.2. Laboratory cell
An aluminium prismatic cell (500 cm^3) placed inside a water bath at 36 °C was used in order to make laboratory scale measurements (figure 1). Bottom and side walls were in contact with the water bath. The top of the cell was open and in contact with the ambient air. Milk was thermostatized inside the cell before the addition of rennet. A transducer (3.5MHz, 5MHz and 10MHz) was placed through one of the side walls and a 10 mm aluminium plate was glued to the opposite wall to avoid overlapping of the reflected signal. The ultrasonic measurement consisted in emitting a short ultrasonic pulse with the transducer and recording the first reflected echo. Variations in the time of flight and the pulse energy were calculated for this echo. Temperature in the centre of the cell was measured during coagulation.

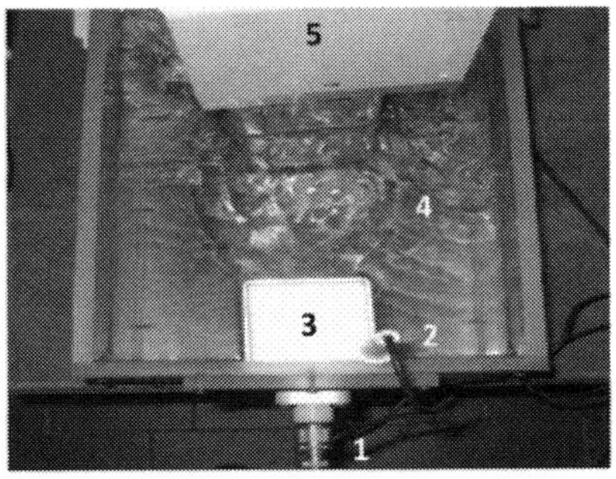

Figure 1. Laboratory chamber (1) Ultrasonic transducer. (2) Temperature sensor. (3) Coagulating milk. (4) Thermostatized water bath. (5) Temperature controller.

2.3. Submergible cell
An aluminium cell capable of being submerged in a coagulation tank was made in order to make measurements independent from border conditions. It consisted in a prism with two open sides (figure 2). A transducer was placed through one of the walls and the opposite wall was constructed thick enough to avoid overlapping of the reflected signals. Coagulation was made in a 5 L container placed inside a thermostatized water tank (36 °C). The cell was submerged inside the milk and the system was left to thermostatise before the addition of rennet. Ultrasonic measurements were performed as described for the laboratory cell assays (c.f. 2.2).

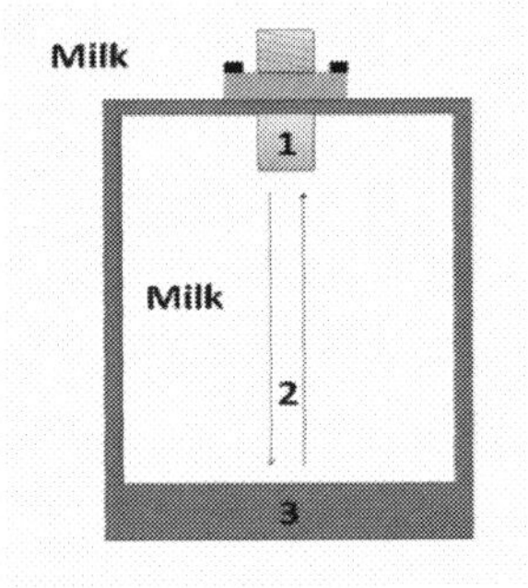

Figure 2. Submergible cell. (1) Ultrasonic transducer. (2) Ultrasonic path. (3) Enlarged front wall.

3. Results and discussion

3.1. Laboratory cell
Significant variations in the time of flight were recorded but they showed a high correlation with the temperature variations inside the cell. Figure 3 shows the results obtained for three repetitions made in

similar conditions. Changes in room temperature lead to changes in the time of flight but no significant differences in the structure of the coagulum were obtained. These results show limitations of the time of flight measurements when room temperature cannot be controlled. An alternative to avoid this limitation at industrial scale could be to measure in the centre of the coagulation tank where conditions could be considered independent from room temperature.
In these experiments energy showed a non-repeteative and unpredictable evolution.

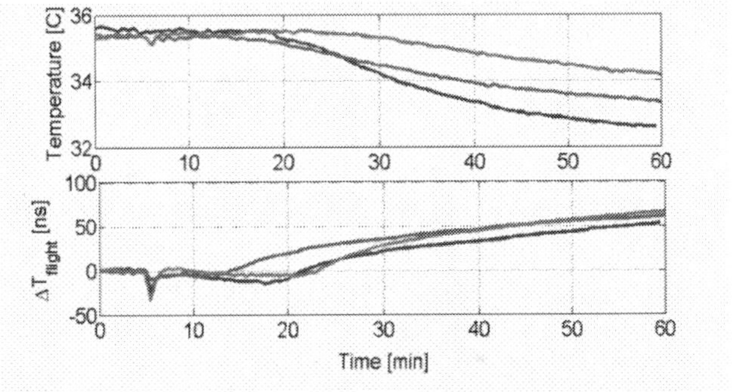

Figure 3. Temperature and time of flight variation for three repetitions. Rennet was added at minute 5.

3.2. Submergible cell

As expected, temperature where the submergible cell was placed remained practically constant. Variations in the time of flight during coagulations were unsignificant, showing that the changes observed in the laboratory cell experiments were due to temperature variations and not to the coagulation process.
Once again energy showed a non-repeteative behaviour during coagulation, so a more detailed study of its evolution was made using this set-up. Energy decreased even before addition of the rennet. To find the cause of this decrease, experiments with no rennet addition were conducted. It was found that cleaning of the transducer surface led to an increase in the energy. Figure 4 shows the evolution of the echo energy when the surface of the transducer was cleaned every five minutes. This decrease was caused by the formation of a thin layer of milk skin on the transducer which was removed when cleaning the transducer's surface. Removal of this layer when agitating after rennet addition could be the cause of the unpredictable changes in energy observed during coagulation experiments in both coagulation cells.

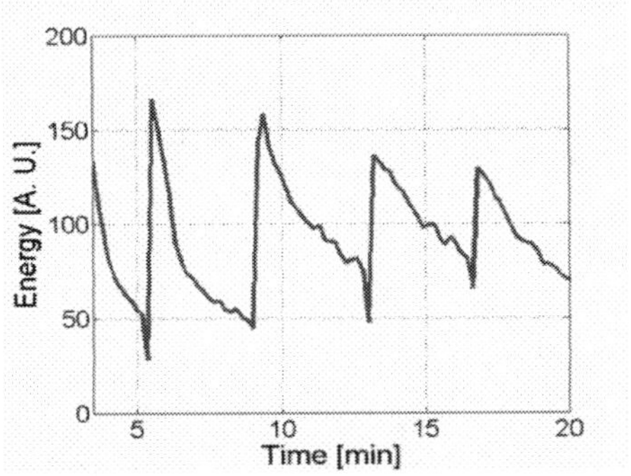

Figure 4. Evolution of the first echo energy using only milk. Every five minutes the transducer surface is cleaned using a small brush. Note the effect of the layer of milk skin.

4. Conclusions

Even though laboratory scale experiments suggested potential of time of flight measurements for coagulation monitoring, a strong dependence with temperature was found. At relatively constant temperature conditions, time of flight changes were insignificant. From this we can conclude that the laboratory scale measurements conducted in this work responded to temperature changes and not to structure variations caused by coagulation. Regarding energy evolution measurements application, limitations were also found. Significant changes were observed when milk was not coagulating. These changes were caused by a thin layer of milk skin formed on the transducer surface. This phenomenon cannot be controlled and is not repetitive; therefore monitoring energy variations of the first echo is not a suitable technique for industrial application. Further research is needed in order to look for an ultrasonic method that could be used at industrial scale, sensitive to changes in the structure caused by coagulation.

References

[1] Dolatowski Z J, Stadnik J and Stasiak D 2007 *Acta Sci. Pol., Technol. Aliment.* **6**(3) 89
[2] Elvira L, Resa P and Montero de Espinosa F 2002 Ultrasonic propagation and thermal changes during milk gelation processes *Proc. XXXIII Congreso Nacional de Acústica–Tecniacústica 2002* (Sevilla, Spain, 16-20 September 2002) ed Sociedad Española de Acústica
[3] Koc A and Ozer B 2008 *Food Res. Int.* **41**(7) 745
[4] Gunasekaran S and Ay C 1996 *J. Food Process Eng.* **19** 63
[5] Kudryashov E D, Hunt N T, Arikainen E O and Buckin V A 2001 *J. Dairy Sci.* **84** 375
[6] Fox F, Guinee T, Cogan T and McSweeney P 2000 *Fundamentals of Cheese Science* (Gaithersburg, Maryland, USA :Aspen Publishers, Inc.)

IOP Conf. Series: Materials Science and Engineering **42** (2012) 012042 doi:10.1088/1757-899X/42/1/012042

Ultrasonic analysis to discriminate bread dough of different types of flour

J García-Álvarez[1,3], C M Rosell[2], M J García-Hernández[1], J A Chávez[1], A Turó[1] and J Salazar[1]

[1]Grup Sistemes Sensors, Departament Enginyeria Electrònica, Universitat Politècnica de Catalunya, Jordi Girona 1-3, 08034 Barcelona, Spain
[2]Cereal Group, Food Science Department. Institute of Agrochemistry and Food Technology (IATA-CSIC), PO Box 73, Burjasot-46100, Valencia, Spain

javier.garcia-alvarez@upc.edu

Abstract. Many varieties of bread are prepared using flour coming from wheat. However, there are other types of flours milled from rice, legumes and some fruits and vegetables that are also suitable for baking purposes, used alone or in combination with wheat flour. The type of flour employed strongly influences the dough consistency, which is a relevant property for determining the dough potential for breadmaking purposes. Traditional methods for dough testing are relatively expensive, time-consuming, off-line and often require skilled operators. In this work, ultrasonic analysis are performed in order to obtain acoustic properties of bread dough samples prepared using two different types of flour, wheat flour and rice flour. The dough acoustic properties can be related to its viscoelastic characteristics, which in turn determine the dough feasibility for baking. The main advantages of the ultrasonic dough testing can be, among others, its low cost, fast, hygienic and on-line performance. The obtained results point out the potential of the ultrasonic analysis to discriminate doughs of different types of flour.

1. Introduction

Most flour used for baking purposes comes from wheat. This type of flour can produce dough with a viscoelastic matrix, named gluten, capable of retaining gas and producing thus bread or pastry within the common quality parameters. Moreover, flour milled from corn, rice or some fruits and vegetables can also be used in baking. The doughs obtained from these flours lacks a natural viscoelastic matrix, having generally lower consistency and capability to retain air bubbles, and therefore can be less adequate to meet the general breadmaking requirements. However, these flours are highly appreciated by the people who cannot tolerate gluten, known as celiacs. Moreover, within a specific type of flour, a further classification into different varieties can be done regarding to the final product that the flour is more apt to produce.

The classification of flours into different varieties can be done regarding to strictly flour characteristics, such as the ash and protein content, and also considering the viscoelastic properties of the resultant bread dough. The viscoelastic properties of bread dough can be determined by means fundamental and empirical rheology, which have been extensively reported in the literature [1, 2].

[3] To whom any correspondence should be addressed.

Ultrasound can also provide relevant information of the flour properties through the analysis of the consistency of the produced bread dough. Acoustic properties of materials are related to their elasticity, consistency and other physical properties. Thus, ultrasound can be regarded as a potential analysis in order to differentiate flour types and varieties. In addition, ultrasound systems are usually fast, relatively cheap and hygienic. Furthermore, previous works shown the feasibility of the ultrasound techniques in order to acoustically determine some relevant properties of both wheat flour [3] and gluten-free [4] dough samples.

In this work, the capability of ultrasound to discriminate different types and varieties of flour has been analysed. That capability of discrimination can be related to the sensitivity of ultrasound to the consistency of the dough samples, which strongly depends on the flour characteristics. One of the types of flour studied were gluten free and the other non-gluten free: rice flour and wheat flour respectively. Within these two types, a set of flour samples intended for different purposes and with diverse physical and quality attributes were tested.

2. Materials and methods

The experimental set-up is depicted in figure 1. It consists of a signal generator (HP33120A for the generation of a 3 cycles of sine burst 100 kHz excitation signal), a power amplifier (AG series amplifier, T&C Power Conversion, Inc. Rochester, NY), two ultrasonic transducers with a resonant frequency of 100 kHz (Panametrics-Olympus, Japan) and a digital oscilloscope (LeCroy LT344). The operation frequency of the ultrasound experiments is the same than the resonant frequency of the shear transducers, of 100 kHz. That relatively low frequency, in the low kHz range, is employed in order to reach the smaller possible attenuation. An additional amplifier with 60 dB gain (Panametrics-Olympus, Japan) is connected to the receiver transducer so as to amplify the generally weak received signals. A dough sample is placed between the two carefully aligned transducers. The attenuation and velocity of the ultrasound wave is obtained as it is shown in [3].

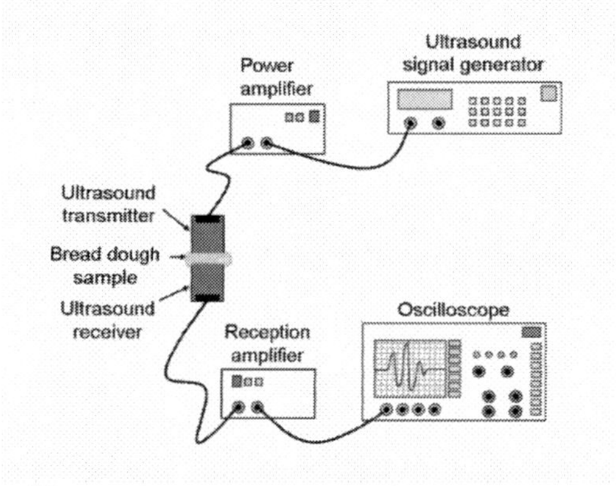

Figure 1. Ultrasonic experimental set-up.

Moreover, making very accurate measurements was very difficult due to the time-dependent nature of the dough. In order to minimise time-dependence effects on measurements, once the dough sample was placed between both transducers, measurements were always done after a same time (1 minute). Furthermore, fluctuations in the measured parameters were observed within the samples prepared by the same mixing procedures. It was found that sample manipulation during the measurements had a significant effect on repeatability.

The recipe and the preparation procedure of the dough samples are described in [3] for the wheat flour dough samples and in [4] for the gluten-free dough samples.

3. Experimental results

The sensitivity of the ultrasonic technique to the consistency of dough samples prepared from different types of flours was first analyzed. The ultrasound attenuation and velocity in wheat flour samples with diverse water content, from 50% to 65%, are shown in figure 2. As can be seen, ultrasound attenuation strongly increases and velocity decreases with water content. The dough presents lower consistency and a more liquid-like behavior as the water content increases, leading to a more difficult propagation of the ultrasonic shear waves, which do not propagate through liquids. That arduous propagation of shear waves in relatively low consistency materials generally results in high values of attenuation and low values of velocity.

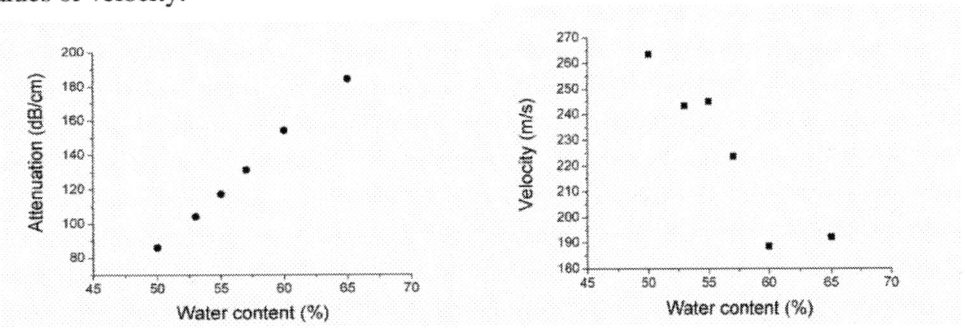

Figure 2. Ultrasound attenuation in dB/cm (left) and velocity in m/s (right) with water content in %.

The ultrasound attenuation and velocity were also measured in gluten-free dough samples with diverse level of consistency, attained through the addition of different amounts of soybean protein (SP) to a combination of rice flour and transglutaminase. The addition of SP increase the dough consistency, which results in decreasing values of attenuation and increasing values of ultrasound velocity with SP content, as it is shown in figure 3. These results agree with the previously commented difficulties of shear waves to propagate through low consistency samples. Moreover, it should be noted that the attenuation values obtained in rice flour samples are generally higher and the velocity values are lower than those measured in wheat flour samples, which agree with the usual lower consistency of rice flour doughs compared to wheat flour doughs, due to the absence of a natural protein viscoelastic matrix in the gluten-free doughs.

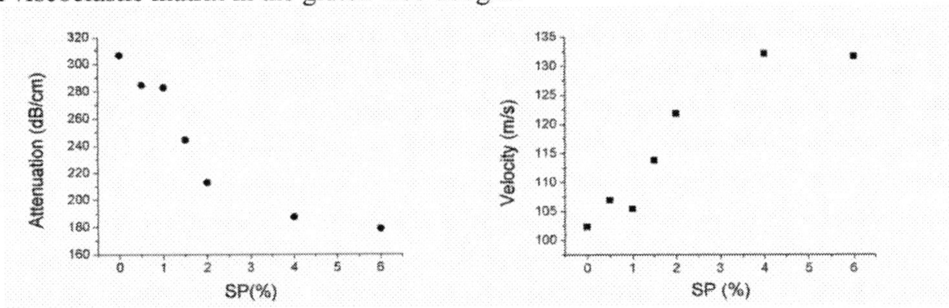

Figure 3. Ultrasound attenuation in dB/cm (left) and velocity in m/s (right) with soybean protein (SP) in %.

In order to further study the capability of ultrasound to discriminate flour types and varieties, the individual values of velocity and attenuation for all dough samples are plotted against each other in figure 4. As can be seen, the flour samples can be clearly separated in two groups, with some diagonal disposition. On one hand, rice flours tend to be situated at the top left corner while wheat flours tend to be situated at the opposite corner. According to this chart, rice flour dough, with comparatively high attenuation and low velocity values, should have lower consistency than wheat flour dough, which is usually right as above commented. Moreover, the same trend can be observed within the group of samples of a specific type of flour. For rice flour, the samples with less SP content, and therefore lower consistency, tend to be placed in the top left corner, with high attenuation and low velocity values, while the samples with more SP content, and consequently higher consistency, are placed in the opposite corner of the group. Likewise, the wheat flour dough with more water content and therefore less consistency are in the top left angle of its group, and the less hydrated dough, with consequently higher consistency, are in the contrary angle. Therefore, the attenuation/velocity ratio could be used as an indicator of the dough consistency, and thus also of the type of flour used.

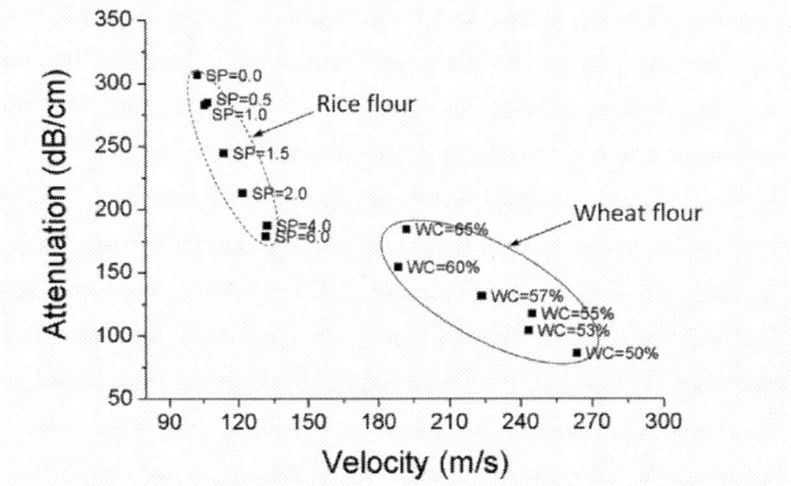

Figure 4. Ultrasound measurements for wheat flour dough with different water content (from 50% to 65%) and rice flour dough with different SP addition (from 0% to 6%).

4. Conclusions

The ultrasonic measurements performed in the dough samples can be clearly separated in two groups depending on the type of flour used, wheat flour or rice flour. Moreover, a general arrangement of the samples within a group of a specific type of flour in function of their consistency can be observed. Therefore, ultrasonic techniques can be seen as a potential low cost tool for flour type and variety discrimination, complementary to conventional techniques, for wheat and gluten-free flour types.

Acknowledgments

This work was financially supported by Ministerio de Ciencia e Innovación, Project (DPI2009-14468-C02-01), Spain.

References

[1] Watanabe E, Bell A E and Brockway B E 1992 Rheological studies on wheat flour dough. I. Empirical and fundamental methods *Food Control* **3** 97-101

[2] Watanabe E, Bell A E and Brockway B E 1992 Rheological studies on wheat flour dough. II. Relationships between frequency, storage modulus and torque amplitude *Food Control* **3** 102-4

[3] Garcia-Alvarez J, Salazar J and Rosell C M 2011 Ultrasonic study of wheat flour properties *Ultrasonics* **51** 223-8

[4] Rosell C M, Marco C, Garcia-Alvarez J and Salazar J 2011 Rheological Properties of Rice–Soybean Protein Composite Flours Assessed by Mixolab and Ultrasound *Journal of Food Process Engineering* **34** 1838-59

IOP Conf. Series: Materials Science and Engineering **42** (2012) 012043 doi:10.1088/1757-899X/42/1/012043

Ultrasonic characterization of pork meat salting

J V García-Pérez, M De Prados, N Pérez-Muelas, J A Cárcel and J Benedito [1]
Grupo de Análisis y Simulación de Procesos Alimentarios (ASPA), Departamento de Tecnología de los Alimentos, Universitat Politècnica de València, 46022, Valencia, Spain

E-mail: jjbenedi@tal.upv.es

Abstract. Salting process plays a key role in the preservation and quality of dry-cured meat products. Therefore, an adequate monitoring of salt content during salting is necessary to reach high quality products. Thus, the main objective of this work was to test the ability of low intensity ultrasound to monitor the salting process of pork meat. Cylindrical samples (diameter 36 mm, height 60±10 mm) of Biceps femoris were salted (brine 20% NaCl, w/w) at 2 °C for 1, 2, 4 and 7 days. During salting and at each experimental time, three cylinders were taken in order to measure the ultrasonic velocity at 2 °C. Afterwards, the cylinders were split in three sections (height 20 mm), measuring again the ultrasonic velocity and determining the salt and the moisture content by AOAC standards. In the whole cylinders, moisture content was reduced from 763 (g/kg sample) in fresh samples to 723 (g/kg sample) in samples salted for 7 days, while the maximum salt gain was 37.3 (g/kg sample). Although, moisture and salt contents up to 673 and 118 (g/kg sample) were reached in the sections of meat cylinders, respectively. During salting, the ultrasonic velocity increased due to salt gain and water loss. Thus, significant ($p<0.05$) linear relationships were found between the ultrasonic velocity and the salt (R^2=0.975) and moisture (R^2=0.863) contents. In addition, the change of the ultrasonic velocity with the increase of the salt content showed a good agreement with the Kinsler equation. Therefore, low intensity ultrasound emerges as a potential technique to monitor, in a non destructive way, the meat salting processes carried out in the food industry.

1. Introduction

Salting is one of the most ancients preservation methods used in food stuffs [1] because of the fact that salt is a multifunctional ingredient that affects both quality and safety. One of the main salted products is the dry-cured meat. Dry-cured ham is one of the most valuable in Spanish meat market [2]. During salting, salt content increases (by 3-5 %) and water content decreases (by 9 %) and the product loses weight [3]. There is a great variability in the final salt content due to among other factors, the different structure and composition of raw meat. Moreover, there is also a higher consumer's demand of dry-cured products with low salt content due to changes in sensory preferences and the increase public concern as regards the high intake of sodium in the diet [4]. These facts lead to the industry to seek for an adequate monitoring of salting process, especially in a non-destructive way.

The number of applications of low intensity ultrasound in the food industry has mostly increased during the last decade [5]. The main reason for this increase arises from the fact that ultrasonic technologies are non-destructive, rapid, low-cost and easy to be automated on-line [6]. Ultrasonic has

[1] To whom any correspondence should be addressed.

been used to determine the physicochemical properties of many foods [7]. In the meat industry, ultrasound has been used to determine the carcass yields of Iberian pigs, the intramuscular fat content in meat and carcasses [8] and to estimate the moisture content of dry-cured meat products [9]. Furthermore, ultrasonic velocity has been also used to assess the quality of fresh back-fat from Iberian pigs [10]. Thus, the main objective of this work is to test the ability of low intensity ultrasound to monitor the salting process of pork meat.

2. Materials and methods

2.1 Sample Preparation

Salting experiments were carried out using 12 meat cylinders of 36 mm in diameter and 60±10 mm in length obtained from Biceps Femoris of white pork. Meat cylinders were placed in a copper cylinder in order to avoid the mass transfer through radial direction. Only one side of the cylinder was immersed in the brine solution, while the other side was covered by plastic film to avoid air drying. Thus, from a theoretical point of view, it could be considered that meat cylinders behave as infinite slabs (L=60±10 mm) and the mass transfer is simplified as one-dimensional. Salting was carried out using a brine solution of 20 % (NaCl, w/w) and temperature was hold at 2 °C. Three cylinders were removed from the brine at different times (24, 48, 96 and 168 hours) and the ultrasonic velocity was measured in the whole cylinder. Afterwards, the cylinders were sliced in 3 parts of 20±3 mm thickness, being measured the ultrasonic velocity and the salt and moisture content in each section.

2.2 Ultrasonic velocity measurements

Figure 1 shows the experimental set-up used in this work, it mainly consisted of a couple of narrow-band ultrasonic transducers (5MHz, 0.75" diameter, V308 model, Panametrics NAT, Walthman, MA, USA), a pulser-receiver (Panametrics, Model 5058PR, Waltham, USA) and a digital oscilloscope (Tektronix, TDS5034, Digital phosphor oscilloscope. Tektronix inc. Bearverton, Oregon. USA). A custom digital height gage was designed and built, and linked to the computer by a RS232 interface to measure the sample thickness. The ultrasonic velocity in the sample was computed from the time of flight [11] and the thickness provided by the height gage. The ultrasonic velocity was determined for all samples (the whole cylinders and their three sections) at 2 °C in a temperature-controlled chamber.

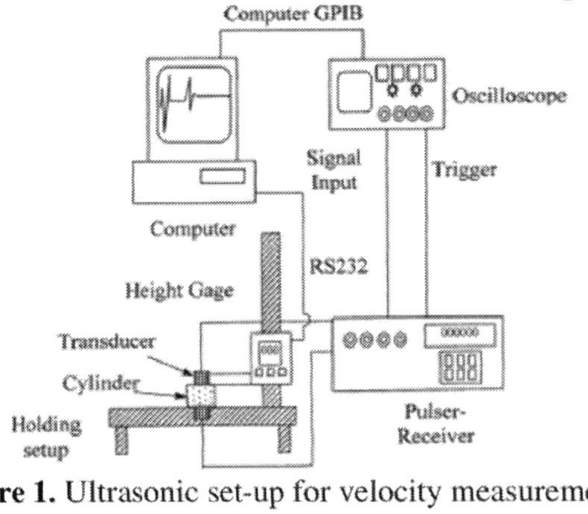

Figure 1. Ultrasonic set-up for velocity measurements.

2.3 Chemical composition

Salt and moisture content were measured according to AOAC standards (AOAC, 1997). In all cases, the measurement was carried out, at least, in triplicate.

3. Results and discussion

3.1 Influence of salting time on composition and ultrasonic velocity

Figure 2 shows the evolution of moisture and salt content during salting in the whole cylinders. As can be observed, samples lost water and gained salt during salting. The average water content decreased from 763±37 to 723±78 (g/kg sample). While the salt content increased from 1.0±0.3 to 37.1 ±8.2 (g/kg sample). In the sections of meat cylinders, moisture and salt contents up to 673 (g/kg sample) and 118 (g/kg sample) were reached, respectively

The influence of salting time on the ultrasonic velocity is shown in figure 3. The ultrasonic velocity increased during salting. This increase is ascribed to two superimposed effects, the decrease of moisture and the gain of salt, which increase the solid content in the liquid phase. Similar results have been obtained for the ripening of several types of cheese, where the ultrasonic velocity increases due to the decrease of water content and the increase of solids [12].

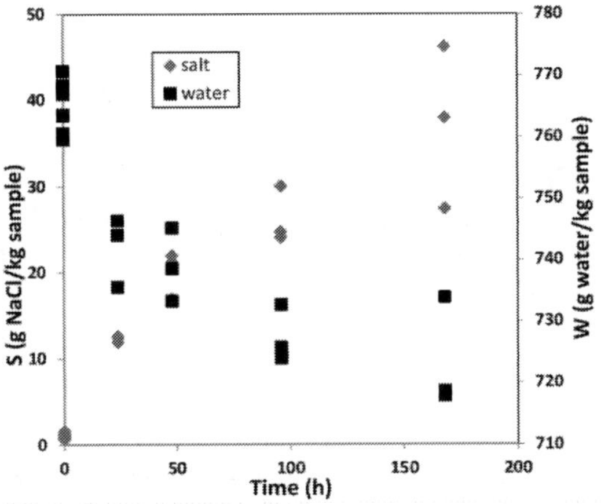

Figure 2. Evolution of moisture and salt content of whole cylinders (37 mm in diameter y 60±10 mm in length) of Biceps Femoris (BF) during salting 2 °C.

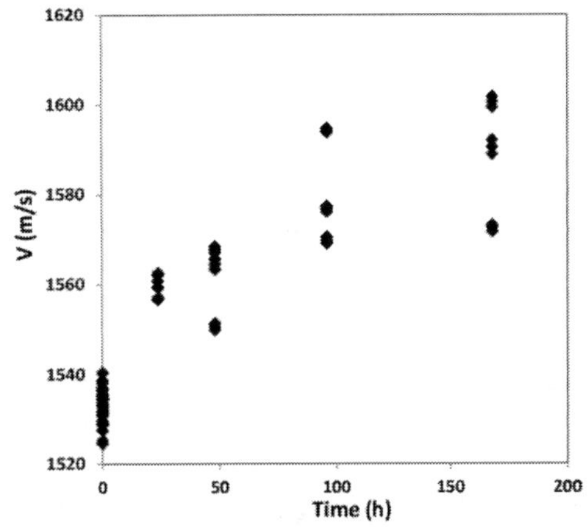

Figure 3. Evolution of ultrasonic velocity in whole cylinders (37 mm in diameter y 60±10 mm in length) of Biceps Femoris (BF) during salting 2 °C.

3.2 Influence of salt and moisture content on the ultrasonic velocity

Figure 4 and 5 show, for the cylinder's sections, the influence of salt and moisture content on the ultrasonic velocity. The ultrasonic velocity increased with the reduction of moisture content and the gain of salt, being established significant ($p<0.05$) lineal relationships in both cases. The slope of the linear relationship determines how the ultrasonic velocity changes due to the change of moisture or salt. Thus, in the case of salt, the slope indicates that the ultrasonic velocity increases 1.46 m/s per g/kg of salt gain, while in the moisture content, the slope means that the ultrasonic velocity decreases 1.62 m/s per g/kg of water lost. During salting, salt and moisture transfer are superimposed effects, thus, further works will be addressed to separate and correctly discriminate the influence of moisture and salt on ultrasonic velocity. Figure 5 also shows the experimental equation of Kinsler (equation (1)), which represents the influence of salt content on the ultrasonic velocity in a water solution [13]. It is observed in figure 5 that, regardless the different structure of meat and water solution, the slope of both linear relationships are similar. The results obtained in this work show that the ultrasound velocity measurements could be used to monitor the pork meat salting for quality control purposes.

$$V = 1449 + 4.6T - 0.055T^2 + 0.0003T^3 + (1.39 - 0.012T)(C - 35) \quad (1)$$

Where V is velocity (m/s), T is temperature (°C) and C is the salt concentration (g/kg).

IOP Conf. Series: Materials Science and Engineering **42** (2012) 012043 doi:10.1088/1757-899X/42/1/012043

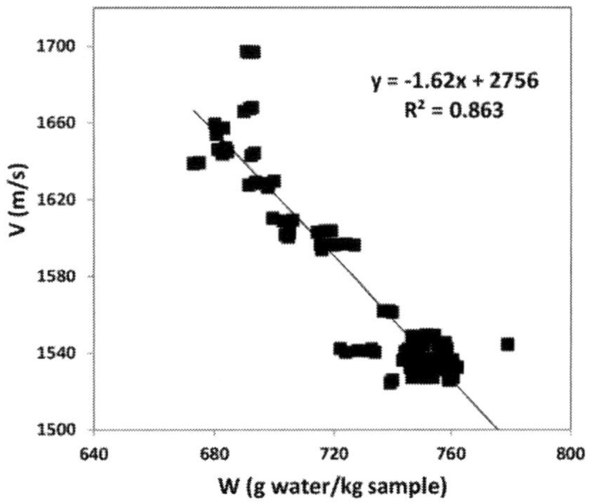

Figure 4. Relationship between ultrasonic velocity and the moisture content of Biceps Femoris (BF) cylindrical sections.

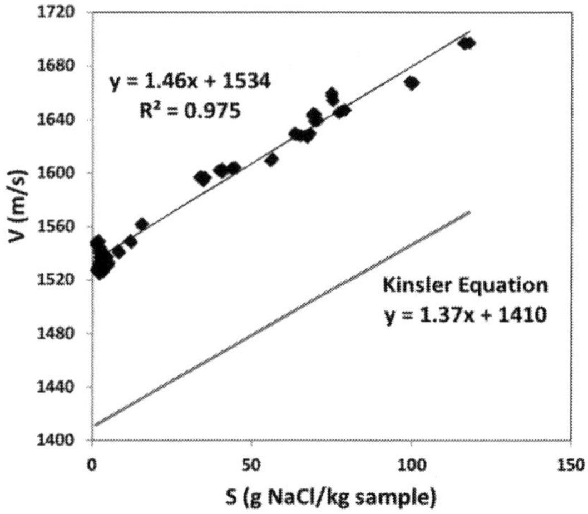

Figure 5. Relationship between ultrasonic velocity and the salt content of Biceps Femoris (BF) for cylindrical sections and Kinsler equation.

4. Conclusion

The ultrasonic velocity increased in the samples during salting due to the gain of salt and the loss of water. Moreover, significant linear relationships were established between the salt and moisture content and the ultrasonic velocity. According to these results, the application of ultrasound to monitor the change of salt and moisture during salting could be of great interest for industrial purposes.

Acknowledgements

The authors acknowledge financial support from the "Ministerio de Economía y Competitividad" in Spain (Project RTA 2010-00029-C04-02).

References

[1] Binkerd E F and Kolari O E 1975 *Food and Cosmetics Toxicology* **13** 655-661
[2] Gou P, Comaposada J and Arnau J 2002 *Meat Science* **61** (1) 25-31
[3] Bello J 2008 *Jamón Curado Aspectos Científicos y Tecnológicos. Perspectivas desde la Unión Europea* (Spain: Diaz de Santos) pp 206-216
[4] World Health Organization/Food and Agriculture Organization 2003 *Diet, Nutrition and the Prevention of Chronic Disease* Technical Report Series 916 (Geneva: World Health Organization)
[5] Benedito J, Carcel J A, Gonzalez R and Mulet A 2002 *Ultrasonics* **40** 19-23
[6] D J McClements 1997 *Critical Reviews Food Science Nutrition* **37** 1
[7] Coupland J N 2004 *Food Research International* **37** 537-543
[8] Mörlein D, Rosner F, Brand S, Jenderka K-V and Wiche M 2005 *Meat Science* **69** 187-199
[9] Llul P, Simal S, Femenia A, Benedito J and Rosello C 2002 *J. Food Eng.* **52** 323-330
[10] Niñoles L 2007 *Contribución a la Caracterización de la Calidad de Productos del Cerdo mediante Ultrasonidos de Señal* Doctoral Thesis (Valencia: UPV)
[11] Benedito J, Carcel J, Roselló C and Mulet A 2001 *Meat Science* **57** 365-370
[12] Benedito J J 1998 *Contribución a la Caracterización de Quesos mediante el Uso de Ultrasonidos de Señal* Doctoral Thesis (Valencia: UPV)
[13] Kinsler L E, Frey A R, Coppens A B and Sanders J V 1982 *Fundamentals of Acoustics* (New York: John Wiley and Sons)

IOP Conf. Series: Materials Science and Engineering **42** (2012) 012044 doi:10.1088/1757-899X/42/1/012044

An experimental device for accurate ultrasounds measurements in liquid foods at high pressure

E Hidalgo-Baltasar[1, 2], M Taravillo[1, 2], V G Baonza[1, 2], P D Sanz[1, 3] and B Guignon[1, 3, 4]

[1]MALTA-Consolider Team
[2]Departamento de Química Física I, Facultad de Ciencias Químicas, Universidad Complutense de Madrid, 28040 Madrid, Spain.
[3]Departamento de Procesos, Instituto de Ciencia y Tecnología de Alimentos y Nutrición ICTAN - CSIC, C/ José Antonio Nováis 10, 28040 Madrid, Spain

E-mail: bguignon@ictan.csic.es

Abstract. The use of high hydrostatic pressure to ensure safe and high-quality product has markedly increased in the food industry during the last decade. Ultrasonic sensors can be employed to control such processes in an equivalent way as they are currently used in processes carried out at room pressure. However, their installation, calibration and use are particularly challenging in the context of a high pressure environment. Besides, data about acoustic properties of food under pressure and even for water are quite scarce in the pressure range of interest for food treatment (namely, above 200 MPa). The objective of this work was to establish a methodology to determine the speed of sound in foods under pressure. An ultrasonic sensor using the multiple reflections method was adapted to a lab-scale HHP equipment to determine the speed of sound in water between 253.15 and 348.15 K, and at pressures up to 700 MPa. The experimental speed-of-sound data were compared to the data calculated from the equation of state of water (IAPWS-95 formulation). From this analysis, the way to calibrate cell path was validated. After this calibration procedure, the speed of sound could be determined in liquid foods by using this sensor with a relative uncertainty between (0.22 and 0.32) % at a confidence level of 95 % over the whole pressure domain.

1. Introduction

High hydrostatic pressure (HHP) processes are spreading throughout the food industry as a successful preservation technology [1]. Pressure plays an antagonist effect to temperature on biochemical reactions providing a protection to thermolabile components even at relatively high temperature. Thanks to this interesting characteristic, pressure appears as a possible additional parameter for conventional food processing. It has been combined with different temperature ranges getting significant improvements over the corresponding traditional processes regarding food quality [2].

Ultrasonic sensors were efficiently introduced in the food industry to control processes at atmospheric pressure [3]. They can also be useful to control HHP processes. This kind of sensors are a priori good candidates for HHP processes due to their pressure resistance, waterproof feature, non-invasive character, versatility of applications and their fast response to changes in the matter. In order

[4] To whom any correspondence should be addressed.

IOP Conf. Series: Materials Science and Engineering **42** (2012) 012044 doi:10.1088/1757-899X/42/1/012044

to be able to implement such sensors in HHP process equipment, the first step is to test them at lab-scale. Therefore, the objective of this study was to establish a methodology to determine accurately the speed-of-sound in foods under pressure.

2. Experiments

2.1. Samples
Speed of sound was measured in deionized water (pure water type I, Milli-Q system, Millipore, Billerica, MA, USA), pasteurized skim milk (Puleva, Spain), orange juice (12°Brix, Nufri, Spain) and orange juice concentrate (40°Brix, Nufri, Spain). Food samples were gently mixed and about 17 mL were poured in the cylindrical sample holder. The sample was degassed for 3-5 min in an ultrasonic bath (model 3000515, J.P. Selecta S.A., Barcelona, Spain) just before starting each experiment.

2.2. HHP equipment and ultrasonic cell
High pressure experiments were carried out in lab-scale equipment U111 (Unipress, Warsaw, Poland). The ultrasonic cell was based on the multireflection method and employed two piezoelectric ceramics (Pz26, 10 mm diameter, 1 mm thick, main resonance frequency 2 MHz, Ferroperm, Denmark), a pulse generator (model 5072PR, Panametrics, Tech Instruments Inc., Waltham, MA, USA) and an oscilloscope (TDS 5032B digital phosphor oscilloscope 350 Hz, 5 GSample/s, Tektronix, Beaverton, OR, USA). The ultrasonic cell used electric feed-through made in the high pressure plug to connect the piezoelectric ceramics immersed in the sample with the rest of the ultrasonic measurement setup. The complete description of the experimental setup can be found in [4].

2.3. Measurement procedure
The ultrasonic cell is introduced in the stainless steel sample holder containing the liquid sample. A layer of silicone oil is added in order to isolate electric connections from the sample. The container is then screwed home on the high pressure plug. This assembly is placed on the HHP vessel and thus hermetically closed it. The vessel is immersed in the bath and let equilibrate to the target temperature. Once it is reached, the pressure is increased by 20 MPa steps up to 700 MPa. At each step, the sample is kept at least 5 min: this was the time necessary to eliminate compression heat.

2.4. Signal analysis and speed-of-sound determination
The recorded signals were analyzed using the OriginPro 8.0 package (OriginLab Corporation, Northampton, MA, USA). Each signal was imported as a text file and plotted. The Fast Fourier Transform function was applied with a band pass 0-3 MHz. The time-of-flight τ was determined as the elapsed time between the same peaks identified in two successive echoes. The speed-of-sound w was calculated as twice the distance between the two piezoelectric pieces and then divided by the time-of-flight. This distance d is fixed by the stainless steel screws that hold the structure. Under HHP, the screws are compressed and the distance is slightly reduced. Its deformation Δd is calculated assuming elastic deformation under hydrostatic conditions at each pressure as:

$$\Delta d = d_{atm} \cdot p \frac{(1-2\mu)}{E} \tag{1}$$

where d_{atm} is the distance at atmospheric pressure, p the pressure, μ Poisson coefficient (= 0.3, for stainless steel) and E Young modulus (= 193 GPa, for stainless steel).

3. Results and discussion

3.1. Cell path calibration and calculation

The time-of-flight measured in pure water is used to calibrate the distance between the piezoelectric pieces (i.e. the cell path) together with the very accurate data of speed-of-sound calculated from the equation of state of water IAPWS-95 formulation [5] in the same *p, T* conditions. Thus:

$$d_{calib} = \frac{\tau}{2} \cdot w_{calc}(p,T) \tag{2}$$

When this is done at atmospheric pressure, the relative deviations between our measurements and the accurate data calculated from IAPWS-95 formulation are as observed in figure 1(a).

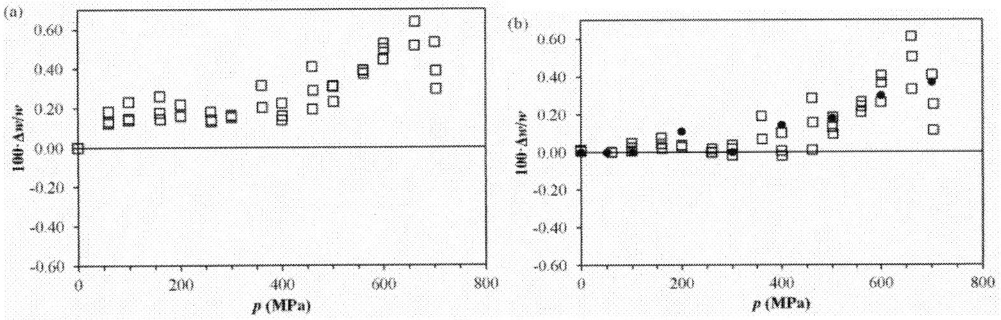

Figure 1. Fractional deviations $\Delta w = w(\exp) - w(\mathrm{calc})$ of the experimental speed of sound $w(\exp)$ from values $w(\mathrm{calc})$ calculated using the IAPWS-95 formulation, as a function of pressure at 298 K. (a) using the distance calibrated with data at atmospheric pressure, (b) using the distance calibrated with the data at 60 MPa. (□) This work, (●) Data at 294 K from Vance et al. 2010 [6].

A systematic deviation is noticed at high pressure which is likely due to some uncertainty sources in the measurement of speed-of-sound at atmospheric pressure. Among the possible sources of uncertainty, air microbubbles were designated as the responsible for this systematic deviation. Indeed, some air bubbles might remain entrapped when submerging the ultrasonic cell in the sample. When pressure is increased, air naturally dissolved in the water and speed-of-sound measurements are not disturbed anymore. Thus the calibration of the cell path has to be made at a given moderate pressure. We selected a pressure level of 60 MPa for that purpose because available data are accurate enough and because air is no longer expected to distort measurement. The distance is calculated as a function of pressure according to (equation (1)) and using the calibrated distance under HHP (equation (2)). As a result, the speed-of-sound at atmospheric pressure was preferably measured separately with the apparatus DSA5000M (Anton-Paar, Graz, Austria). Once the calibration procedure is corrected, our measurements are in excellent agreement with data calculated from IAPWS-95 formulation with a maximal fractional deviation of 0.08 % at pressure below 200 MPa where data are not extrapolated (Figure 1(b)). They also correlate well with recent experimental data at 294 K [6].

For speeds-of-sound determined at temperatures below 273 K, data at the first pressure of measurement were used for cell path calibration. At subsequent pressures, cell path was calculated as above.

3.2. Speed-of-sound in water above 200 MPa between 253 and 348 K

The speed-of-sound in water increases non-linearly with pressure as it has been already stated by other authors. Compared to the IAPWS-95 formulation, the relative maximal difference between our experimental data and the calculated values is 0.3 % while data are not extrapolated, otherwise it can reaches 1 % below 288 K and near 600 MPa. Estimated uncertainties of our measurements at a confidence level of 95 % range between (0.22 and 0.32) % over the whole pressure and temperature range studied.

3.3. Speed-of-sound in liquid foods at 298 K.

Speed of sound in pasteurized skim milk, orange juice (12°Brix) and orange juice concentrate (40°Brix) was measured at 298 K and pressures to 700 MPa. The mean values of three experiments are plotted in figure 2. Compared to water, speed-of-sound in liquid food is higher due to the presence of solutes. However, the same monotonic increasing trend with pressure is observed as for water.

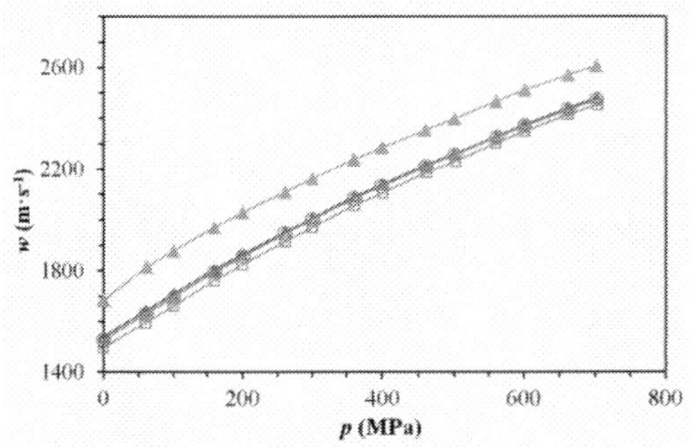

Figure 2. Speed of sound in liquid foods at 298 K as a function of pressure. (□) Water, (○) Pasteurized skim milk, (Δ) Orange juice, (▲) Orange juice concentrated. Lines are only guides to the eyes. Standard deviations between experiments are on the order of line thickness.

4. Conclusion

The developed ultrasonic sensor for HHP process control was shown to give speed-of-sound values in water with a low uncertainty (below 0.32 %). Data are now known for some liquid foods at 298 K up to 700 MPa. Although sensor implementation is challenging at industrial scale, technology progresses are expected which will allow for finding solutions to the current technical difficulties encountered with large HHP vessels. This work provides an interesting basis to develop such high pressure ultrasonic sensors.

Acknowledgments

This work has been supported by MICINN through the projects CSD2007-00045 and CTQ2009-14596-C02-01 and by Comunidad de Madrid through the project QUIMAPRES S2009/PPQ-1551. E.H.B. acknowledges a FPI grant from Spanish MICINN.

References

[1] Bermúdez-Aguirre D and Barbosa-Cánovas G V 2011 An update on high hydrostatic pressure, from the laboratory to industrial applications *Food Eng. Rev.* **3** 44-61
[2] Rastogi N K, Raghavarao K S M S, Balasubramaniam V M, Niranjan K and Knorr D 2007 Opportunities and challenges in high pressure processing of foods *Crit. Rev. Food Sci. Nutr.* **47** 69–112
[3] Povey M J W 1989 Ultrasonics in food engineering. Part II: Applications *J. Food Eng.* **9** 1-20.
[4] Hidalgo Baltasar E, Taravillo M, Baonza V G, Sanz P D and Guignon B 2011 Speed of Sound in Liquid Water from (253.15 to 348.15) K and Pressures from (0.1 to 700) MPa *J. Chem. Eng. Data* **56** 4800–4807
[5] Wagner W and Pruss A 2002 The IAPWS Formulation 1995 for the thermodynamic properties of ordinary water substance for general and scientific use *J. Phys. Chem. Ref. Data* **31** 387-535
[6] Vance S and Brown J M 2010 Sound velocities and thermodynamic properties of water to 700 MPa and -10 to 100°C *J. Acoust. Soc. Am.* **127** 174-180

Ultrasonic preliminary measurements of oenological malolactic fermentation parameters in red wine

DF Novoa-Díaz[1], A Puig-Pujol[2], J García-Álvarez[1], J A Chávez[1], A Turó[1], S Mínguez[2], MJ García-Hernández[1], E Bertran[2] and J Salazar[1]

[1] Grup Sistemes Sensors. Departament Enginyeria Electrònica. Universitat Politècnica de Catalunya, Jordi Girona 1-3, 08034 Barcelona, Spain.
[2] Secció d'Investigació Enològica. Estació de Viticultura i Enologia. Institut Català de la Vinya i el Vi (INCAVI), Plaça Àgora, 2. Polígon Industrial Domenys II. 08720 Vilafranca del Penedès, Spain.

E-mail: danielfnovoa@yahoo.com

Abstract. In the winemaking process, the malolactic fermentation is an essential process in the production of high quality red wines which concerns the conversion of malate into lactate. In this work, the ultrasonic velocity through wine samples with different concentrations of malate and lactate was measured using the pulse echo technique with 1 MHz tone burst signals. The evolution of these concentrations during malolactic fermentation was taken into account in order to determine the ratio between concentrations of malate and lactate of the different samples. These preliminary results have revealed that the ultrasonic velocity increases during the conversion of malate to lactate. In addition, measurements have been conducted to quantify the influence of variations in turbidity and temperature on test samples. Therefore, these results show the possibility of using ultrasonic velocity measurements for on-line monitoring the malolactic fermentation of red wine and may help to improve and contribute to the development of the winemaking process.

1. Introduction

There are two fermentation processes that occur during the production of most red wines. The first is known as alcoholic fermentation and is achieved through the use of yeast [1,2,3,4]. The second is known as malolactic fermentation (MLF) and occurs when lactic bacteria transform malic acid into lactic acid, giving rise to changes in a number of wine's properties – including those of an acoustic nature [5,6]. MLF is often unpredictable and sometimes does not occur owing to a variety of factors hindering bacterial growth. In light of this latter process, this study intends to establish a sound basic knowledge on which to facilitate the introduction of ultrasound technology into the monitoring and subsequent control of MLF in red wine.

To achieve this objective, the study is divided into two stages: the first is centred on an initial study consisting of laboratory tests in which the influence of ultrasound on temperature, turbidity and concentrations of malic and lactic acid were analysed; and the second involves applying the technology in a semi-industrial environment where MLF in three fermentation tanks is monitored. In this instance, the results consisted of the information obtained by measuring the velocity of ultrasonic waves travelling through the wine and the corresponding comparison and correlation of the data against those of the conventional measurements (typically enzymatic methods) used by winemakers to control MLF.

2. Materials and methods

2.1. Wine samples

To conduct the different laboratory tests we used a hydroalcoholic solution (12% alcohol) for temperature test, a base wine for turbidity test and finally a batch of base wines (of the Merlot variety), along with DL-Malic Acid (Panreac Química S.A.U., Barcelona, Spain, 99-100%) and L(+)-Lactic Acid (Panreac, 85%) we prepared a series of samples to analyse. The second stage was carried out using wine at the MLF stage in three stainless-steel tanks, two of which contained wine of the Tempranillo variety and the third wine, Merlot variety.

2.2. Technique used to measure ultrasonic velocity

Ultrasonic measurements were based on the pulse-echo technique, in combination with the configuration shown in figure 1(a). The system was composed of an ultrasonic transducer (emitter-receiver) attached to a buffer rod (a cylinder of neutral plastic), the other end of which is placed in contact with the wine sample, while an acoustic reflector is placed at the bottom of the container at a fixed distance, d. Two echoes are thus obtained, E1 and E2 (figure 1(b)), where the time interval between both echoes corresponds to the time taken for the ultrasonic wave to travel through the wine sample (Time of Flight, TOF), and from this data it is then possible to calculate ultrasonic velocity through wine ($v_w = 2d/TOF$).

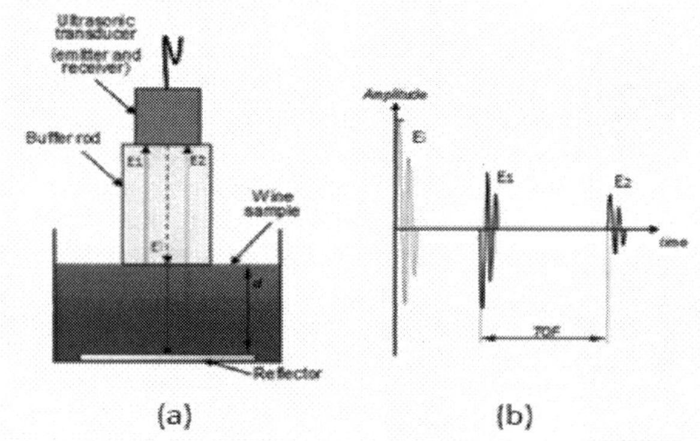

Figure 1. (a) Configuration used to measure ultrasonic velocity and (b) Example of transmitted (E_i) and received (E_1 and E_2) signals by the ultrasonic transducer.

The measurement system was complemented with a signal generator (Agilent 33522A) that supplied the excitation waves to the transducer (a ten-cycle burst of 1 MHz frequency and 10V amplitude sine oscillation) and a digital oscilloscope (Tektronix TDS2024B) to acquire and display the signals. A piezoelectric transducer with a resonance frequency of 1 MHz (Panametrics NTD, Japan) was used. Finally, an application created using MATLAB was employed to process the information gathered and obtain the data of interest to this study.

3. Results

As figure 2 shows, in a wide range of temperatures within which MLF fermentation tends to occur (22°C) in red wine, ultrasonic velocity displays an increasingly linear behaviour that is strongly dependent on temperature variations. Thus, during experiments, it will be absolutely necessary to compensate the relative increase or decrease in velocity due to temperature variations since temperature cannot be rigorously kept constant, as tends to be the case in wineries.

Results shown in figure 3 clearly demonstrate the linear reduction in ultrasonic velocity when there are major variations in turbidity (several hundred NTUs). Consequently, for processes involving substantial changes in turbidity – such as alcoholic fermentation – it is important to take into account the effects of these changes on the acoustic properties of the wine. However, as turbidity varies very little in MLF, the influence of this parameter on the ultrasonic measurements obtained can be considered negligible for most of the time.

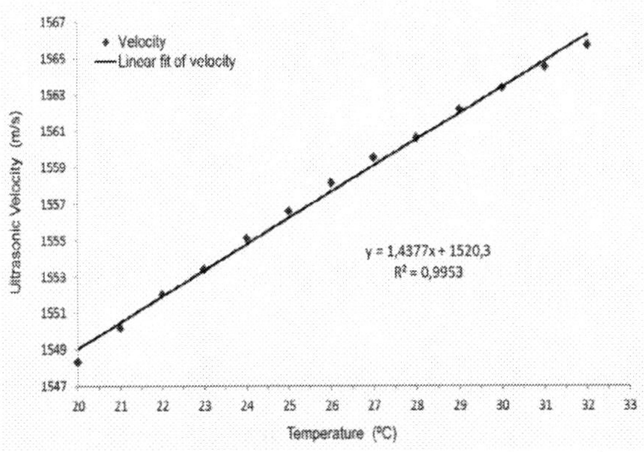

Figure 2. Ultrasonic velocity with wine temperature.

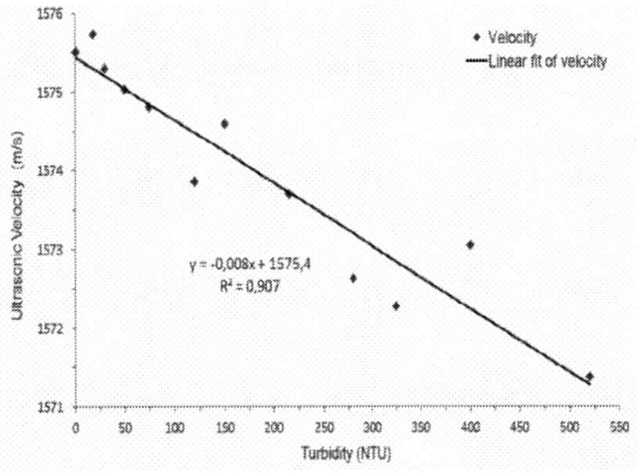

Figure 3. Ultrasonic velocity with wine turbidity.

Working at a controlled temperature of 22ºC ±0.1ºC, figure 4 shows a relative decrease rate of -0.5588 m/s per g/L in the velocity of ultrasound waves when only the concentration of malic acid increases and a relative decrease rate of -0.2423 m/s per g/L if only the concentration of lactic acid increases.

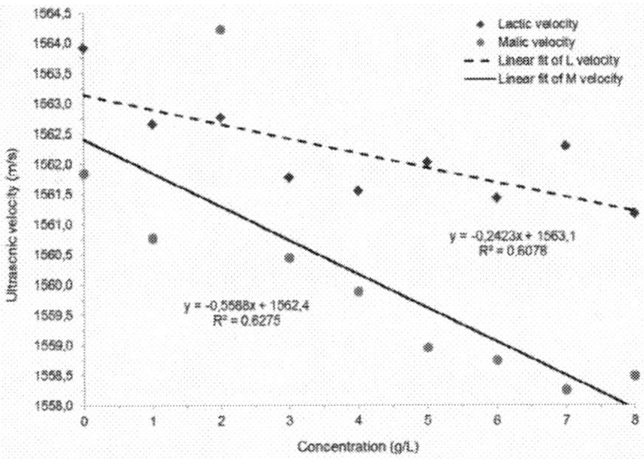

Figure 4. Ultrasonic velocity with malic and lactic acids concentrations.

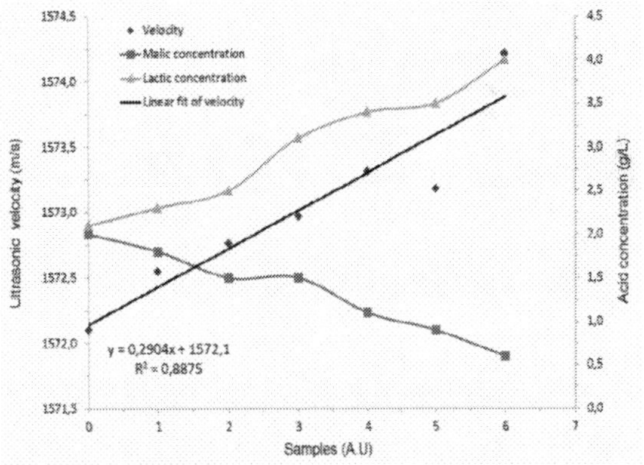

Figure 5. FML monitoring with traditional and ultrasonic methods.

Given that the malic acid concentration decreases while the lactic acid concentration increases during MLF, and taking into account the results shown in figure 4, when using ultrasonic technology it is expected that monitoring will reveal an approximate relative increase rate of 0.3165 m/s per g/L transformed (0.5588-0.2423) in ultrasonic velocity through wine while MLF progresses. This concurs well with the relative increase rate of 0.2904 shown in figure 5, which correspond to the monitoring of MLF in one of the three aforementioned fermentation tanks. This means that during a MLF the ultrasonic velocity can increase in a range of 2 to 4 m/s.

Finally, table 1 shows the calculation of Pearson's Correlation Coefficient and the P value from the conventional variables and the ultrasonic velocity recorded during the monitoring of MLF in the three fermentation tanks over 18 consecutive days. As can be observed, the correlation notably improves when the corresponding adjustment for temperature is made to the measured ultrasonic velocity; this ensures that parameters such as the concentrations of malic and lactic acid and the CFU logarithm have a high level of correlation with the ultrasonic measurements.

Table 1. Correlations among malic and lactic acid concentrations, CFU logarithm and ultrasonic velocity with (v_cT) and without (v) temperature compensation.

Tanks and wine variety	Ultrasonic velocity (m/s)	Malic (g/L)	Lactic (g/l)	Log(UFC/ml)
Tank 1 Tempranillo	v	-0,622 (0,073)	0,596 (0,090)	0,533 (0,140)
	v_cT	-0,964 (0,000)	0,966 (0,000)	0,872 (0,002)
Tank 2 Merlot	v	-0,262 (0,496)	0,232 (0,548)	0,262 (0,496)
	v_cT	-0,903 (0,001)	0,953 (0,000)	0,805 (0,009)
Tank 3 Tempranillo	v	-0,478 (0,193)	0,243 (0,529)	0,447 (0,228)
	v_cT	-0,926 (0,000)	0,897 (0,001)	0,917 (0,001)

4. Conclusions

Preliminary ultrasonic measurements have been performed in a batch of several red wine samples with different concentrations of malate and lactate and the MLF of the red wine has been monitored. These preliminary results have revealed that the ultrasonic velocity increases during the conversion of malate to lactate and that appropriate temperature compensation is necessary for best results. Therefore, these results show the possibility of using ultrasonic velocity measurements for on-line monitoring the MLF of red wine and may help to improve and contribute to the development of the winemaking process.

Acknowledgements

This work was supported by the Spanish Research Support Scheme of the Ministerio de Economía y Competitividad, project reference DPI2009-14468-C02-01.

References

[1] Lamberti N, Ardia L, Albanese D and Di Matteo M 2009 *Ultrasonics* **49** 94-97
[2] Resa P, Elvira L, Montero de Espinosa F, González R and Barcenilla J 2009 *Bioproc. Biosys. Eng.* **32** 321-31
[3] Resa P, Elvira L and Montero de Espinosa F 2004 *Food Res. Internal.* **37** 587-94
[4] Vatandas M, Koc A B, and Koc C 2007 *Europ. Food Res and Tech.* **225** 525-32
[5] Resa P, Bolumar T, Elvira L, Pérez G and Montero de Espinosa F 2007 *J. Food Eng.* **78** 1083-91
[6] Resa P, Elvira L, Montero de Espinosa F, Bolumar T and Pérez G 2004 *IEEE Internal. Ultras. Ferro. and Freq.* 2322-25

IOP Conf. Series: Materials Science and Engineering **42** (2012) 012046 doi:10.1088/1757-899X/42/1/012046

Apparent negative mass in QCM sensors due to punctual rigid loading

P Castro, P Resa and L Elvira

Centro de Acústica Aplicada y Evaluación No Destructiva (CAEND), UPM-CSIC
C/Serrano 144, 28006 Madrid, Spain
Phone number: (+34) 915 61 88 06 (259)
Fax number: (+34) 914 11 76 51

E-mail: pedro.castro@caend.upm-csic.es

Abstract. Quartz Crystal Microbalances (QCM) are highly sensitive piezoelectric sensors able to detect very small loads attached to them. These devices are widely employed in many applications including process control and industrial and environmental monitoring. Mass loading is usually related to frequency shift by the well-known Sauerbrey's equation, valid for thin rigid homogeneous films. However, a significant deviation from this equation can occur when the mass is not uniformly distributed over the surface. Whereas the effects of a thin film on a QCM have been thoroughly studied, there are relatively few results on punctual loads, even though particles are usually deposited randomly and non-uniformly on the resonator surface. In this work, we have studied the effect of punctual rigid loading on the resonant frequency shift of a QCM sensor, both experimentally and using finite element method (FEM). The FEM numerical analysis was done using COMSOL software, 3D modeling a linear elastic piezoelectric solid and introducing the properties of an AT-cut quartz crystal. It is shown that a punctual rigid mass deposition on the surface of a QCM sensor can lead to positive shifts of resonance frequency, contrary to Sauerbrey's equation.

1. Introduction

Quartz Crystal Microbalances (QCM) are extremely sensitive mass devices (< 10 ng cm^{-2}), the measuring principle of which is based on the relationship between the attached mass and the resonant parameters (basically, resonant frequency and bandwidth). This type of sensors consists of a thin piezoelectric plate with metal electrodes deposited on each side. Due to the piezoelectric effect, when an AC electrical field is applied perpendicular to the plate, a shear wave is generated that propagates through the crystal inducing its oscillation. QCM are commonly made up of AT-cut quartz crystals, which provide high quality factors and very good stability near room temperature [1]. Although the effects of a thin uniform film on a QCM have been thoroughly studied, there is relatively little understanding of the effects of discrete particles on the surface of a QCM sensor [2], even though particles can be deposited randomly and non-uniformly in many cases.

Assuming an homogeneous film rigidly attached to the QCM surface, the Sauerbrey expression [3] is commonly used to relate mass loading and resonance frequency shift, Δf :

$$\Delta f = -\frac{f_0 \rho_f h_f}{\rho_q h_q} \qquad (1)$$

where f_0 is the resonant frequency, h_q is the thickness of the quartz crystal, h_f is the thickness of the attached uniform film, and ρ_q and ρ_f are the quartz and film density, respectively. Therefore, according to equation (1), it can be seen that uniform mass deposition on the surface of a QCM sensor leads to a negative shift of the resonance frequency.

However, Dultsev and Kolosovsky [4] analyzed theoretically the effect of a point load on the surface of a QCM resonator with a simple one-dimensional model based on longitudinal oscillatory motions in thin plates. It was mathematically demonstrated that, in contrast to the case of uniform mass loading, a point load on a membrane causes a positive frequency shift, Δf :

$$\frac{\Delta f}{f} = +\frac{m}{M}\cdot\sin^2\left(\pi x_0 / L\right) \quad (2)$$

where m is a single nano-size load, M, the mass of the working strip of the resonator, x_0, the position of the load, and L, the length of the resonator.

In the present work, the effect of punctual rigid loading on a QCM sensor has been studied both numerically, using Finite Element Modeling (FEM) analysis, and experimentally, with a fiberglass needle. This constitutes a first step to understand the effect of non uniform mass distributions upon the resonance characteristics of QCM sensors.

2. Methodology

A 3D FEM model (COMSOL Multiphysics) has been employed to evaluate the influence of punctual rigid loading on the resonant frequency of an electrode circular plate (diameter, 5.1054 mm), vibrating in the fundamental thickness-shear mode at 8 MHz (thickness, 0.208 mm), which takes into account both the crystal anisotropy and the piezoelectric effect. First, the matrices of natural quartz are set up and then, the coordinate system is rotated about the y axis by 35.25° to obtain the AT-cut [5]. For the mesh of the model, a total of 91.684 elements were used, resulting in 515.620 degrees of freedom. The force, F_i, of the particle at a point was defined by an opposing inertial force [6]:

$$F_i = -m \cdot u_i \cdot \omega^2 \quad (3)$$

where m, u_i and ω are the mass of the particle, the displacement and the radial frequency, respectively. As a result, a force is applied in all directions as a load boundary condition. In this way, the effects of the mass position and changes in mass are studied and the thickness of the attached mass is neglected.

On the other hand, the experimental frequency shifts of a QCM sensor under punctual loading have been measured using an Agilent 4294A Precision Impedance Analyzer, ranging from 40 Hz to 110 MHz. 8 MHz AT-cut quartz crystal (diameter, 13.665 mm), coated with optically flat polished gold electrodes on both sides (sensing gold electrode area, 0.2 cm^2), were purchased from International Crystal Manufacturing Co, Inc (ICM, Oklahoma, USA). The crystals were cleaned before use by rinsing with absolute ethanol, then rinsed with distilled water and dried in air. To simulate a "point of mass" on the QCM surface, the resonator was pressed with a fiberglass needle with a diameter less than 0.12 mm (so as not to damage the surface of the sensor). In brief, a "punctual load" in air at 25 °C has been simulated. For this purpose, a system was designed to achieve a mass point capable of moving over the sensor surface. The strategy was similar to other references [7], but with a needle of fiberglass instead of elastic small spheres.

3. Results and discussion

Firstly, the effect of punctual loading on the resonator has been simulated using 3D FEM analysis. The resonant frequency shift as a function of the position of the load is depicted in figure 1. The distance from the center of the disc varies from 0 mm (center) to 2.553 mm (edge), and the resonant frequency shift was calculated at these positions always for the same weight (1000 ng). As expected, the sensor is more sensitive when the punctual rigid load is near to central positions rather than located closer to the edge. Frequency shifts are always positive ("apparent" negative mass).

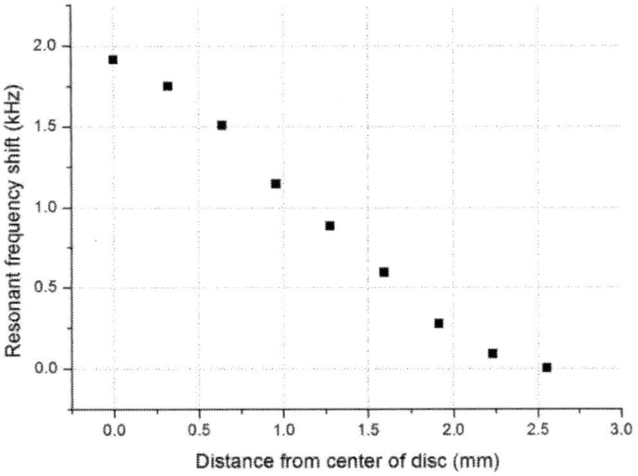

Figure 1. Dependence of the resonant frequency shift with the position of the punctual load (x axis).

Changes in the resonant frequency shift as a function of the weight are shown in figure 2. It can be observed that these frequency shifts are always positive for every mass. In this case, the mass is located at the point (x=0.1 mm; y=0.1 mm), but a similar behavior was obtained at other positions (taking into account the dependence of sensitivity with position). Above a mass value of about 0.1 mg, the resonant frequency shift becomes stable.

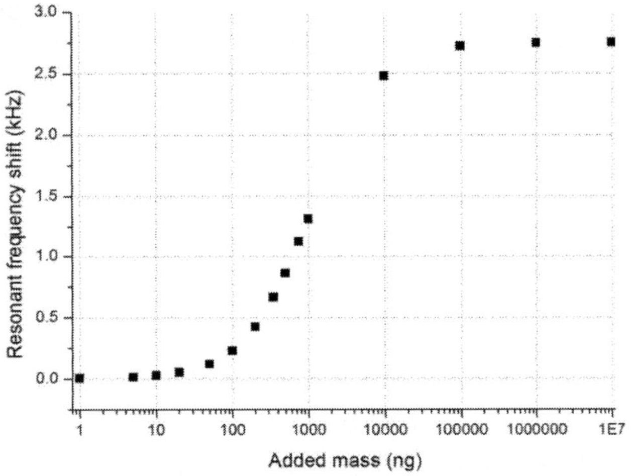

Figure 2. Resonant frequency shift depending on the weight of a discrete particle at a specific position (x=0.1 mm; y=0.1 mm).

In figure 3, the experimental results obtained are presented. In agreement with the FEM method, positive resonant frequency shifts can be observed. In particular, a positive shift of about +250 Hz is obtained when a punctual load is applied close to the center of the resonator surface.

These "positive" (versus "negative") resonant frequency shifts are in agreement with equation (2) and have also been reported in previous works that assume mechanical models [4] or consider elastic attached particles [2].

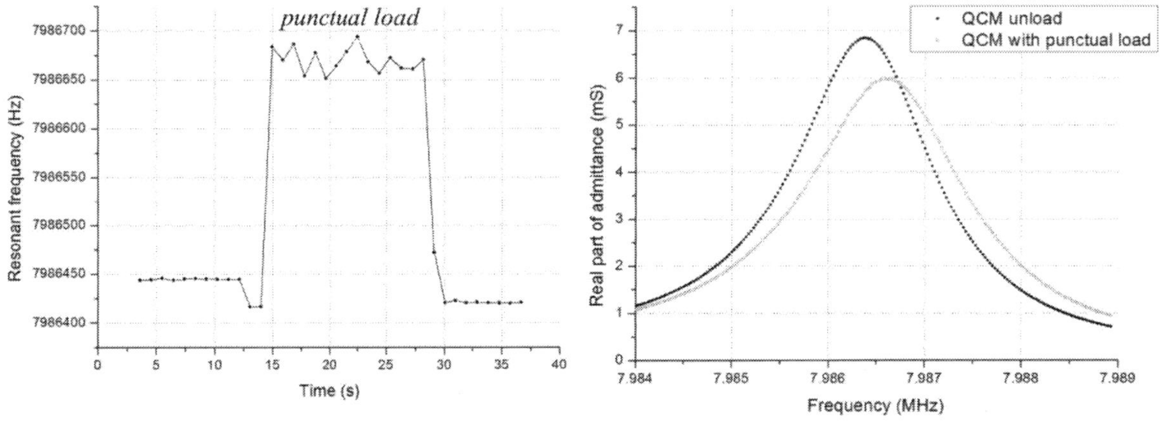

Figure 3. Experimental resonant frequency shift under punctual loading (pressed with a fiberglass needle). On the left: time-course plot of the resonant frequency. On the right: resonant frequency with and without loading.

4. Conclusions

In this paper, the effect of a punctual rigid mass on the surface of a QCM sensor has been studied using both 3D FEM analysis and experimental measurements. While the Sauerbrey equation predicts a decrease in the resonant frequency for thin uniform rigid films, we have shown that the resonant frequency increases (apparent negative mass) when a punctual rigid load is applied.

These results are in agreement with previous theoretical approaches and can significantly contribute to the interpretation of QCM data in some industrial and natural processes, where the mass distributions are highly non-uniform.

Acknowledgements

This work has been supported by the Spanish Ministry of Science and Innovation (CICYT DPI2010-17716), a CSIC intramural project (ref. 201150E045) and the JAE-CSIC Program/European Social Fund (JAEPREDOC2011-058 & JAEDOC2008-065).

References

[1] Bechmann R, Ballato A D and Lukaszek T J 1962 Higher-Order Temperature Coefficients of the Elastic Stiffinesses and Compliances of Alpha-Quartz *Proceedings of the IRE* **50**(8) 1812-22
[2] Zhang C, Liu N, Yang J and Chen W 2011 Thickness-shear vibration of AT-cut quartz plates carrying finite-size particles with rotational degree of freedom and rotatory inertia [Correspondence] *IEEE Trans. Ultrason. Ferroelectr. Freq. Control* **58**(3) 666-670
[3] Sauerbrey G 1959 Use of quartz crystal vibrator for weighting thin films on a microbalance *Z Phys.* **155** 206-222
[4] Dultsev F N and Kolosovsky E A 2009 Identifying a single biological nano-sized particle using a quartz crystal microbalance A mathematical model *Sensors Actuators B: Chem.* **143** 17-24
[5] ANSI/IEEE Standard 176-1987 1988 *IEEE Standard on piezoelectricity* (New York: The Institute of Electrical and Electronics Engineers)
[6] Puccio D and EerNisse E P 2008 Finite element analysis of thickness shear crystals with surface irregularities *2008 IEEE International Frequency Control Symp.* (Honolulu, 19-21 May 2008) (IEE Conference Publications) pp 25-30
[7] Berg S, Johannsmann D and Ruths M 2002 Frequency response of quartz crystal shear-resonator during an adhesive, elastic contact in a surface forces apparatus *Journal of Applied Physics* **92** 6905-10

IOP Conf. Series: Materials Science and Engineering **42** (2012) 012047 doi:10.1088/1757-899X/42/1/012047

Excitation signal's influence on ultrasonic transit time flow meter's performance

L Svilainis, P Kabisius, A Aleksandrovas, and A Chaziachmetovas

Kaunas University of Technology, Department of Signal Processing, Studentu 50-340, LT-51368, Kaunas, Lithuania

E-mail: linas.svilainis@ktu.lt

Abstract. Ultrasonic flow meter performance was analyzed. Ultrasound transit time was used for flow rate estimation. Time of flight was measured using cross correlation processing. Simultaneous channels excitation was used. Ultrasonic signals were excited using low voltage (4V pp) signal generator, received signals were amplified 30 dB and simultaneously acquired by 100 Ms/s 10 bit analog-to-digit converters. Subsample delay estimation was used. Flow rate was varied from 10 l/h to 200 l/h. Measurement channel diameter 8 mm was used. It is complicated to obtain the unbiased reference signal for correlation processing. Various combinations of signals travelling in measurement channel were used for cross-correlation processing. Performance of correlation function and time of flight estimator variance were studied. Variable gain amplifier usually is used for signal dynamic range matching to A/D converter input. Gain influence on time of flight was a subject to study. It has been concluded that gain control introduces systematic errors in time of flight estimator. Influence of the temperature of electronics (pulser, receiver, A/D converter, reference clock etc.) and ultrasonic transducers on the delay estimator was studied. It was concluded that the major temperature-related systematic error comes from the pulser. Performance of the meter was studied when narrowband and spread spectrum signals were used for ultrasound excitation across temperature and flow rate range. It has been concluded that spread spectrum signal allows for better zero flow stability over temperature and lower time of flight variation.

1. Introduction

With the rapid development of transport fleet management systems [1] number of parameters acquired is increasing. Such systems find application in sea and surface logistics, solid and liquid waste collection process management and other areas. New generation of the systems is not just transmitting the position information over communication networks but also relays real time vehicle parameters, including temperature, engine mode and fuel consumption [2]. It is desirable to have the ability to monitor the real time fuel consumption using flow meter. Volume displacement meters are large and do not cover the expected flow range. Differential pressure meters are not suitable due to pressure loss, poor accuracy, low turndown ratio and risk of clogging [3]. Thermal mass flow meters [4] fail due to inertness of temperature sensors and thermal propagation speeds in liquids. Coriolis meters [5] measure mass flow directly and can provide high accuracy, but are subject to vibrations. Then, the ultrasonic flow meter [6] seems the best solution. Ultrasonic transit time flow meter is under development for such purpose. Prototype measurement chamber with channel diameter 8 mm was developed. Expected flow rate is 10 l/h to 200 l/h, which corresponds to 0.05…1.1 m/s. It is quite

IOP Conf. Series: Materials Science and Engineering **42** (2012) 012047 doi:10.1088/1757-899X/42/1/012047

small value in comparison to 1500 m/s ultrasound velocity. This puts strict limitation on time of flight (ToF) measurement accuracy. Investigation of the time of flight variation caused by electronics temperature and excitation signal type is presented.

2. Transit time flow measurement system

We consider the ultrasonic pulse transit time technique: it has been well studied, allow for full duplex of measurement channel with least complications. Measurement chamber is arranged in such way that ultrasonic signal path is aligned with flow (figure 1). Then ultrasonic signal sent against the flow will take longer time ToF_{up} than the signal send down the flow [6].

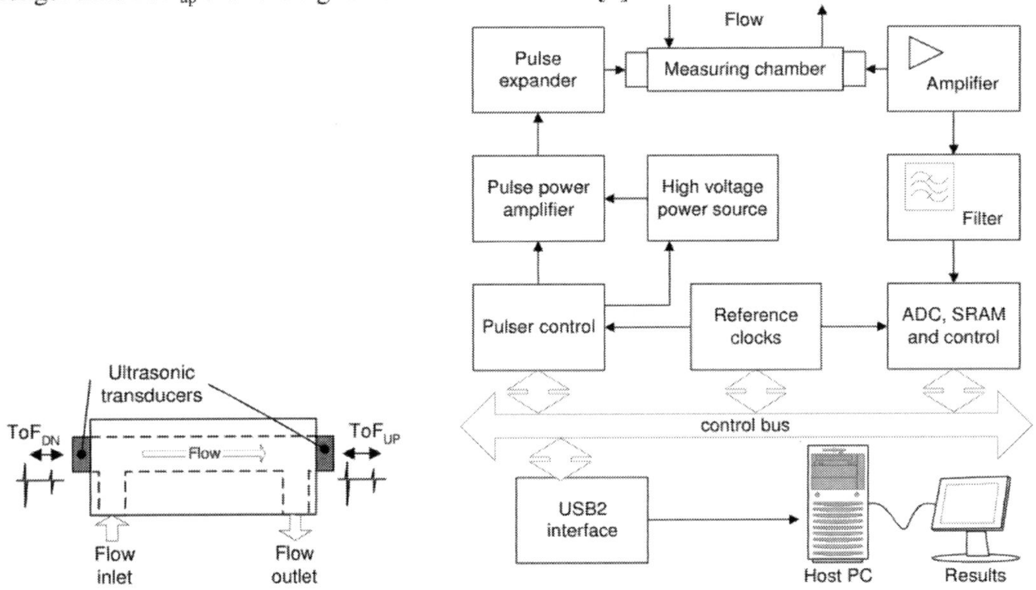

Figure 1. Measurement chamber (left) and electronics structure (right) used in experiment.

The experimental system structure is presented in figure 1 right. It contains pulser control logic (arbitrary pulse train with sampling rate 100 MHz can be uploaded), pulser output stage powered by programmable high (0 V…300 V) or low (0.1 V to 5 V) voltage power supply, pulse expander, low noise preamplifier with signal protection (+/- 0.5 V) and filters (0.5 MHz…30 MHz 3-rd order Butterworth), dual channel, 100 MHz sampling frequency 10 bit flash ADC with buffer memory and glue logic. Control and acquisition is accomplished by host PC using high speed USB2 interface.

3. Experimental results

Three types of investigation were carried out: i) electronics temperature influence; ii) signal type influence of ToF measurement. ToF was used as performance measure. The correlation peak position was used as the ToF estimator [7]. Parabolic interpolation was used [8] for the ToF subsample estimation.

3.1. Electronics influence

To evaluate the errors induced by the signal acquisition electronics the dry channel experiment has been carried out. Transducers were replaced by direct connection via 20 dB, 30 dB or 40 dB attenuator. Signals recorded were filtered using matched filter: first signal from the record was stored as reference and the rest used for cross correlation processing.

Variable gain amplifier AD8331 was used to matching the signal dynamic range to A/D input. Gain variation influence on ToF was studied. Investigation revealed that gain control introduces systematic errors in ToF estimator: 500 ps bias was noted when gain varied 10 dB. Gain control was excluded from the rest of experiments.

Pulser logic stages temperature influence has been studied by connecting the attenuated logic output directly to A/D input. Thanks to low voltages used, slew rate variance on ToF was insignificant: 2.5 ps/°C for bias and 6 ps (temperature invariant) for random errors.

Pulser output stages temperature influence study has been carried out by feeding the 40 dB attenuated high voltage output to A/D converter. Here, pulser electronics temperature influences semiconductor switching speed: in case of high voltage output slew rate is high and this gives significant ToF variation: 500 ps/°C for bias and 20 ps/°C for random errors.

Synchronization clock temperature influence was studied by measuring the oscillator frequency drift with temperature. Then drift was converted into time scale distortion: 0.3 ppm/°C drift at 100 mm path (70-100 μs) corresponds to 30 ps/°C ToF bias.

Receiver stages temperature influence is mainly through phase AC response variation and corresponding group delay: 15 ps/°C for bias and 15 ps (temperature invariant) for random errors.

3.2. Signal influence on ToF variation

Channel transmission experiments were carried out using cylindrical 8 mm diameter channel (figure 1, left). Transmission was measured using 0.1…20 MHz (chirp) probing signal (figure 2 left). Investigation indicated that center frequency is around 5 MHz with potential operation region of 3 MHz to 6.3 MHz was estimated. Further experiments were carried out to investigate the signal type influence on ToF. Four types of signals were selected: CW burst, chirp, pulse and step. Two types of the signals had high energy content thanks to their long duration: 30 μs duration spread spectrum (chirp) and CW burst signals. Despite similar energy, their spectral content differs: CW burst is narrowband while chirp bandwidth is large and can be regulated. Chirp signals were selected to cover main operation region 3 MHz to 6.3 MHz and significantly exceeding the transducer passband: 0.5..10 MHz. CW burst signals center frequencies within main operation bandwidth 3 MHz to 6.3 MHz were chosen. Step and pulse signals represent low energy signals due to their short duration. Nine 0.8 Vp-p signals were chosen for excitation: 1) chirp with 0.5…10 MHz bandwidth; 2) chirp 3..6.3 MHz; 3) CW burst 3 MHz; 4) CW burst 5 MHz; 5) CW burst 6.3 MHz; 6) pulse matched for 3 MHz; 7) pulse matched for 5 MHz; 8) pulse matched for 6.3 MHz; 9) positive step with 3ns rise time. Signal spectral content registered at transducer contacts is presented in figure 2 right.

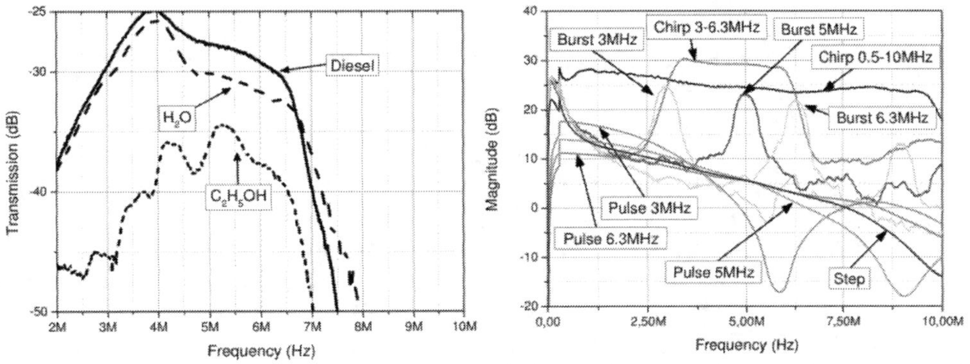

Figure 2. Transmission in measurement chamber (left) and investigation signals' spectrum (right).

Temperature effect interference with signal type was investigated for transducers placed in direct contact. In such way temperature effect of the liquid is neglected. When measurements were started, at measurement number 100 heating of transducers was switched on and maintained until the end of the experiment. Temperature rose from 20 °C at the start of the experiment to 44 °C at the end of the experiment. All nine signals were used in sequential estimation of the ToF over transducers pair, then pause of 1s was introduced. Obtained results are presented in figure 3, separately for high energy signals (left) and low energy signals (right).

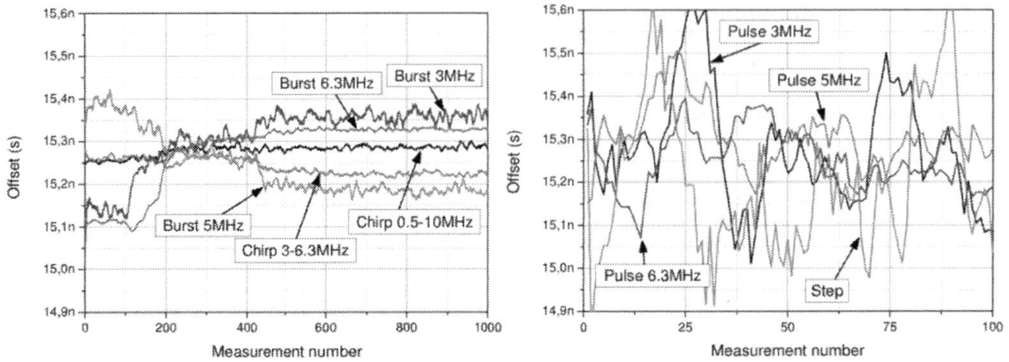

Figure 3. Signal type influence on ToF offset: high energy (left) and low energy (right) signals.

Investigation indicates that best performance is obtained for spread spectrum (chirp) signals: trend of only 50 ps over 20 °C temperature range is observed. Interesting to note that chirp covering low frequency range beyond transducer passband (0.5...10 MHz bandwidth) had opposite polarity trend from chirp operating in main passband. Both random errors and trend was significantly larger for step and pulse signals. This can be explained by stronger dependence of the transmission spectrum on transducer properties.

4. Conclusions

Gain control introduces systematic errors in time of flight estimator. Major temperature-related systematic error comes from the pulser output stages. Frequency range of two transducers' mutual transmission is varying with temperature. Thanks to the presence of wideband signal this effect is minimized. Spread spectrum signals allows for better bias stability over temperature and lower time of flight variation. CW burst performance is comparable to chirp though trend is 3 to 5 times higher.

Acknowledgments

This research was funded by a grant (No. MIP-058/2012) from the Research Council of Lithuania.

References

[1] Gogate C A, Sagaidev B S and Vaidyanathan S G 2004 Monitoring and transmission of heavy vehicle parameters using fixed cellular terminal *Proc. IEEE VTC* (Los Angeles, 26-29 September 2004) vol 6 pp 4100 – 02

[2] Rakha S P, et al. 2010 Development of fuel and emission models for high speed heavy duty trucks, light duty trucks, and light duty vehicles *Proc. IEEE ITSC* (Madeira, 19-22 September 2010) pp 25 – 32

[3] Crabtree M.A 2009 *Industrial Flow Measurement* Masters Thesis (University of Huddersfield)

[4] Terao M, Akutsu T and Tanaka Y 2007 Non-wetted thermal micro flow sensor *Proc. SICE* (Kagawa, Japan, 17-20 September 2007) pp 2084-88

[5] Röck H and Koschmieder F 2008 A new method to detect zero drift and sensitivity of a Coriolis Mass Flow Meter (CMFM) by using phasor control *Proc. ICST* (Taiwan, 30 November-3 December 2008) pp 518-522

[6] Som J N 2009 Research and development in ultrasonic flow measurement *J. Pure and App. Ultrasonics* **31**(3) 115-123

[7] Svilainis L, Dumbrava V, Kitov S and Chaziachmetovas A 2011 The Influence of Digital Domain on Time of Flight Estimation Performance *Proc. Int. Congress on Ultrasonics* (Gdańsk, 5-8 September 2011) (AIP Conf. Proc.) vol 1433 pp 479-482

[8] Svilainis L and Dumbrava V 2008 Analysis of the interpolation techniques for time-of-flight estimation *Ultragarsas* **63** 25-29

Using ultrasonic Doppler velocimetry to investigate the mixing of non-Newtonian fluids

Dineshkumar Patel, Farhad Ein-Mozaffari and Mehrab Mehrvar
Department of Chemical Engineering, Ryerson University
350 Victoria Street, Toronto, Ontario, M5B 2K3, Canada

E-mail: fmozaffa@ryerson.ca

Abstract. Mixing is a critical unit operation, which is widely used in chemical and allied industries. Mixing of non-Newtonian fluids is a challenging task due to the complex rheology exhibited by these fluids. Pseudoplastic fluids with yield stress are an important class of non-Newtonian fluids. In this study, we utilized ultrasonic Doppler velocimetry (UDV) to explore the flow field generated by different impellers in the agitation of xanthan gum solutions and pulp suspensions, which are yield-pseudoplastic fluids.

1. Introduction

Ultrasonic Doppler velocimetry (UDV) is a non-invasive and non-intrusive method and it can be used to measure fluid velocity without obstructing the fluid motion. UDV is appropriate for use in both opaque and transparent systems. These advantages make UDV applicable for studies of all liquids, emulsions, and slurries and make this technique very attractive from an industrial perspective [1]. It measures velocity filed as a function of time and space and is an efficient flow mapping process.

Due to its vast applications, mixing plays a vital role in various chemical and allied industries. Mixing of non-Newtonian fluids (pseudoplastic fluids possessing yield stress) is one of the most challenging tasks. Mixing of yield stress fluids such as pulp suspensions and xanthan gum solutions results in the formation of a zone of intense motion (cavern) in the impeller region surrounded by stagnant or slow moving fluids in the mixing vessel. The presence of such stagnant regions in the mixing vessel rises to low mass and heat transfer rates, which eventually implicates poor product quality and reduces productivity. Therefore, it is necessary to eliminate such stagnant regions by properly designing and operating mixing systems. By knowing the fluid velocity in the mixing vessel one can determine whether stagnant region is present or not. Thus, the aim of this paper is to utilize UDV technique to explore the flow field generated by different impellers in the agitation of yield-pseudoplastic fluids and improve the mixing performance. The principal and methodology of UDV technique are explained in the next section.

2. Ultrasonic Doppler Velocimetry Method

The UDV instrument uses a single probe as both a transmitter and a receiver. An ultrasonic pulse is emitted by probe at a set frequency. This pulse travels through the liquid, and is reflected back towards the probe by solid particles in the suspension or solution. Thus, same probe receives the echo reflected from the particles. The time delay (T_d) between the emitted pulse and received echo can be used to calculate the position of the particle, λ, ($\lambda = cT_d/2$, where c is the speed of sound in the suspension) along the measurement axis [2].

The pulse repetition frequency (PRF) is the frequency of emitted bursts from the probe. The maximum depth, λ_{max}, ($\lambda_{max} = cT_{prf}/2$, where T_{prf} is the PRF period) is a function of the PRF [3]. The velocity can be determined using the following equation:

$$v = \frac{c\Psi}{4\pi\varpi_e(\cos\theta)T_{prf}} = \frac{c\varpi_d}{2\varpi_e(\cos\theta)} \tag{1}$$

where ϖ_e is the frequency emitted from the transducer. The difference between the emitted frequency and the received frequency is called the Doppler shift frequency (ϖ_d). The maximum velocity that can be measured also depends on the PRF. The Nyquist theorem determines the maximum velocity ($v_{\max}$) that can be measured as [4]:

$$v_{\max} = \frac{c}{4\varpi_e T_{prf}} \tag{2}$$

The relationship between the maximum penetration depth and measured velocity can be given by:

$$\lambda_{\max} v_{\max} = \frac{c^2}{8\varpi_e} \tag{3}$$

The above relation shows that maximum velocity measured is inversely proportional to the maximum penetration depth.

3. UDV experimental setup and results

The experimental set-up described by Ein-Mozaffari et al. [5] was used to study the mixing of pulp suspension in a scale-model of an industrial chest. The stock chest was rectangular and equipped with a side-entering impeller. A Maxflo impeller (Chemineer Inc, Dayton, OH), was used for all tests. The tests were made using two fibre types: (i) a hardwood bleached kraft pulp (HW) having a mean length-weighted fibre length of 0.90 mm, and (ii) a softwood bleached kraft pulp (SW) having a mean length-weighted fibre length of 2.47 mm. The mass concentration of the suspension was varied between 2.1-3.3%. An ultrasound Doppler velocimeter (Model DOP2000, Signal Processing, Switzerland) was used to measure velocity profiles within the agitated chest. The single probe was used to measure the axial velocity across the impeller (v_y). The probe was located at an angle (85°) to the chest wall with the velocity profile measured along probe axis (v) projected on the y-axis to calculate the axial velocity across the impeller ($v_y = v\cos(85^o)$).

The mean axial velocity profile measured across the impeller in water is depicted in figure 1. The velocity data have been plotted in dimensionless form ($v_y/\pi ND$, where πND is the impeller tip velocity, N the impeller speed, and D the impeller diameter). As the velocity in front of impeller was so intense, the penetration depth was decreased so that only measurements up to the centre of the impeller were made. It was assumed that the velocity profile was symmetrical, which was then used to calculate the impeller pumping capacity Q_p ($m^3\ s^{-1}$) usually defined as the volumetric flow rate passing through the plane established by the impeller rotation. The dimensionless flow number (Fl) was calculated using following equation:

$$Fl = \frac{Q_p}{ND^3} = \frac{\int_{A_s} v_y dA_s}{ND^3} \tag{4}$$

where A_s (m^2) is the swept surface area. The measurement of the flow number for the Maxflo impeller in water using the UDV data (0.64 ± 0.04) gave an identical value to that provided by the manufacture (0.643). As listed in table 1, flow numbers were also measured for HW (2.1%, 2.7%, 3.3%) and SW (2.1%). As the fibre mass concentration increases, the number of fibre/ fibre interaction increases which in turn increases the suspension yield stress. Table 1 shows that the flow number decreased as the suspension concentration and yield stress increased.

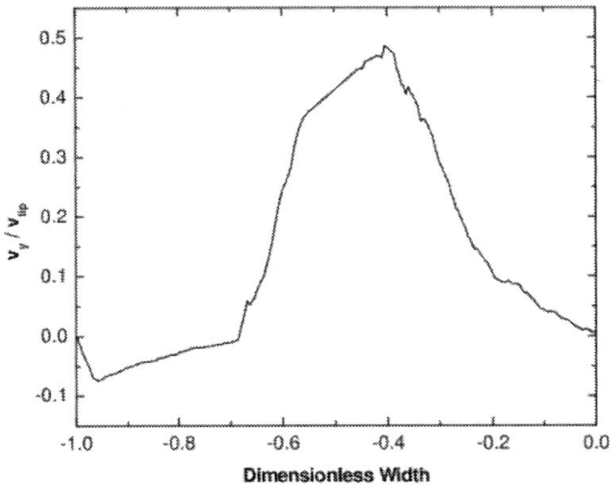

Figure 2. Dimensionless axial velocity ($v_y/\pi ND$) in front of the impeller in water.

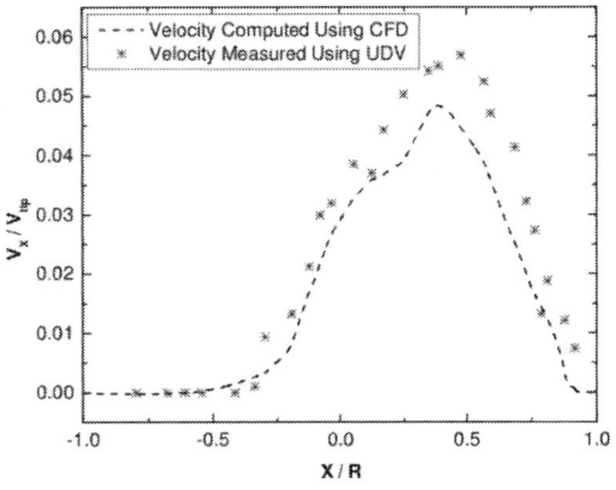

Figure 1. Calculated and measured velocities (below the suspension surface).

Table 1. Impeller flow number

Fibre type	Fibre mass concentration (%)	Suspension yield stress (Pa)	Flow number
HW	2.1	23	0.588
HW	2.7	59	0.513
HW	3.3	99	0.431
SW	2.1	89	0.447

Table 2. Rheological parameters of xanthan gum

Mass concentration (%)	Yield stress (Pa)	Consistency index, (Pa.s^n)	Power law index
0.5	1.79	3	0.11
1.0	5.25	8	0.12
1.5	7.45	14	0.14
3.5	20.6	33.1	0.18

UDV technique was used to study the mixing of pulp suspension in a cylindrical tank equipped with a side-entering impeller (an axial-flow Lightnin A310 impeller). The details on experimental set-up and procedure can be found in Saeed et al. [6]. The yield stress of the 3.3% SW pulp suspension was set to 475. To validate the CFD model, a comparison of the velocity data from the UDV measurements with the velocities computed using CFD was done (figure 2). In this figure, $v_{tip} = \pi ND$ is the impeller tip velocity. The graph shows that computed velocities agree with the measured data.

To investigate the mixing of xanthan gum solution in a cylindrical flat-bottom tank, the experimental setup explained in Ein-Mozaffari and Upreti [7] was used. Three axial-flow impellers were utilized in the experiments: pitch blade turbine (PBT), marine propeller, and A310 impeller, each having diameter equal to $0.44T$. Xanthan gum is a pseudoplastic fluid with yield stress and its rheology can be described by the Herschel-Bulkley model [8]:

$$\tau = \tau_y + K(\dot{\gamma})^n \quad (5)$$

where τ is the shear stress, τ_y is the yield stress, K is the consistency index, $\dot{\gamma}$ is the shear rate and n is the power-law index. Table 2 summarizes the rheological parameters of xanthan gum solutions. The UDV measured data were used to validate the CFD model for velocity field. The flow number for the axial flow impellers was estimated using equation (4). The axial velocities obtained using CFD and UDV were used to calculate flow numbers for pitched blade turbine, marine propeller, and A310 impellers. The flow numbers for marine propeller, pitched blade turbine, and A310 impeller were: 0.42, 0.47, and 0.37, respectively (CFD); and 0.39, 0.43, and 0.38, respectively (UDV). It shows that computed values agreed well with measured data.

The mixing of xanthan gum solutions (1.5% and 3.5%) with helical ribbon impeller were also investigated in a flat-bottom cylindrical vessel using UDV (see details in reference [9]). The rheological parameters of xanthan gum solutions are provided in table 2. The UDV measured data for velocity field were applied to validate the CFD model. A comparison of the velocities measured using UDV with velocities computed using CFD is depicted in figure 3. It shows that computed velocities agreed well with the measured data. The velocity profiles shown in figure 3 were used to verify that the net axial flow was zero [10]:

$$Q = 2\pi \int_{R_s}^{R_i} r v_z \mathrm{d}r = -2\pi \int_{R_i}^{R_w} r v_z \mathrm{d}r = 2\pi^2 N R_b R_w^2 \int_{Rs}^{Ri} \frac{v_z}{\Omega R_b} \mathrm{d}(r / R_w)^2 \qquad (6)$$

where R_s, R_i, R_b, R_w, v_z, N, Ω, and Q are the radius of impeller shaft, the radial position at zero axial velocity, the impeller outer radius, the vessel radius, the axial velocity, the impeller rotational speed, the angular velocity, and the volumetric flow rate, respectively.

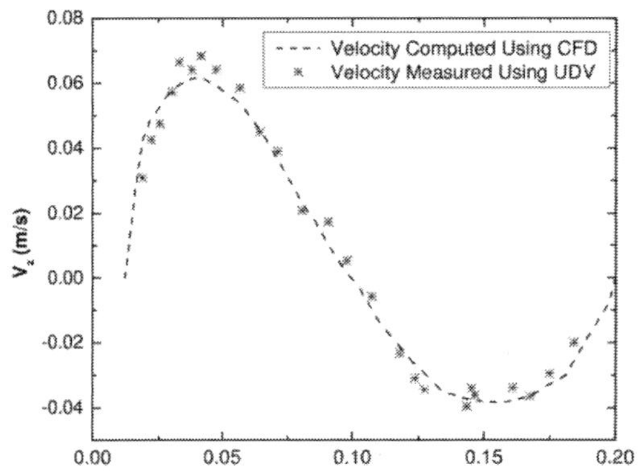

Figure 3. Calculated and measured axial velocities.

4. Conclusions

The ability of UDV to work with opaque fluids makes it a suitable tool for measuring flow velocity in yield- pseudoplastic fluids mixing operations. Using measured velocity profiles across the impeller, the effect of fibre type on the impeller flow number was quantified. It was found that as suspension yield stress increased (by increasing fibre mass concentration) impeller flow number decreased. The CFD calculations picked up the features of the flow field, and the computed velocities agreed well with the measured data. The axial velocity obtained using CFD and UDV were used to calculate the flow numbers for marine propeller, PBT, and A310 impellers. Thus, UDV is an efficient technique to investigate the mixing of opaque non-Newtonian fluids possessing yield stress.

References

[1] McClements D J, Povery M J W and Jury M 1990 *Ultrason.* **28** 266
[2] Takeda Y 1991 *Nucl. Eng. Des.* **126** 277
[3] Petitjeans P, Wesfreid J E and Attiach J C 1997 *Exp. Fluids* **22** 351
[4] Bouillard J, Alban B, Jacques P and Xuereb C 2001 *Chem. Eng. Sci.* **56** 747
[5] Ein-Mozaffari F, Bennington C P J, Dumont G A and Buckingham D 2007 *Chem. Eng. Res. Des.* **85** (A5) 591
[6] Saeed S, Ein-Mozaffari F and Upreti S R 2007 *Ind. Eng. Chem. Res.* **46** 2172
[7] Ein-Mozaffari F and Upreti S R 2009 *Chem. Eng. Res. Des.* **87** 515
[8] Patel D, Ein-Mozaffari F and Mehrvar M 2011 *Ind. Eng. Chem. Res.* **50** (15) 9377
[9] Ihejirika I and Ein-Mozaffari F 2007 *Chem. Eng. Technol.* **30** (5) 606
[10] Carreau P J, Patterson I and Yap C Y 1976 *Can. J. Chem. Eng.* **54** 135

International Symposium on Ultrasound in the Control of Industrial Processes (UCIP 2012) IOP Publishing
IOP Conf. Series: Materials Science and Engineering **42** (2012) 012049 doi:10.1088/1757-899X/42/1/012049

Effect of temperature on thermoacoustic properties of olive oil in alcohol mixtures

R S Shriwas[1], O P Chimankar[1,4], P V Tabhane[1], S P Dange[2] and Y D Tembhurkar[3]

[1]Department of Physics, RTM Nagpur University, Campus, Nagpur-440033, INDIA
[2]Department of Physics, Sindhu Mahavidyalaya, Nagpur
[3]Department of Physics, S.K.Porwal College, Kamptee, Dist-Nagpur, India

E-mail: opchimankar28@gmail.com
E-mail: ranjeeta.shriwas@gmail.com

Abstract. The ultrasonic studies in liquids are great use in understanding the nature and strength of molecular interactions. Recently ultrasonic is the rapidly growing field research, which has been used in the food industry for both analysis and modification of food products. This paper presents ultrasonic velocity, density, adiabatic compressibility in olive oil with alcohol at different concentration that has been measured in the temperature range from 283.15K to 298.15K. The Moelwyn-Hughes parameter has been utilised to establish some simple relations between the available volumes, Bayer's non- linearity parameter, internal pressure, van der Waal's constant, Debye temperature etc. a relationship among the isobaric, isothermal and isochoric thermo-acoustical parameter have been studied and analyzed in the case olive oil with alcohol mixture. The present treatments had the distinct advantages of the thermo-acoustic parameters in the particular mixture.

1. Introduction

Food engineering related to different oils with organic solvents for various purification and treatment procedures such as deacidification, winterization etc. In extraction and refining processes, vegetable oils are mixed with the organic solvents. The enthalpy of mixing and the viscosity of the organic solvents in vegetable oil mixtures are important in the design of process equipment for the oil and seed processing industry. The mixing thermodynamics properties behaviour as a function of temperature or pressure has been investigated for several decades. But Biological macromolecules have received little attention, which is wonderful when one considers their role in food processing industry and in biotechnology processes. In order to understand more precisely to alkanol contributions to the thermodynamics of olive oil mixtures, new measurement of physical properties at several temperatures for short aliphatic alcohols + olive oil mixtures are presented [1-3]. The speed of sound in binary liquid mixtures is a reliable property for various analyses in the oil based industries [4-6]. In recent years, Dukhin and Goetz acoustic techniques have been applied to food industry [7-9].

As less work has been made on the various thermoacoustic properties dependent on thermal expansion coefficient in different oils with organic solvents, the present study is an attempt to the study of intermolecular interactions in the binary mixtures of olive oil and aliphatic alcohols like

[4] To whom any correspondence should be addressed.

IOP Conf. Series: Materials Science and Engineering **42** (2012) 012049 doi:10.1088/1757-899X/42/1/012049

ethanol, 1-propanol, 2-propanol and 1-butanol by measuring Moelwyn-Hughes parameter, Bayer's non- linearity parameter, internal pressure, van der Waal's constant, Debye temperature, Gruneisen parameter etc. of the olive oil mixtures with alcohol at temperature 283.15K-298.15K at intervals of 5K.

2. Theoretical Aspect of the work

The experimental data of ultrasonic velocity and density of the olive oil in ethanol, 1-propanol, 2-propanol and 1-butanol are used for the present study from the literature [10].

This paper presents Moelwyn-Hughes parameter, Bayer's non- linearity parameter, internal pressure, van der Waal's constant, Debye temperature, Gruneisen parameter etc. in olive oil with alcohol at different concentration that has been measured in the temperature range from 283.15K to 298.15K. The results of thermo acoustic parameters in olive oil with aliphatic alcohols of the present work are shown by graphical representation.

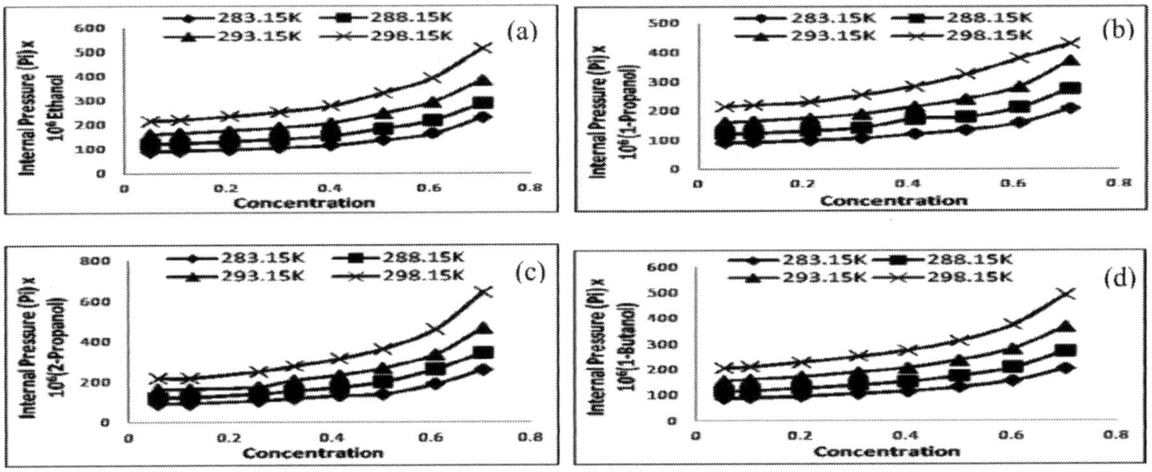

Figure 1. Internal Pressure (P_i) for: (a) Ethanol (b) 1-Propanol (c) 2-Propanol (d) 1-Butanol at temperature 283.15K-298.15K.

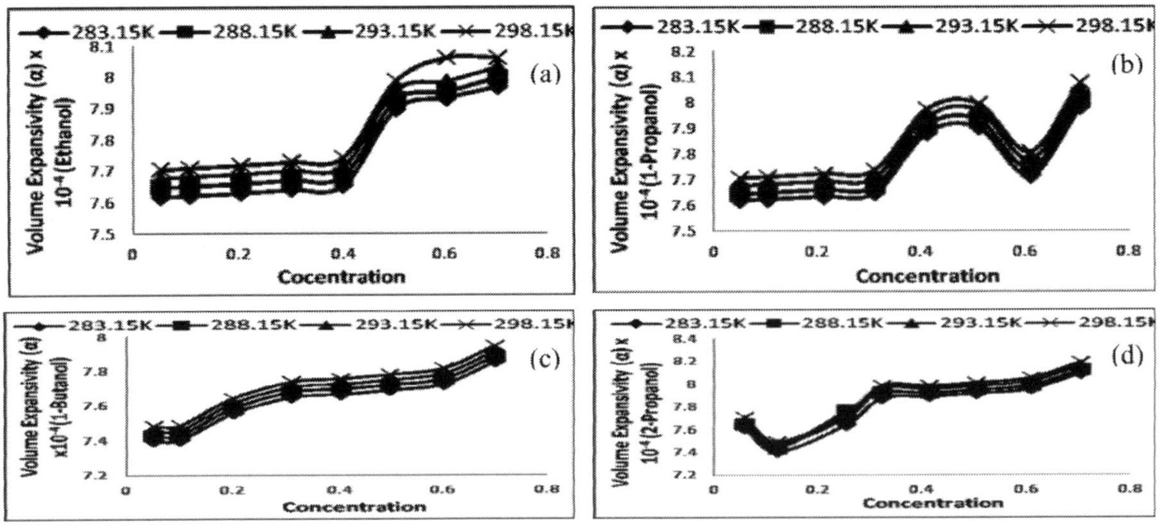

Figure 2. Volume expansivity (α) for: (a) Ethanol (b) 1-Propanol (c) 2-Propanol (d) 1-Butanol at temperature 283.15K-298.15K.

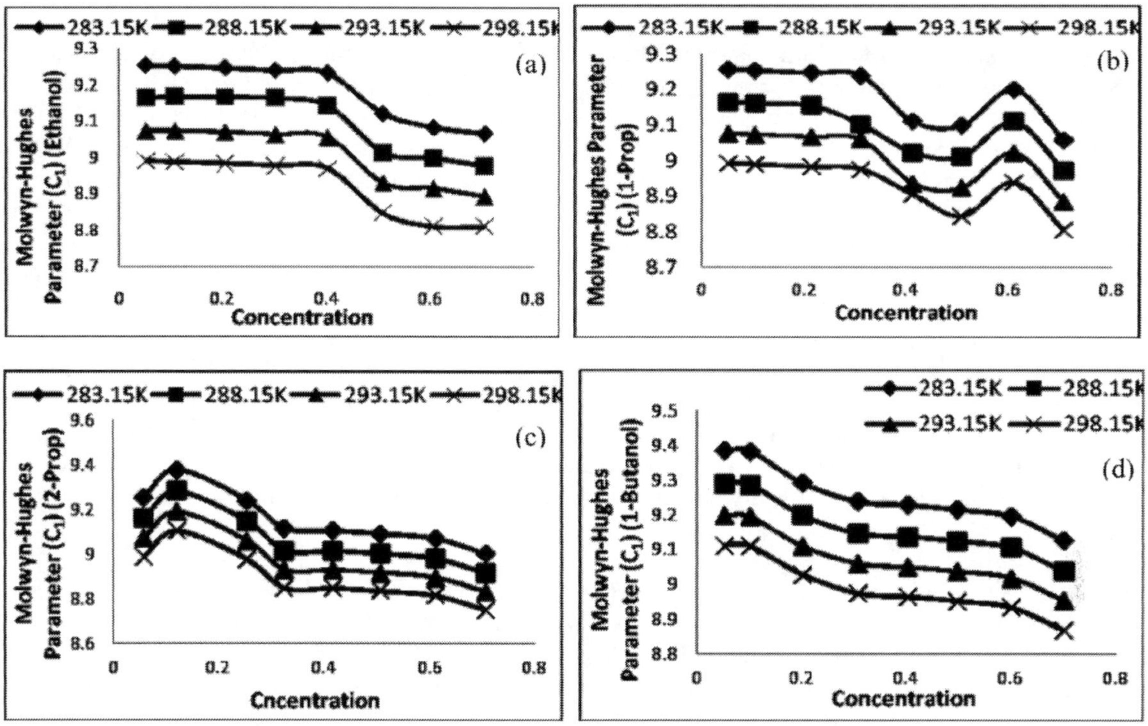

Figure 3. Molwyn-Hughes parameter (C_1) for: (a) Ethanol (b) 1-Propanol (c) 2-Propanol (d) 1-Butanol at temperature 283.15K-298.15K.

3. Results and Discussion

The internal pressure (P_i) (figure 1) shows an increasing trend which indicates a slight interaction between the solute and solvent molecules for increasing compositions and a reinforcement of solvent-solvent interaction. Low value of internal pressure shows the degree of association in the molecules at high temperature. The internal pressure (P_i) is the resultant of attractive and repulsive forces between the molecules. It measures the molecular cohesion and the instantaneous volume derived of the olive oil. The increase P_i with increase in mass concentration indicates the orientation of the solvent molecules around the ions. This may due to the influence of electrostatic field of ions i.e. solutions become harder to compress. This also shows associating tendency of the molecules in olive oil with aliphatic alcohols.

Volume expansivity (α) is calculated from density measurement the corresponding curves (figure 2) showing nonlinear trend for all oil mixture at different temperature. The abnormal behaviour of olive oil with 1-propanol at higher concentration of alcohol and 2-propanol at lower concentration of alcohol due to hydrophobic interaction at oil rich and alcohol rich region.. These measurements are closely related to changes in the solvent layer around triglyceride molecules [11]. At diluted mixtures of oil a typical tridimentional cage like structure is formed, the solution being organized predominantly by intense alcohol-alcohol interactions. Under these conditions, linear chains or three dimensional rings of alcohols have been proposed to exist [12].

The variation of Molwyn-Hughes parameter (C_1) (figure 3) of olive oil with aliphatic alcohols shows continuous non-linear decreasing trend except 1-propanol and 2-propanol at different concentration. In all the systems, the variation of C_1 and α shows the opposite behaviour with the increase concentration of aliphatic alcohol in olive oil and at various temperature i.e. when C_1 increases, volume expansivity decreases and vice-versa. This may be due to the associating tendency of the liquid mixture.

The solvent 1-Propanol and 2-propanol showed a higher nonideality than the corresponding linear aliphatic isomer. The minimum value is shown at high solvent composition. For two cases (1-propanol, 2-propanol) phase separation occurs, a greater immiscibility being observed at triglyceride rich and alcohol rich region of the functional series due to higher polar activity of hydroxyl functional group. All the alcohol mixtures show total miscibility, the same decreasing negative trend being observed for an increasing molar weight, although the values are similar to those observed in alcohol mixtures. These type of mixtures shows a high non-ideal behaviour which is severely affected by two phenomena; associative trend by means of hydrogen bonds (self- interaction of alcohol molecules), polar or dispersive interaction (dispersive among aliphatic chains), and the other is the huge difference in molecular volume between alcohol (Polar Solvent) and triglyceride (non-polar flexible chain). This leads to steric hindrance effect between residual ends. This fact is reflected by a general expansive effect in mixtures except for dilution of olive oil in propanol isomers due to their aliphatic chain and active hydroxyl group. This surprising capability of 1-propanol and 2-propanol leads to a slight contraction in diluted oil and phase splitting when the concentration of alcohol is increased. Some authors [10, 11, and 14] have previously reported the same behaviour from the derived ultrasonic velocity and adiabatic compressibility at different concentration and temperature.

4. Conclusions

The observed trend of changes in all thermoacoustical properties shows alcohol-alcohol and triglyceride-triglyceride interaction in the olive oil - alcohol mixtures, which leads to no chemical reaction at the time of residual ends. These types of interactions among hydroxyl solvents, and triglyceride molecules provide promising expectations for their application in winterization and purification of various oils in oil and seed processing industry.

Acknowledgment

The one of the author (OPC) is greatful to University grant commission, New Delhi for providing financial support to this work through Major research project letter F.No.39-456/2010 (SR).

References

[1] Gonzalez C, Resa J M, Lanz J, Rui A and Gutierrez J I 1996 J. *of Chemical and Engineering Data* **41** 796-798
[2] Gonzalez C, Resa J M, Lanz J, Ruiz A and Gutierrez J I 1997 *J. of Chemical and Engineering Data* **42** 339-341
[3] Gonzalez C, Iglesias M, Lanz J and Resa J M 1999 *Thermochimica Acta* **328** 277-296
[4] Gonzalez C, Resa J M, Lanz J and Fanega M A 2002a *Journal of Chemical and Engineering Data* **47** 1116-1122
[5] Gonzalez C, Resa J M, Lanz J and Fanega M A 2002b *Journal of Chemical and Engineering Data* 79 536-542
[6] Gonzalez C, Resa J M, Lanz J and Fanega M A 2002c *Journal of Chemical and Engineering Data* **79** 543-548
[7] Dukhin A S and Goetz P J 1996 *Longmuir* **12**(21) 4987-4997
[8] Dukhin A S and Goetz P J 1996 *Longmuir* **12**(21) 4998-5004
[9] Dukhin A S and Goetz P J 1996 *Longmuir* **12**(21) 4334-4344
[10] Gonzalez C, Iglesias M, Lanz J, Marino G, Orge B and Resa J M 2001 *J. of Food Engineering* **50** 29-40
[11] Parke S A and Birch G G, *Food Chemistry* **67** 241-246
[12] Frank H S and Ives D J G 2002 *Quarterly Reviews* **1** 44-48
[13] Resa J M, Gonzalez C, Fanega M A, Ortiz de Landaluce S and Lanz J 2002 *J. of Food Engineering* **51** 113-118
[14] Gonzalez C, Resa J M, Lanz J and Iglesias M 2006 *Journal of Food Engineering* **77** 152-161

IOP Conf. Series: Materials Science and Engineering **42** (2012) 012050 doi:10.1088/1757-899X/42/1/012050

Non-destructive monitoring of curing process in precast concrete

S Aparicio[1], J Ranz[1], R Fernández[2], V Albert[2], J V Fuente[2] and M G Hernández[1]

[1]Centre of Applied Acoustic and Non Destructive Evaluation, CAEND, CSIC-UPM, Madrid, Spain,
[2]Technological Institute of Construction, AIDICO, Valencia, Spain

E-Mail: sofia.aparicio@caend.upm-csic.es

Abstract. Currently, the use of precast concrete elements has gained importance because it offers many advantages over site-cast concrete. A disadvantage of site-cast concrete is that its properties vary according to the manufacturing method, the environment and even the operator who carried out the mixing, pouring and implementation of the concrete. Precast concrete elements are manufactured in a controlled environment (typically referred to as a precast plant) and this reduces the shrinkage and creep. One of the key properties of precast concrete is the capability to gain compressive strength rapidly under the appropriate conditions. The compressive strength determines if the precast can be stripped from the form or manipulated. This parameter is measured using destructive testing over cylindrical or cubic samples. The quality control of precast is derived from the fracture suffered by these elements, resulting in a "pass or fail" evaluation. In most cases, the solution to this problem is to allow the material to cure for a few hours until it acquires sufficient strength to handle the precast element. The focus of this paper is the description of the research project "CUREND". This project aims to design a non-destructive methodology to monitor the curing process in precast concrete. The monitoring will be performed using wireless sensor networks.

1. Introduction

The current criteria followed in precast plants to determine if a precast element is ready for processing (cutting or removal) is the value of its compressive strength. This strength is determined from two cylindrical specimens obtained from the same mix. The quality control is done using a series of precast rupture experiments, which results in a "pass or fail" evaluation, the last test is performed after 28 days as a reference of the product when it reaches the maximum strength. Moreover, curing times are usually calculated by previous experience (measurements) in preceding seasons, knowing that in summer these times are shorter than in other seasons. The problem arises when the constituent materials (raw materials), such as cement, sand or gravel, and/or the climatic conditions within a time or season change. The solution to this problem is often to allow the material to cure for a longer time on the track until it acquires the strength needed to process the precast element.

The project "CUREND" is a public-private cooperation between research organizations and companies within the program INNPACTO. Participating institutions include the Center for Applied Acoustic and Non-destructive Evaluation, the Technological Center of the Construction AIDICO, the PRECAMP company and the precast companies: CALDERON S.L., HERMO S.L. and FORJADOS SECUSA. The new research project aims to develop a remote monitoring system for a non-destructive quality control of the strength properties of precast concrete throughout the curing process in production plants. This research benefits from previous knowledge acquired by the participating groups. The monitoring system will largely solve the quality control in precast concrete plants. Monitoring is done using wireless sensor networks that measure the temperature, humidity and ultrasonic velocity. Different models will be applied to interpret the measurements and to predict the strength gain of concrete throughout the curing process and the properties of the final product. To predict these final properties it is necessary to estimate in advance the degree of curing.

The degree of curing of concrete can be estimated from the product of the internal temperature in the material and the curing time or from the hydration rate of cement. The concept of maturity assumed that the strength gain in cementitious materials is a combined effect of temperature and curing time. The maturity method measures the temperature variation during the hydration of the material to obtain a correlation with the increasing of the strength [1], [2]. The procedure to obtain the maturity-strength relationship was described in [3]. The major limitation of this method is this relationship varies in the real process compared to the laboratory. The reason is that the dosage of the cementitious material, the temperature and the humidity are better controlled in the laboratory environment. Different functions are used to relate the hydration of cement with time. The linear or parabolic models developed by Knudsen [4] are widely used. These models differ in the kinetics of hydration. The linear model assumes that the hydration of cement particles is linear in time while the parabolic model considers the hydration as a quadratic function of time. The real conditions of hydration are generally within these two cases.

On the other hand, there are numerous studies that attempt to determine the relationship between concrete strength and different ultrasonic parameters that include the ultrasonic velocity and the reflection coefficient [5], [6].

2. Experimental Description

The methodology proposed in this project is developed in three phases. In the first phase the characteristics of each material from different companies are studied and their behavior under optimum curing conditions is evaluated. In the second phase, the monitoring is performed on concrete specimens manufactured and cured in the production line. This phase allows to assess the performance of equipment and instrumentation to be developed for operation in the harsh conditions of manufacturing where they are exposed to water, dust, cutting machinery, welding, etc. The final phase consists on monitoring of prefabricated lines at different points of the production line.

In the first phase, experiments were performed with laboratory instrumentation, in the second phase the temperature and humidity (T/RH) were monitored with wireless networks and ultrasonic velocity with conventional wired equipment. In the third phase, all sensors will be incorporated into the wireless networks. In this work, the first results obtained in the first and second phases of the project are shown.

In the first phase, concrete specimens were monitored with the material from each precast company under optimum curing conditions (20°C and 98% of relative humidity) and performing assays rupture at short intervals of time. In the second phase, currently underway, the monitoring of the specimens is repeated at the curing conditions of each prefabricated line. This monitoring over several months will allow to validate the developed instrumentation and to have an extensive database of the material behavior to further study the influence of environmental conditions in the curing.

The monitoring of T/RH was performed using SHT-7 sensors and the ultrasonic velocity was measured with a Pundit equipment and using longitudinal transducers of 54 kHz. The sensors are placed in measuring cells designed by the research team [7]. The monitoring was done during the first 50 hours of curing, taking measurements every 2 minutes. The objective was to determine the strength in the early ages of curing allowing a rapid and efficient handling of the precast element and to estimate their properties at 28 days of curing. This provides a quality control of the material. Furthermore, strength tests were performed at different stages of curing. The system, the methodology and the experimental design for monitoring the curing of cementitious materials were patented by the research team [7].

3. Experimental Results

A relation of the results obtained during the monitoring of the curing process is presented. Because there is a confidentiality agreement between the participating entities some of the scales values were omitted.

Figure 1 shows the evolution of the temperature inside the concrete and the ambient temperature during the curing process. In the first phase, blue trace, the temperature increase corresponding to the hydration reactions is shown. In the other experiments the effect of hydration is masked by the influence of ambient temperature which modifies the thermal history of the material.

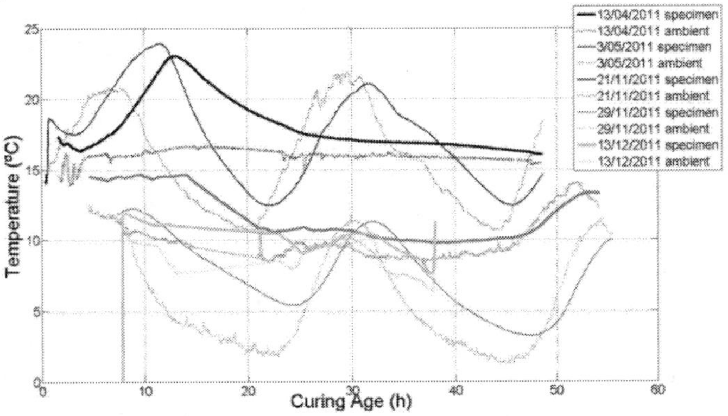

Figure 1. Temperature evolution (inside and ambient) during the curing process.

Figure 2 shows the behavior of compressive strength (left) and ultrasonic velocity (right) measured and estimated during the first 120 hours of curing. The strength and the ultrasonic velocity are adjusted using the method of Knudsen [4]. The time interval corresponding to a given value of the ultrasonic velocity is obtained and the strength values are determined, figure 3.

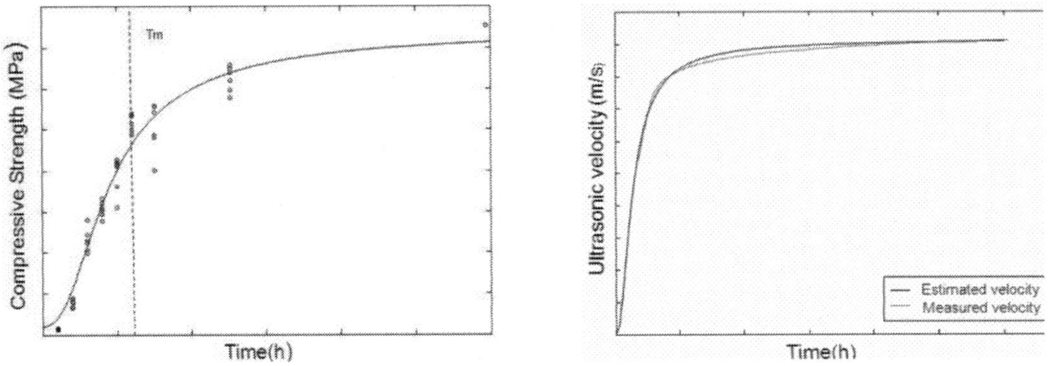

Figure 2. Compressive strength (left) and ultrasonic velocity (right) during the curing process.

Figure 3 shows the estimation of compressive strength (solid line) based on the ultrasonic velocity measurements and the measured values (circles) for a given experiment. The estimation obtained from the developed methodology provides a reasonable fit of the measured values. These results are still being analyzed to improve the prediction model using other parameters such as the wave reflection coefficient during the curing process.

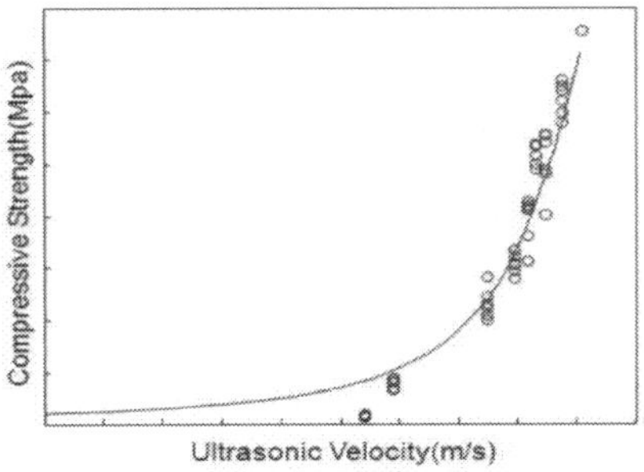

Figure 3. Estimation of compressive strength based on the ultrasonic velocity measurements.

4. Conclusions

The CUREND project develops a methodology and a non-destructive instrumentation for monitoring the curing of precast concrete plants, which will allow a better control of the quality of the final product. The results obtained in the early stages of the project have shown the feasibility of the mechanical and electronic instrumentation developed for the precast concrete plants. The parameters T/RH and ultrasonic velocity for the estimation of the compressive strength of concrete during the curing process were validated. The instrumentation and methodology were recently patented.

Acknowledgements

The authors wish to thank the Ministry of Science and Innovation the grant project BIA 2009-14395-C04-01 and IPT-020000-2010-26 and the FPI grant awarded to J. Ranz, as well as the support provided by the company PRECAMP and the precast companies: CALDERON S.L., HERMO S.L. and FORJADOS SECUSA to carry out this work. Dr. S. Aparicio thanks the funding of the postdoctoral program JAE-Doc from CSIC.

References

[1] Carino N J. 1984 *J. Cem. Conc. Agg.* **6** 61-73
[2] Pane I and Hansen W 2002 *ACI Mater. J.* **99** 534–542
[3] ASTM C 1074 1998 *2000 Annual Book of ASTM Standards* vol 04.02 (West Conshohocken, PA: ASTM)
[4] Knudsen T. 1984 *Cem. Concr. Res.* **14**, 622-630.
[5] Trtnik G, Franci K and Turk G 2009 *Ultrasonics* **49** 53-60
[6] Reinhardt H W and Grosse C U 2004 *Constr Build Mater* **18** 145–154
[7] Aparicio S *et al* 2011 *Procedimiento y sistema inalámbrico de medida del grado de fraguado y endurecimiento de materiales cementicios para la predicción de resistencias mecánicas* Patent nº 201131005

International Symposium on Ultrasound in the Control of Industrial Processes (UCIP 2012) IOP Publishing
IOP Conf. Series: Materials Science and Engineering **42** (2012) 012051 doi:10.1088/1757-899X/42/1/012051

Numerical simulation of ultrasonic tomography inspections of highly heterogeneous materials

M Molero[1,2], L Medina[1], D Lluveras[2], MAG Izquierdo[3] and JJ Anaya[2]

[1]Departamento de Física, Facultad de Ciencias, UNAM, México D.F., México
[2]Centro de Acústica Aplicada y Evaluación No Destructiva, CAEND, (CSIC-UPM), España
[3]Departamento S.S.R., E.T.S.I. Telecomunicación (UPM), Madrid España.

E-mail: miguel.molero@csic.es

Abstract. This paper deals with the simulation of ultrasonic transmission tomography systems in water-immersed to nondestructively inspect highly heterogeneous materials with fractures. The time-domain Elastodynamic Finite Integration Technique (EFIT) was employed for all numerical simulations because is able to reliably simulate this type of ultrasonic problems. The EFIT code was implemented using OpenCL and PyOpenCL. Several ultrasonic tomography inspection setups were numerically simulated under different conditions varying the number of ultrasonic sources and their size and number and different operation schemes. Sinograms of concrete scenarios were computed and compared for each configuration, using homogeneous materials with similar fracture types and experimentally validated.

1. Introduction

Ultrasonic tomography imaging is an emerging and promising nondestructive evaluation technology for assessment of quality and integrity of concrete structures. This technique reconstructs cross-sectional images from different projections through an object under test, providing visual information about its structural condition and deterioration state. Ultrasonic imaging techniques have shown a spectacular success in the biomedical field, however the non-destructive evaluation in civil engineering (NDE-CE) applications still remain to be investigated. Ultrasonic tomography inspections in highly heterogeneous materials like concrete requires to study a complex wave propagation consisting of reflections, dispersion, scattering, and mode conversion masking relevant information on the material structural properties. This paper deals with the simulation of ultrasonic transmission tomography systems in immersion, to provide a practical tool for optimal design of ultrasonic inspection systems and reconstruction techniques for highly heterogeneous materials. The numerical simulations were performed with the Elastodynamic Finite Integration Technique (EFIT). The EFIT code is implemented using the new open industry standard for parallel programming (OpenCL) and the open-source toolkit PyOpenCL.

2. Elastodynamic Finite Integration Technique

The Elastodynamic Finite Integration Technique (EFIT) has been widely used as a numerical NDE tool to simulate the wave propagation in several heterogeneous composite material scenarios [1–4]. This time-domain numerical approach is able to automatically handle boundary conditions and different acoustic phenomena such as reflections, dispersion, scattering, and mode conversion. The basis of this technique relies on the field component to be calculated, is placed at the centre of an appropriate control volume and the integration of the corresponding equation is calculated over this cell. Given the bi-dimensional linear elastodynamics equations in terms of their velocity (v_x,v_y) and stress (τ_{xx}, τ_{yy},τ_{xy}) components, as:

$$\rho\dot{v}_x = \frac{\partial\tau_{xx}}{\partial x} + \frac{\partial\tau_{xy}}{\partial y} + f_x; \quad \rho\dot{v}_y = \frac{\partial\tau_{xy}}{\partial x} + \frac{\partial\tau_{yy}}{\partial y} + f_y \tag{1}$$

$$\dot{\tau}_{xx} = (\lambda + 2\mu)\frac{\partial v_x}{\partial x} + \lambda\frac{\partial v_y}{\partial y} + g_{xx}; \dot{\tau}_{yy} = (\lambda + 2\mu)\frac{\partial v_y}{\partial y} + \lambda\frac{\partial v_x}{\partial x} + g_{yy}; \dot{\tau}_{xy} = \mu\left(\frac{\partial v_x}{\partial y} + \frac{\partial v_y}{\partial x}\right) + g_{xy} \tag{2}$$

where the elastic medium properties are given by the Lamé constants (λ, μ) and the mass density (ρ). $f_{(\circledast)}$ and $g_{(\circledast)}$ are the force and the stress sources components. Then the discretised versions of $\dot{v}_x$, $\dot{\tau}_{xx}$ and $\dot{\tau}_{xy}$ respect to the n-cell and regular grid (e.g. $\Delta x = \Delta y$) are:

$$v^n_{x,i,j} = v^{n-1}_{x,i,j} + \frac{\Delta t}{\Delta x}B_x\left(\tau^{n-1/2}_{xx,i+1,j} - \tau^{n-1/2}_{xx,i,j,} + \tau^{n-1/2}_{xy,i,j,} - \tau^{n-1/2}_{xy,i,j-1}\right) + \Delta t B_x f_{x,i,j} \tag{3}$$

$$\tau^{n+1/2}_{xx,i,j} = \tau^{n-1/2}_{xx,i,j} + \frac{\Delta t}{\Delta x}\left\{(\lambda + 2\mu)\left[v^n_{x,i,j} - v^n_{x,i-1,j}\right] + \lambda\left[v^n_{y,i,j} - v^n_{y,i,j-1}\right]\right\} + \Delta t g_{xx,i,j} \tag{4}$$

$$\tau^{n+1/2}_{xy,i,j} = \tau^{n-1/2}_{xy,i,j} + \frac{\Delta t}{\Delta x}\mu_{xy}\left[v^n_{x,i,j+1} - v^n_{x,i,j} + v^n_{y,i+1,j} - v^n_{y,i,j}\right] + \Delta t g_{xy,i,j} \tag{5}$$

where B_x, B_y are effective buoyancies, and μ_{xy} is the effective rigidity. To guarantee numerical stability of the EFIT code, the spatial and time steps must be at least $\Delta x \leq C_{min}/(10F_{max})$ and $\Delta t \leq (1/\sqrt{2})\Delta x/C_{max}$, where C_{min} and C_{max} correspond to the lowest and highest velocities in the heterogeneous medium, and F_{max} is the highest frequency component in the source signal with operating frequency F_c (e.g. $F_{max} = 2F_c$).

3. [PY]OpenCL EFIT implementation

To handle the intensive computation calculations involved in our simulations, we implemented the EFIT code using the new open industry standard for parallel programming of heterogeneous systems, OpenCL (Open Computing Language) [5]. The OpenCL framework provides an open, royalty-free and neutral-vendor parallel programming solution. It delivers a high degree of portability across different computing devices such as multi-core CPU's and GPU's. OpenCL includes a language based on C99 for writing functions called kernels which will be executed on OpenCL devices, and application programming interfaces (APIs) used to define and control the platforms. To simply our OpenCL EFIT implementation, we used an open-source toolkit called PyOpenCL allowing us the access of the OpenCL APIs via Python programming [6]. The PyOpenCL toolkit offers an attractive solution to implement parallel numerical codes using the high-level programming environment of Python needing much less explicit code to define and control the OpenCL programming model.

4. Numerical simulation of ultrasonic tomography inspections

Numerical simulations of ultrasonic wave propagation in heterogeneous media were performed with the [Py]OpenCL EFIT code implementation. A bi-dimensional geometric model of the concrete microstructure representing the three-phase material model (cement matrix, aggregates and air voids) was used. This geometric model was built using basic geometric forms distributed into a homogeneous cement matrix. Aggregates and air voids were modeled as infinite ellipse cylinders and circular cylinders, respectively. The inspection setup consists of one transducer as an emitter and eleven receiver transducer radially located around the concrete specimens with 30° separation. The circular voids representing cracks are water-saturated. The emitter transducer drives a raised cosine signal with an operating frequency of 500 kHz. Both emitter and receiver transducers were 25.4 mm in length. Figure 1 shows numerical results (snapshots and receiver traces) for three different scenarios using the parallel EFIT code: plain circular concrete specimens, see figure 1(a), and two different concrete specimens with circular voids embedded in their microstructures, see figures 1(b) and 1(c).

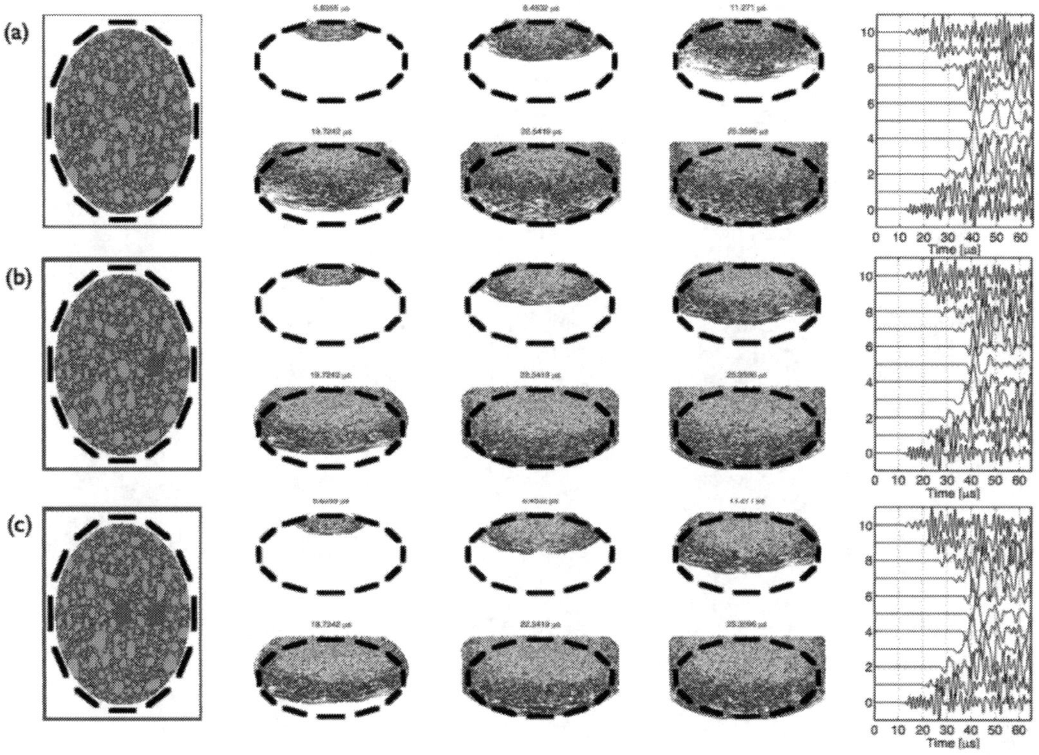

Figure 1. Concrete models, snapshots and ultrasonic traces from eleven receivers.

In order to verify our parallel EFIT implementation, Figure 2 shows a comparison between ultrasonic experimental traces measured from an aluminum cylinder probe and numerical simulations by EFIT. Both experimental and numerical ultrasonic inspection setups consist of a single emitter transducer and several receivers located around the specimen in the range [-135° ; 135°] with a single 1° step and with respect to the source. Figure 2 shows a very good agreements between experimental data and numerical simulations.

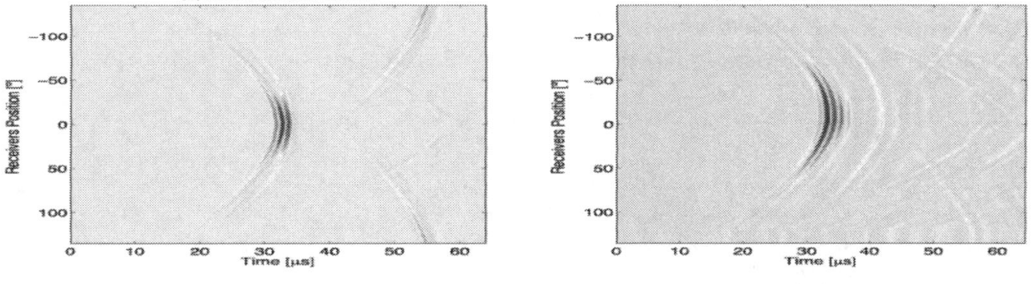

Figure 2. Comparison between numerical simulations and experimental data.

To numerically test our parallel EFIT code for a complete tomography setup, several amplitude sinograms were computed for different homogeneous and heterogeneous material scenarios. The amplitude sinograms were calculated with the maximum value of the ultrasonic signal corresponding to each projection of every single transducers acting as an emitter and the rest as receivers. Figure 3 shows the comparison between experimental data and the numerical simulations. The experimental data were measured using four specimens: two cylinder aluminum probes and two cylinder concrete specimens in which one or two cylinder voids were made to simulate cracks. In figures 3(a) and 3(b) corresponding to the aluminum scenarios (homogeneous media with voids) it can be observed that the simulated sinograms agree with the experimentally-acquired data. The simulated sinograms corresponding to the concrete scenarios, see figures 3(c) and 3(d) (highly heterogeneous media with voids) show a little discrepancy with the experimental ones. This probably due to the used geometric concrete model since simple geometrical shapes for aggregates and air voids were used.

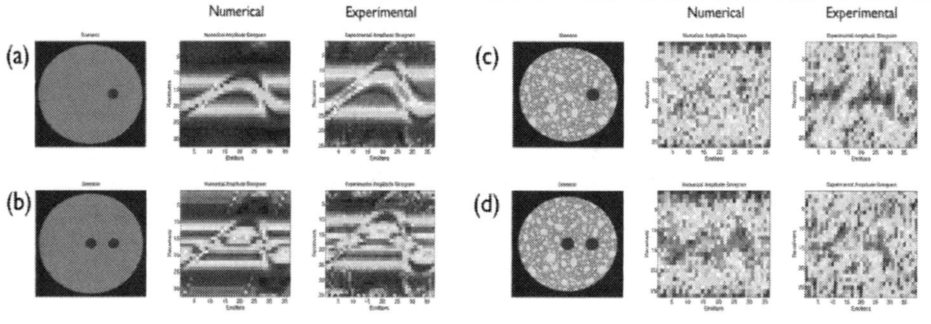

Figure 3. Comparison between numerical simulations using our parallel EFIT and experimental data of an aluminum cylindrical probe. The transducer length was 12.7 mm and had a 10° separation between them. The source signals operated at 500 kHz.

5. Concluding Remarks

The parallel EFIT code was successfully implemented to overcome the limitations imposed by numerical simulations demanding a great deal of computational requirements. Numerical results showed a very good agreement between experimental and numerical data for ultrasonic tomography inspections of highly heterogeneous materials. This approach may serve as a reliable numerical tool to previously study materials commonly used in civil engineering.

Acknowledgments

The financial support of the Spanish Science and Innovation Ministry (Project BIA 2009-14395-C04-01) is greatly acknowledged. L.M. and M.M. acknowledge financial support from CONACYT-131937 and DGAPA-PAPIIT-IT118811-3 projects. M.M. thanks CONACYT for a posdoc stipend and the support of the department of education of the Community of Madrid. D.L. thanks the CSIC JAE-PREDOC program.

References

[1] Fellinger P, Marklein R, Langenberg KJ and Klaholz S 1995 *Wave Motion* **1** 47
[2] Schubert F, Peiffer A, Koehler B and Sanderson T 1998 *J. Acoust. Soc. Am* **5** 2604
[3] Schubert F and Koehler B 2001 *J. Comput. Acoust.* **4** 1543
[4] Schubert F 2004 *Ultrasonics* **42** 221
[5] OpenCL `http://www.khronos.org/opencl/`
[6] Klöckner A, Pinto N, Lee Y, Catanzaro B, Ivanov P and Fasih A 2012 *Parallel Comput.* **38** 157

Characterization of the state of a droplet at a micro-textured silicon wafer using a finite difference time-domain (FDTD) modeling method

N Saad[1, 2], B Merheb[2], G Nassar [1], P Campistron [1], J Carlier [1], M Ajaka [2] and B Nongaillard [1]

[1]IEMN UMR CNRS 8520, Université de Valenciennes, Le Mont Houy 59313, France
[2]LPA, Université Libanaise, P.O.Box 90656, Jdeidet, Liban

E-mail : Nadine.Saad@univ-valenciennes.fr

Abstract. In this study, we introduce a finite difference time domain method to study the propagation and reflection of an acoustic wave on smooth and micro-textured silicon surfaces in interaction with droplets in different states. This will enable numerical investigations of interfaces composed of periodically distributed well-defined pillars. One type of transducer was modeled generating longitudinal wave. Three configurations were studied: the Cassie state, the Wenzel state and a composite state for which the droplet collapsed into the middle height of the pillars. After analysis of the displacement along y direction in the silicon wafer, we were able to show that a longitudinal wave is sensitive to the detection of the state of the droplet. The first experimental results made it possible to show a good agreement between modeling and experiments.

1. Introduction

Superhydrophobic materials hold substantial capacity for potential applications extending from self-cleaning surfaces, completely water impermeable textiles to low cost energy displacement of liquids in lab-on-chip devices. Superhydrophobic surfaces displaying high contact angle and low contact angle hysteresis also reveal a self-cleaning features and low drag for fluid flow. Lotus leaves and bird feathers are one of many surfaces that are known to be superhydrophobic due to their rough surface. So in order to modify the hydrophobic property of the surface, it is possible, not just modify its chemical composition but also its hierarchical structures by changing surface roughness.

On such surfaces a drop of water can experience two different states: suspended on a composite surface of material and air pockets (Cassie-Baxter state) or following all the contours of the rough surface (Wenzel state).

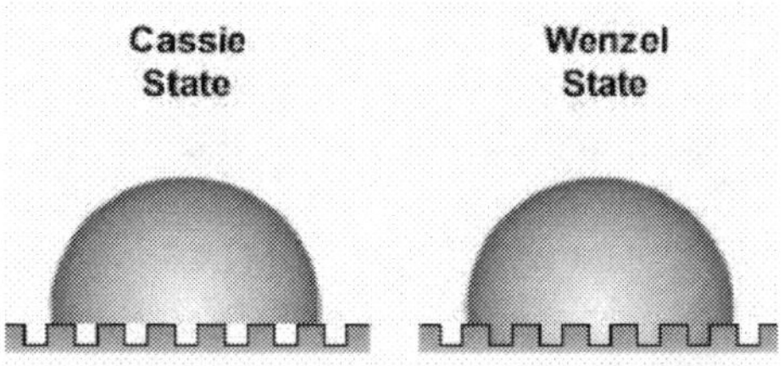

Figure 1. Etats Cassie et Wenzel.

IOP Conf. Series: Materials Science and Engineering **42** (2012) 012052 doi:10.1088/1757-899X/42/1/012052

This subject is receiving much attention; several models based on theoretical calculations were developed in order to estimate the critical values (e.g. line density criterion, surface roughness height [1], drop's surface energy criteria [2], etc..) for which one can predict in advance whether the drop will collapse or be suspended by pillars. Experimentally, current measurements are done by optical means, which often are limited by technological barriers. Ultrasound, beside acoustic, are generally used either to detect and characterize defects in structures [3], or to highlight the levels of adhesion [4]. While a consistent method for characterizing textured interfaces is missing, the aim of the present work and thus its originality lies in determining the potential of acoustics to characterize the state of a droplet on microtextured surfaces.

This paper is arranged as follows. Section II presents a finite difference time domain (FDTD) model design for acoustic wave propagation through a microtextured structure in interaction with a fluid. A brief overview of the early experimental work as well as initials results are reported in section III. Finally, the last section summarizes the main findings and conclusions of this joint study.

2. Methods and Models

Figure 2 illustrates a 2D cut of the structure of different models developed in this work. In the simulation, the two-dimensional model is assumed infinite in the Z direction and finite in X and Y Cartesian coordinates (OXYZ). Every structure consists of three different regions. The first one is composed of an anisotropic single crystal silicon microtextured with different patterns. Its physical characteristics are those of an anisotropic monocrystalline silicon of mass density ρ = 2330 kg.m^{-3}, longitudinal speed C_l = 8433 ms^{-1} and transverse speed C_t = 5843 ms^{-1}. Pillars of width `a`, height `h` and pitch `b`, are considered parallel to the Y axis, that is the direction of wave propagation. The second region is water of mass density ρ = 1000 kg m^{-3} and longitudinal speed C_l = 1490 ms^{-1} (Ct = 0 ms^{-1}).

The third region is a homogeneous region formed of air modeled as a fluid of mass density ρ = 1.3 kg.m^{-3}, in which the speed is C_l = 340 ms^{-1}.

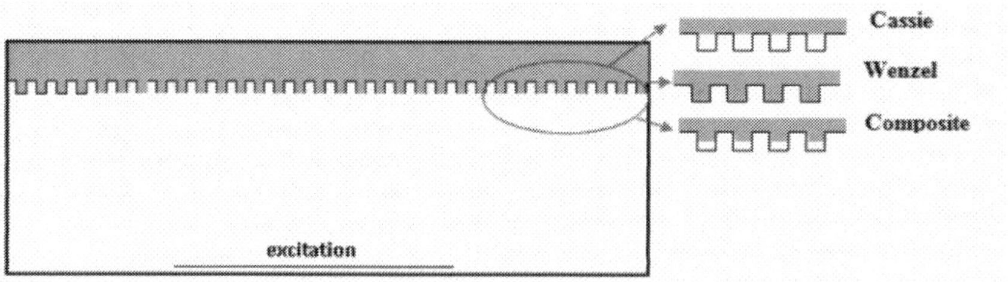

Figure 2. 2D cross section of the FDTD model structure.

In many fields of application, the study of the propagation waves in complex media has long been an important area of interest. Sigalas et al. [5] described in details the FDTD method for acoustic and elastic wave propagation in two dimensional composites. Based on this later study, our model of acoustic propagation has been developed and results will be directly reported below.

The first set of simulations was carried out with the objective of studying the reflection of an acoustic wave on silicon surfaces. Figure 3 illustrates the results of comparing a smooth surface to a textured surface plotted in terms of the absolute values of the y displacements along the transducer in function of time. The textured surface displays the following parameters: pillar width a= 15 µm, pillar height h= 20 µm and pitch b= 15 µm (where the pitch defines the inter pillars distance).

Figure 3 shows the result of the first simulation: The ratio between the amplitudes of the first reflected echo on the smooth surface and that of the textured one is 50% which matches with the ratio of the reflecting surface. Concerning the second echo appearing on the curve corresponding to the textured surface, it is delayed of about 4.6 nanoseconds with the respect to the first one. The propagation delay corresponds to the acoustic propagation Δt in the pillar (c_l is the speed of the longitudinal wave in y direction).

IOP Conf. Series: Materials Science and Engineering **42** (2012) 012052 doi:10.1088/1757-899X/42/1/012052

$$\Delta t = \frac{2 \cdot h}{c_l} = \frac{2.20.10^{-6}}{8433} = 4{,}7.10^{-9} s = 4{,}7ns \quad (1)$$

This means that the first echo derived from the reception of the wave signal reflected on the bottom of the grooves and the second one from the reception of the wave signal reflected on the top of the pillars.

The same simulation was achieved for a second textured structure having different height and the same pillar width and pitch. The result proves our initial finding pointing that time spacing between echoes enables the height of the pillars to be assessed.

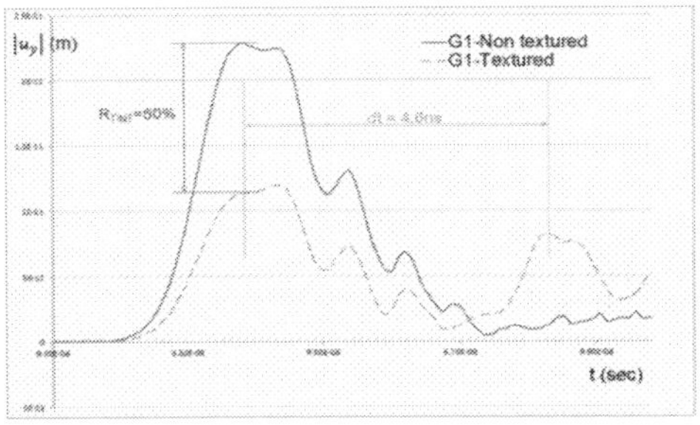

Figure 3. Absolute values of y displacement for a smooth and textured surface in function of time.

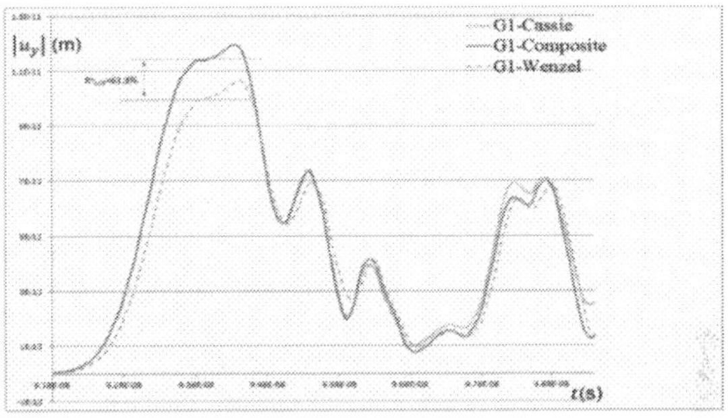

Figure 4. Absolute values of y displacement for a textured surface of height 2h in function of time.

The second set of simulations determines the wetting behavior of the above-mentioned first textured structure. Three cases were studied: The Wenzel state, the Cassie state and an intermediate state for which liquid penetrate into the middle height of the pillars.

We can clearly see for these cases the decreasing of the amplitude of the second echo with the increasing of the contact area between the solid and water. This phenomenon is explained by the fact that with the decreasing of solid/liquid interfacial area, the liquid loading is less important which means that reflection is becoming more important. For example, for the first echoes, the ratio of the amplitudes between Cassie and Wenzel is approximately 80% which is the reflection coefficient for a solid-water interface.

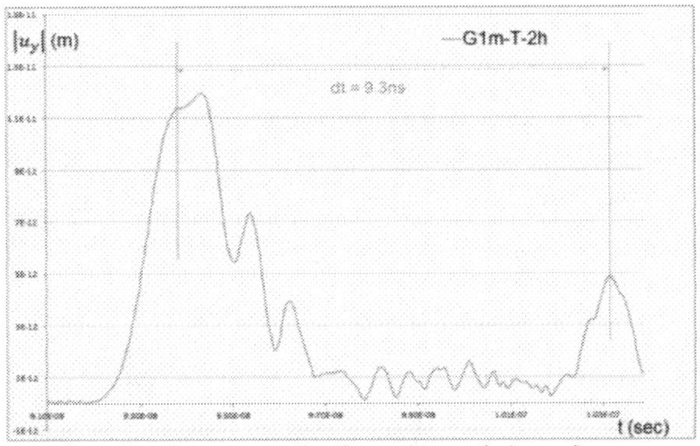

Figure 5. Absolute values of y displacement in function of time for a textured surface in interaction with different state of a droplet (Cassie, Composite and Wenzel).

3. Experimental Method

We limit the experimental study on a microtextured Silicon sample patterned with an array of pillars of diameter 15 μm, height 20 μm and pitch 30 μm [6]. The ultrasonic emission source used in the experiment is an 800 MHz ZnO transducer deposited by radio frequency magnetron sputtering on the rear face of the sample as presented previously [7]. Experimental measurements were carried out by measuring the S_{11} scattering parameter identified as being the ratio of the complex amplitudes of the reflected to the incident voltage thanks to a Suss Microtech prober coupled with a Hewlett Packard 8753 Vector Network Analyser [8]. The impulse response of the system is obtained by inverse Fourier transforming the frequency domain received signal. Initial results clearly show the two echoes obtained from modeling. The time delay between them matches the propagation time in pillars. A difference in amplitude between the echoes for the different states of the droplet is even verified.

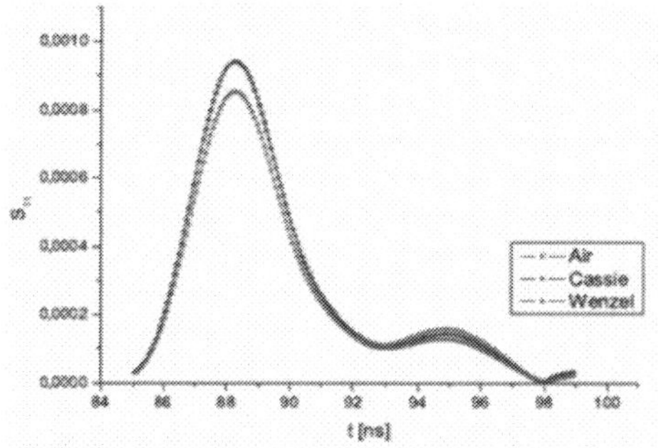

Figure 6. Experimental results for the absolute values of y displacement in function of time for a textured surface in interaction with different state of a droplet (Cassie, Composite and Wenzel).

4. Conclusion

We have presented high frequency ultrasonic method able to characterize textured surfaces. This method can be useful to evaluate pillar's height. Initial experimental results are demonstrated to be in good agreement with the simulation showing promising and exciting results of our approach.

References

[1] Extrand C W 2004 Criteria for ultralyophobic surfaces *Langmuir* **20** (12) 5013-18

[2] PAtankar NA 2010 Consolidation of hydrophobic transition criteria by using an approximate energy minimization approach *Langmuir* **26** (11) 8941-45

[3] Rus G, Wooh SC and Gallego R 2007 Processing of ultrasonic array signals for characterizing defects. Part II: experimental work *IEEE Trans Ultrason. Ferroelectr. Freq. Control.* **54**(10) 2139-45

[4] Sukmana D D and Ihara I 2006 Novel noncontact method for characterizing surface profile by air-coupled ultrasound scattering *AZojomo* **2**

[5] Kafesaki M, Sigalas M M and Garcia N 2001 *Photonic Crystals and Light Localization in the 21st Century* ed C M Soukoulis (Dordrecht, The Netherlands: Kluwer) p 69

[6] R Dufour,M Harnois, Y Coffinier, V Thomy, R Boukherroub and V Senez 2001 Engineering sticky superhomniphobic surfaces on transparent and flexible PDMS substrate *Langmuir* **26** (22) 17242-47

[7] Wang S, Campistron P, Carlier J, Callens-Debavelaere D, Nongaillard B, Ndiguene A, Nassar G, Soyer C and Zhao X 2009 SU-8-based nanocomposites for acoustical matching layer *IEEE Trans. Ultrason. Ferroelectr. Freq. Control* **56** (7) 1483-89

[8] Campistron P, Carlier J, Saad N, Gao J, Toubal M, Dupont L, Nassar G and Nongaillard B 2011 High frequency ultrasound, a tool for elastic properties measurement of thin films fabricated on silicon *Advanced Materials Research* **324** 277-281

IOP Conf. Series: Materials Science and Engineering **42** (2012) 012053 doi:10.1088/1757-899X/42/1/012053

A novel ultrasonic method to quantify pressure in bolted joints

A Sisman[1,2], J Martinez[2] and R Guldiken[2]

[1]Marmara University, Department of Electrical and Electronical Engineering, Goztepe, 34722 Istanbul, Turkey.
[2]University of South Florida, Department of Mechanical Engineering, 4202 E Fowler Ave. Tampa, FL, 33620, USA.

E-mail: alper.sisman@marmara.edu.tr

Abstract. In structures most of fasteners are not critical for rigidity, however significant damage can occur if majority of uncritical fasteners fail. Furthermore, the maintenance is not feasible due to their vast quantity. In this study, we explored a new method to significantly improve the maintenance capability for these fasteners. The method distantly detects the preload between the bolt and the surface. We use a linear transducer array that generates acoustic beam using surface acoustic waves (SAWs) transmitted by array elements. The bolt boundary reflects SAW and array elements receive the echo signals. The signals are processed in the beamformer to generate the boundary's image which quantifies the preload to enable a strong estimation about the health of the bolted joint. We experimentally investigated the proposed method by using used a 5-MHz, 50-elements synthetic array. We measured the preload for five different tightening levels. As expected, two saturation regions (for too tightening and too loosening cases) and a linear region are observed (between them). The electronically steering of the array enables to make simultaneous preload measurements of many bolted joints in the same surface. The distant pressure measurement sensor could be integrated to a data acquisition system to track the health of the joints over time.

1. Introduction

The threaded fasteners are one of the more versatile methods to assembly structural components, especially in steel structures such as pipelines, bridges and platforms. In steel building structures threaded fasteners are used for two purposes: High strength fasteners require careful design and comprehensive care for maintenance. The second group of fasteners has less importance in structural rigidity, however significant damage can occur if majority of the fasteners fail. Furthermore, the maintenance of these fasteners is not feasible due to their vast quantity.

A very common cause of structure failure is the self-loosening of the bolted joints [1,2]. Direct and indirect shear loads result in a "slip" at the thread-plate and head-plate interfaces [3]. In addition to the shear loads, the bolt tension also tends to loose due to the generation of loosening moment related to the helical shape of the thread. Another factor that contribute significantly to the loosening process is the elastics deformation of the fasteners and clamped members; member deformation causes "slip" in the fastener heads and increases the loosening moment in the threads [3]. The clamping force (CF) is the key quantity that indicates the good operation of the joint. The CF is too low in loose bolts; in addition low CF indicates the acceleration in the natural loosening process [3]. The CF is too high in excessive tightened bolts; it may produce damages in the members such as excessive distortion or break [3].

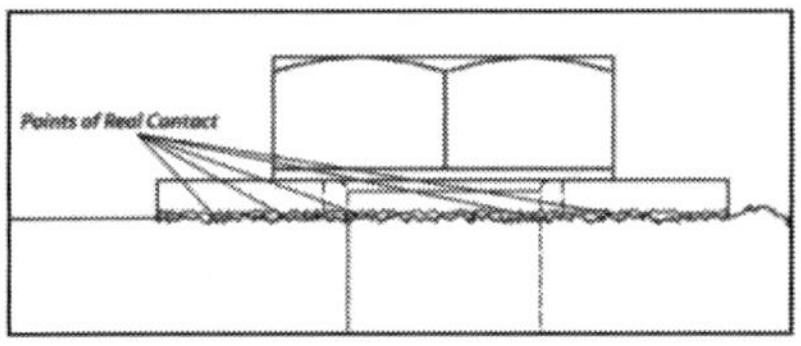

Figure 1. Enlarged representation of the real contact surface of the clamped plate and washer.

It is explained that the clamping force applied to the joint is related to the preload in the threaded fastener by the combined stiffness of the joint and the fastener. The bolts tension is often called bolt preload and it has to be controlled in order to maintain a suitable CF. Several methods are studied in order to quantify the health of bolted joints. Mostly torque control method is applied [2-4]. Other methods are turn of nut control [1], direct preload control [1] and stretch control. The need of previous information about the specific bolt (like its acoustic properties) is common lack for these methods; in addition, existing methods cannot measure distantly and that disables simultaneous measurement.

In this study we investigate a novel method for distant detection of preload between the bolt and the surface. We used an ultrasonic reflection method that doesn't need any previous information about the bolted joint. The method enables simultaneous in *situ* quantification of many bolted joints in the same surface therefore the proposed method significantly improves the maintenance capability.

2. Method

The tension between the bolt and the surface is a function of the contact area formed by the clamped members and the bolt head. The proposed method measures the contact area, but the contact area is not the apparent area we need to determine real area of contact (RAC) [4]. The RAC is the sum of areas of contacted micro scaled surfaces between the joint members as illustrated in figure 1. The interaction between the peak tips explains the formation of the friction force due to molecular interaction and mechanical resistance. At the micro scale level adhesive forces due to electromechanical molecule interaction result in significant friction force component. Vander Waal interaction, ionic, covalent and metallic links between the molecules of both materials generate another opposition force to the movement. In addition to the molecular forces the resistance produced by plastic and elastic deformations of the peaks tips, completes the friction force. The independency of the friction from the geometrical area is a natural cause of the difference between the real and the apparent area of contact. Furthermore, several experiments have shown that the real area of friction is in fact proportional to the normal force when either elastic or plastic deformations are present [4]. Furthermore the torque applied to a bolt (T) is in fact proportional to the preload (F_N) achieved by it. Consequently the RAC is proportional to T and CF in bolted joints, thus: $T\alpha F_N$, $RAC\alpha F_N \Rightarrow RAC\alpha T$.

We use surface acoustic waves (SAWs) to detect the RAC. The SAW like every other type of ultrasonic waves suffers attenuation, reflection and refraction. Hence SAW can be used to get imaging purpose. A phased array that produces SAW beams should be used to get the image of the bolt boundary. The width of bolt boundary is proportional to RAC (figure 2).

The SAW is transmitted to the area of interest and the echo is received using a 1-dimensional (1-D) linear transducer array. Proper delays are applied to each received signal to get the B-Scan image of the interested area. The delays (τ) are calculated by dividing the distance between array element and imaging point by the ultrasound velocity as seen in equation (1) where c is ultrasound velocity, (x_p, z_p) is the coordinate of the point in imaging area and (x_i,0) is the coordinate of array element.

Each image line is obtained by shifting the each received signal according to calculated delay and by summing the shifted signals. Equation (2) shows the expression to find each image line: I(t). Here A_i is the aperture function applied to each receive and transmit element, $s_{i,j}$ is the received signal corresponding to i^{th} transmit and j^{th} receive element, τ_i is refer to transmit and receive delays [5].

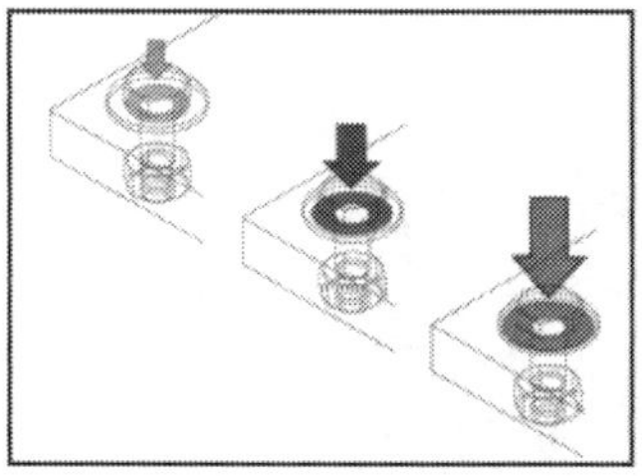

Figure 2. Schematic representation of change in the RAC due to tension increments.

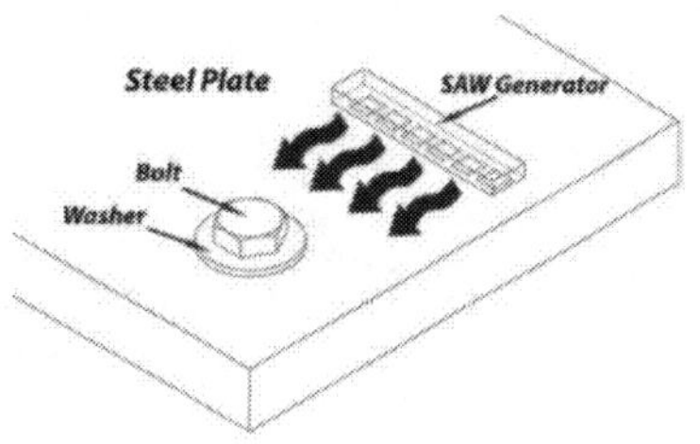

Figure 3. Schematic set up of the proposed methodology.

$$\tau_i = \frac{1}{c}\left(\left(x_i - x_p\right)^2 + {z_p}^2\right)^{1/2} \tag{1}$$

$$I(t) = \sum_{i=1}^{N} A_{tx,i} \sum_{j=1}^{N} A_{rx,j} s_{i,j}(t - \tau_{tx,i} - \tau_{rx,j}) \tag{2}$$

The setup for proposed methodology can be seen in figure 3. SAW generator can receive the echo signals as well. An Nx1 beamformer is fed using the received signals in order to generate B-scan image around the bolt. The distance between the hole and the array is determined using this image. The maximum distance is measured if there is no bolt. In case of the maximum applied torque is the boundary will be closer in an amount equal to width of the washer or the bolt.

3. Experiment results

3.1. Experimental setup

For experimental testing we used synthetic aperture technique instead of phased array imaging to demonstrate the proof of proposed methods. The SAW is generated using a 5-Mhz transducer with a 70 degrees wedge. The transducer is stepped linearly to form a synthetic 1-D array as illustrated in figure 4. We used 50 steps hence the number of elements in the linear array is 50. The pitch is choosen as 0.0254 cm in order to sample the beamspace precisely hence the total length of the array is 1.27 cm. The distance between array and the machined hole boundary is 2.8 cm. The tested joint is made by a 1018 steel plate of 1.27 cm thickness, a 0.635 cm grade 8 bolt and stainless steel washer with 0.5 cm of width.

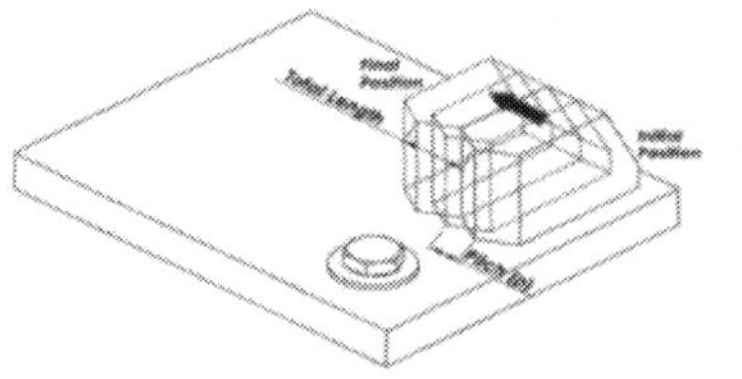

Figure 4. Illustration of forming 1-D linear array.

The SAW velocity in the plate is 2952 m/s. We applied 5 different torque values and measured the boundary distance. The maximum torque exerted to the bolt is 13.78 Nm therefore we have selected 0 Nm (No bolt), 3.61 Nm, 7 Nm, 10.39 Nm and 13.78 Nm.

3.2. The experiment results

The images of the bolt boundary are obtained for each torque value; they are given in figure 5 using the dynamic range 15 dB. We averaged the lines around the bolt boundary to measure the distance between the array and the bolt boundary. The averaged plot is given in figure 6 where the horizontal axis is the depth and the vertical axis is the relative intensity. The distances are measured for each torque value using figure 6.

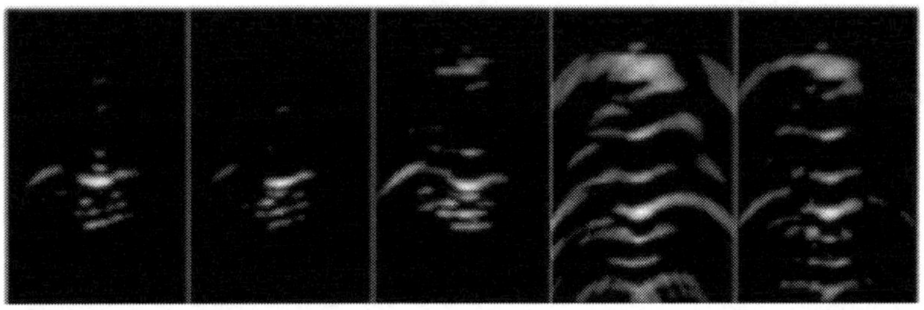

Figure 5. The images of bolt boundary, the array is located at the bottom of the images and the torques are maximum (13.78 Nm) at the rightmost image decreasing to 0 Nm at the leftmost image.

The distance between the boundary and the array is measured 2.8 cm while there is no bolt (the peak of blue plot) as expected. The distance difference corresponding to highest and lowest torque values is 0.12 cm, which is consistent with washer width (The difference between the peaks of blue and cyan). The remaining torque levels are located in between those two. Figure 7 shows the distance of the boundary (peak position) versus torque values and in this figure three regions can be clearly seen: two saturation regions around maximum and minimum torque and a linear region between them. These measurements show the direct relation between the boundary distance and the torque value.

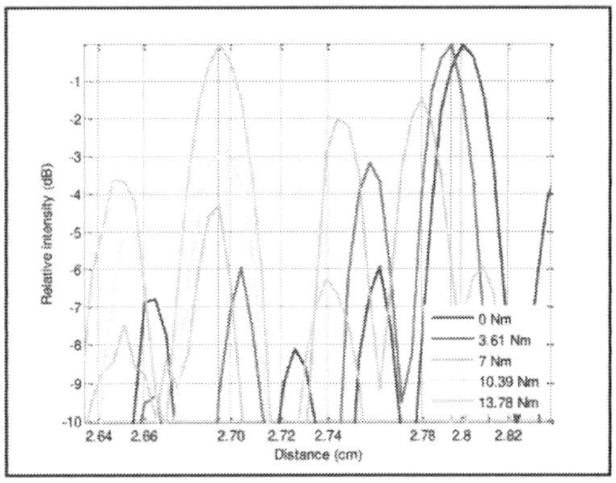

Figure 6. The averaged 1-dimensional plot of the bolt boundary.

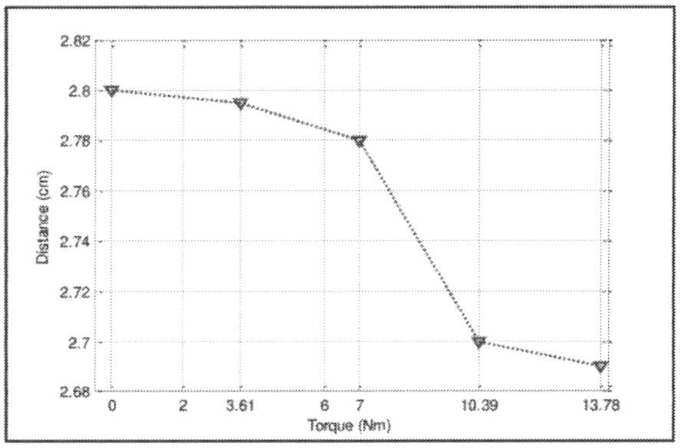

Figure 7. The plot of torque level versus peak position.

4. Conclusion

In this study we demonstrated a novel method to measure the health of the bolted joints. We constructed the images using the SAW echoes from boundary to enable distant measurement. Therefore the method has the potential to make the quantification feasible since the method enables the simultaneous imaging of more than one bolt distantly. The initial results are promising to precisely measure the health of the bolted joints. The design of a phased array will enhance the image quality then the integration with a smart building will be next step.

References

[1] Hess D P 1998 *Handbook of Bolts & Bolted Joints* ed J Bickford and S Nasser (New York: Marcel Dekker)
[2] Goodier J N and Sweeney R J 1945 *Mechanical Engineering* **67** 798-802
[3] Pai N G and Hess D P 2002 Experimental study of loosening of threaded fasteners due to dynamic shear loads *Journal of Sound and Vibration* **253**(3) 585-602
[4] Johnson K L 1995 *Contact mechanics* (Cambridge, London: Cambridge University Press)
[5] Thomenius K E 1996 Evolution of US beamformers *Proc. IEEE Ultrasonics Symp.* (San Antonio, Texas, 3-6 November 1996) vol 2 pp 1615-22

IOP Conf. Series: Materials Science and Engineering **42** (2012) 012054 doi:10.1088/1757-899X/42/1/012054

Continuous monitoring of bacterial biofilm growth using uncoated Thickness-Shear Mode resonators

P Castro[1], P Resa[1]; C Durán[1], J R Maestre[2], M Mateo[2] and L Elvira[1]

[1] Centro de Acústica Aplicada y Evaluación No Destructiva (CAEND), UPM-CSIC
C/Serrano 144, 28 006 Madrid, Spain
Phone number: (+34) 915 61 88 06 (259)
Fax number: (+34) 914 11 76 51
[2] Servicio de Microbiología Clínica. Hospital Central de la Defensa Gómez-Ulla
Glorieta del Ejército, s/n, 28047, Madrid, Spain

Email: pedro.castro@caend.upm-csic.es

Abstract. Quartz Crystal Microbalances (QCM) were used to nondestructively monitor in real time the microbial growth of the bacteria *Staphylococcus epidermidis* (*S. epidermidis*) in a liquid broth. QCM, sometimes referred to as Thickness-Shear Mode (TSM) resonators, are highly sensitive sensors not only able to measure very small mass, but also non-gravimetric contributions of viscoelastic media. These devices can be used as biosensors for bacterial detection and are employed in many applications including their use in the food industry, water and environment monitoring, pharmaceutical sciences and clinical diagnosis. In this work, three strains of *S. epidermidis* (which differ in the ability to produce biofilm) have been continuously monitored using an array of piezoelectric TSM resonators, at 37 °C in a selective culturing media. Microbial growth was followed by measuring the changes in the crystal resonant frequency and bandwidth at several harmonics. It was shown that microbial growth can be monitored in real time using multichannel and multiparametric QCM sensors.

1. Introduction

Thickness-Shear Mode (TSM) sensors, namely Quartz Crystal Microbalances (QCM), are highly sensitive devices not only able to measure the addition of very small mass, but also non-gravimetric contributions of viscoelastic media, which can be operated in air, aqueous conditions and under vacuum. QCM usually consists of a thin AT-cut quartz wafer with two metallic electrodes deposited uniformly onto both sides of the crystal. Due to the piezoelectric effect, when an AC electrical field is applied perpendicular to the plate, a shear wave is generated that propagates through the quartz crystal inducing its oscillation. The Sauerbrey expression [1] is commonly used to relate mass loading and resonant frequency shift, Δf :

$$\Delta f = -C\Delta m \quad (1)$$

where C is the sensitivity factor of the resonator in Hz/(ng/cm^2), and Δm is the change in mass per unit area in ng/cm^2. This expression is only valid for a homogeneous film rigidly attached to the QCM surface. This theory has been extended to liquid applications, and it has been shown that QCM measurements can be performed in liquid media, in which case a decrease in the resonant frequency related to viscosity will be observed:

$$\Delta f = -f_0^{3/2}\left(\eta_l \rho_l / \pi \rho_q \mu_q\right)^{1/2} \quad (2)$$

where f_0 is the resonant frequency, ρ_l is the density of the liquid, η_l is the viscosity of the liquid, ρ_q and μ_q are the density and rigidity modulus of the quartz crystal (respectively).

Uncoated TSM resonators have been used to monitor the growth of bacteria [2]-[4]. For example, in reference [2], the resonant frequency shift of a QCM was measured to monitor for several days the growth of a bacterial biofilm (*Pseudomonas cepacia*). A non-linear behavior was observed when they calibrated the frequency response of the QCM to the bacterial counts measured on the surface (frequency shift from a typical biofilm could not be directly related to the attached mass, as is the case of a thin rigidly attached film).

In the present work, *S. epidermidis* has been chosen as a model organism because of its common occurrence in the environment, the ability to rapidly form a robust biofilm suitable for experimentation, and the large number of experimental studies that have been done with this microorganism. There are several examples of the monitoring of *S. epidermidis* using coated QCM [5]-[8] .

In this study, it was shown that microbial growth and biofilm formation can be monitored in real time using multichannel and multiparametric uncoated QCM sensors. Although only *S. epidermidis* has been studied, the technique could be applied to control other processes of bacterial colonization.

2. Material and methods

2.1. Instrumentation

8 QCM cells made up of an 8 MHz AT-cut quartz crystals, with optically flat polished gold electrodes on both sides, were purchased from International Crystal Manufacturing Co, Inc (ICM, Oklahoma, USA, 13.7 mm blank diameter, 5.1 mm electrode diameter, polished surface finish, 100 Å Ti & 1000 Å Au electrode material). The crystals were cleaned before use by rinsing with absolute ethanol to obtain a clean gold surface. Cleaned crystals were then rinsed with distilled water and dried in air. The crystals were connected to a 4294A Precision Impedance Analyzer ranging from 40 Hz to 110 MHz, through a multiplexer of 8 channels (see figure 1).

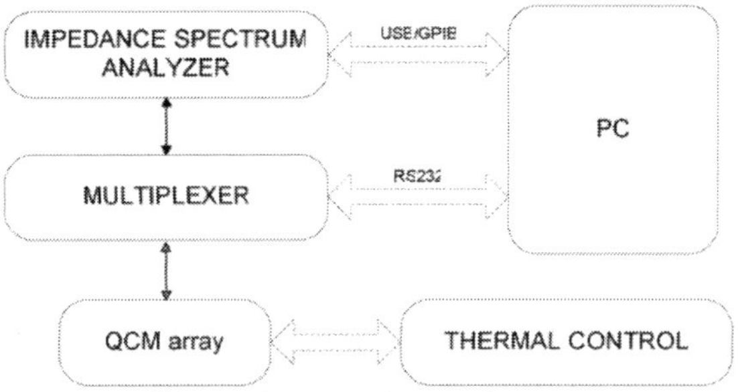

Figure 1. Experimental setup.

The experimental cell (see figure 2) consisted of a 1 mL screw top tube with the quartz wafer silicone glued to the bottom. The electrodes of the piezoelectric transducer were soldered to an RF coaxial cable using copper strips (the side in contact with the liquid was grounded).

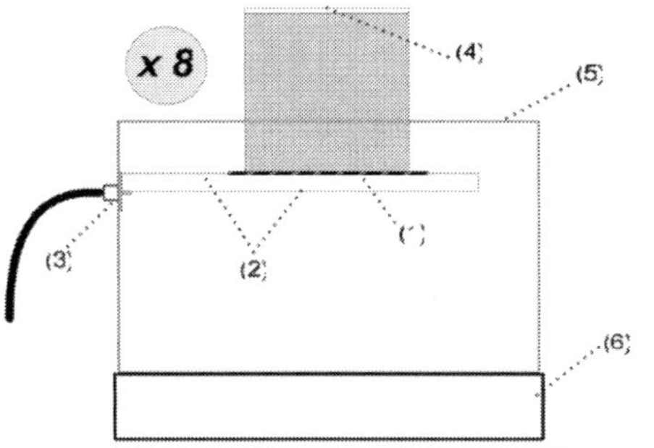

1.- AT-Cut quartz crystal
2.- Electrical connections
3.- Panel waterproof connector
4.- Container of liquid samples with screw-cap
5.- Watertight enclosure
6.- Screw-type closure

Figure 2. Measuring cell.

2.2. *Microbial growth*

Standard bacterial strains of *S. epidermidis* were obtained from *Hospital Central de la Defensa Gómez-Ulla* (Spain). Three strains of *S.epidermidis* have been monitored, which differ in the ability to produce extracelular polymeric substance (EPS): strain 2 (ATT 35984, high biofilm producer), strain 1 (ATCC 35983, medium biofilm producer) and strain 3 (ATCC 12228, not producing biofilm).

Pure bacterial strain cultures were grown in a selective culturing media at 37 ºC for several days (with a shaking incubator that besides controlled temperature). Haemophilus Test Medium, was purchased from Trinova Biochem GmbH (Germany), and the sample volume was 1 mL.

3. Results and discussion

The experimental results obtained are shown in figures 3 and 4. In figure 3, resonance frequency shifts are plotted, while the bandwidth shifts for the same microbial growth are displayed in figure 4. These results were obtained during 4 days for the harmonic n=3. For higher harmonics, similar results were observed (taking into account that the sensitivity depends on frequency).

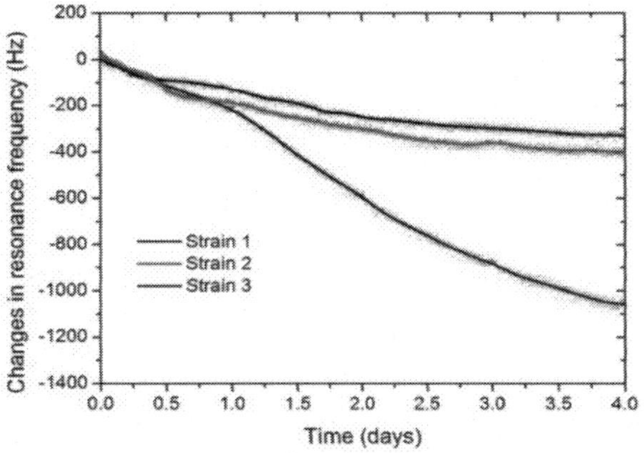

Figure 3. Resonant frequency shifts for 3 different strains of *S. epidermidis* (37ºC, HTM broth).

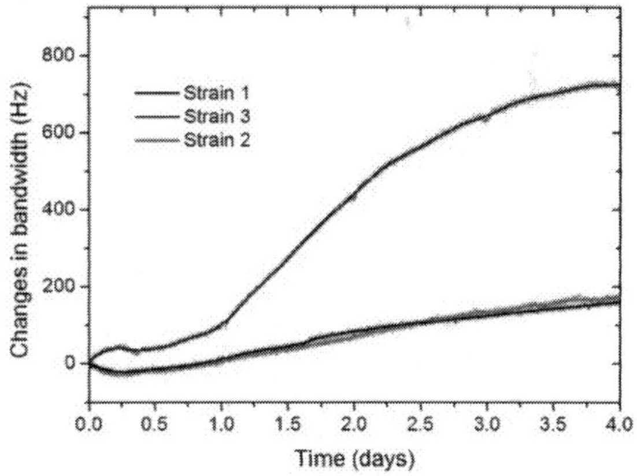

Figure 4. Bandwidth shifts for 3 different strains of *S. epidermidis* (37ºC, HTM broth).

It can be noticed that the highest changes correspond to strain 3, non-producer of EPS, showing that they are not caused by the production of biofilm, but by other metabolic processes. Both strains 1 and 2, producers of EPS, follow a similar behavior. During the first hours, negative changes of bandwidth and negative frequency shifts can be noted, probably due to the formation of biofilm on the surface of the quartz crystal. Both half bandwidth, $\Delta\Gamma$, and resonant frequency shifts, Δf_0 , depend on gravimetric and non-gravimetric effects. Assuming a viscoelastic film in liquid, the ratio in equation (3) is independent on mass:

$$\frac{\Delta\Gamma}{\Delta f_0} = -\frac{\omega\eta J'_f}{1-\omega\eta J''_f} \cong -\omega\eta J'_f \qquad (3)$$

where ω is the radial frequency, η is the viscosity, J' and J'' are the elastic and viscous compliance, respectively. The last equivalency assumes that film viscosity is much higher than liquid viscosity According to equation (3), the sign of the ratio $\Delta\Gamma/\Delta f_0$ will be dependent on the film properties. When the biofilm becomes more rigid, J' decreases and equation (3) becomes less negative. From the first day, the lowest J' is obtained for strain 2 and the highest J' is obtained for strain 3, linking the results to the strain biofilm production. In addition, it is shown that QCM are sensitive to the growth of the strains, which modifies the composition of the culture medium, as substrates are consumed and converted into metabolic end products.

4. Conclusions

TSM resonator measurements are affected by several factors including gravimetric and non-gravimetric effects (biofilm growth, catabolic changes in the liquid, etc.), which can make QCM results difficult to interpret.

Microbial growth and biofilm formation can be monitored using uncoated QCM sensors. This implies: minimal manipulation, small volumes, real-time monitoring, multichannel and multiparametric measurements.

In conclusion, the current QCM array system could be a low-cost, rapid and sensitive method able to continuously monitor the microbial biofilm growth produced by *S. epidermidis* (or other kind of microorganims).

Acknowledgements

This work has been supported by the Spanish Ministry of Science and Innovation (CICYT DPI2010-17716), a CSIC intramural project (ref. 201150E045) and the JAE-CSIC Program/European Social Fund (JAEPREDOC2011-058 & JAEDOC2008-065).

References

[1] Sauerbrey G 1959 Use of quartz crystal vibrator for weighting thin films on a microbalance *Z Phys.* **155** 206-222

[2] Nivens D E, Chambers J Q, Anderson T R and White D C 1993 Long-Term, On-Line Monitoring of Microbial Biofilms Using a Quartz Crystal Microbalance *Anal. Chem.* **65** 65-69

[3] Chen K, Le D, Zhang H, Nie L and Yao S 1996 Model of quartz crystal microbe growth sensor and its application to estimation of microbial populations in mineral waters *Analytica Chimica Acta* **329** 83-89

[4] Reipa V, Almeida J and Cole K D 2006 Long-term monitoring of biofilm growth and disinfection using a quartz crystal microbalance and reflectance measurements *Journal of Microbiological Methods* **66** 449–459

[5] Pavey K D, Ali Z, Olliff C J and Paul, F 1999 Application of the quartz crystal microbalance to the monitoring of staphylococcus epidermidis antigen-antibody agglutination *Journal of Pharmaceutical and Biomedical Analysis* **20**(1-2) 241-245

[6] Xia H, Wang F, Huang Q, Huang, J, Chen M, Wang J and Fu W 2008 Detection of staphylococcus epidermidis by a quartz crystal microbalance nucleic acid biosensor array using au nanoparticle signal amplification *Sensors* **8**(10) 6453-70

[7] Strauss J, Liu Y and Camesano T A 2009 Bacterial adhesion to protein-coated surfaces: An AFM and QCM-D study *Journal of the Minerals* **61**(9) 71-74

[8] Olsson A L J, van der Mei H C, Busscher H J and Sharma P K 2011 Acoustic sensing of the bacterium-substratum interface using QCM-D and the influence of extracellular polymeric substances *Journal of Colloid and Interface Science* **357**(1) 135-138

Non-scanning curvature measurement with a segmented ultrasonic annular array

Michael Lenz[1], **Elfgard Kühnicke**[2]
Technische Universität Dresden, Institut für Festkörperelektronik, D-01062 Dresden, Germany
E-mail: michael.lenz@tu-dresden.de[1], elfgard.kuehnicke@tu-dresden.de[2]

Abstract. The presentation describes an ultrasonic method for the determination of the curvature radius of an object without mechanical scanning. Instead of scanning, a segmented ultrasonic annular array is used, where the phase and amplitude differences on the different receivers are used to determine the curvature radius (spherical reflector) and the lengths of the semi-axes (ellipsoidal reflectors), respectively. Experimental results with spheres as reflectors are presented and the latest results on the determination of the length of the semi-axes with an ellipsoidal reflector are presented.

1. Introduction

Conventional scanning ultrasonic measurements use the time of flight of the ultrasonic wave to determine the distance of an object. In the current presentation, an approach for curvature determination without scanning [1–3] are presented. Instead of scanning, the shape of the reflected sound field is analysed. The approach aims at measurements of spherical or oval inclusions in fluids and solids. A potentially interesting technical application making use of its features is the combined tracking of position and shape of rising bubbles in opaque fluids required for flow research. In medical engineering, a particular interest may lie in the determination of eye lens and retina curvature, where the method is intended to work as a supplement to conventional imaging applications.

2. Approach and Measurement Set-up

The idea of the current approach goes back to the reflection of light. It is known, that a plane electromagnetic wave is converted into a concentric wave when being reflected by a parabolic mirror, where the reflected light has its origin at a virtual focus point F behind the mirror. Vice versa, a concentric wave aiming at this focus point is converted into a plane wave on reflection (Fig. 1, leftmost panel).

The underlying principle is exploited in the ultrasonic domain, here. An ultrasonic array made of curved piezoelectric composite material is used. It has a curvature radius of $R = 70\,\mathrm{mm}$ and a resonance frequency of 3 MHz. The transducer is segmented into four concentric ring-shaped elements, which – in turn – are segmented into four 90° elements, each (Fig. 1, rightmost panel). All elements work as both transmitter and receiver and have been designed to have the same area, as is typical of focusing arrays (see e. g. [4, 5]).

Spheres of different diameters used as calibration objects were immersed into a water basin with a distance of $z = 50\,\mathrm{mm}$ between the sphere surface and the transducer. The spheres have

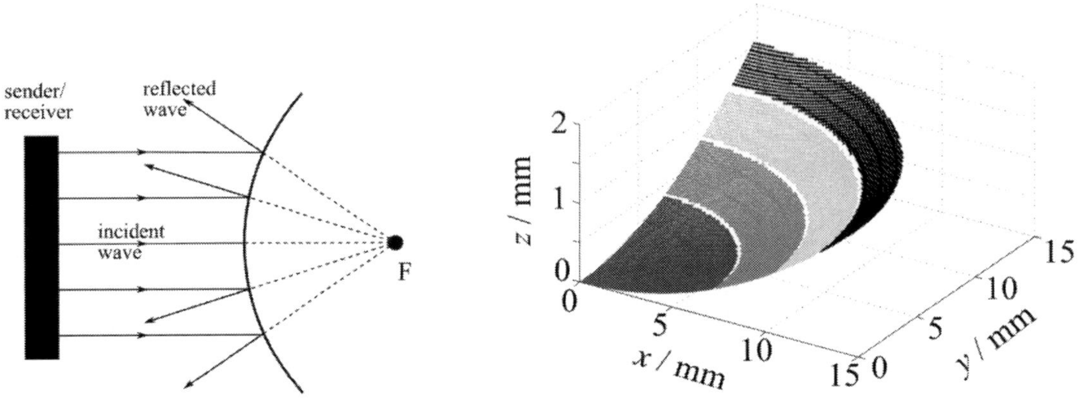

Figure 1. Left panel: Using a parabolic mirror, a plane wave is is converted into a concentric wave having its origin behind the mirror. Right panel: Visualisation of the transducer geometry with a 90° segment comprising four elements shown. The nominal ring radii range from 0 to 7.14 mm, 7.34 mm to 10.23 mm, 10.43 mm to 12.64 mm, and 12.84 mm to 14.69 mm, respectively.

nominal diameters of D = 12 mm, 14 mm, 16 mm, 18 mm, and 22 mm and were positioned in such a way that the acoustical axis passes through the sphere centre. Correct calibration was controlled by comparing the transmit/receive-signals of the elements on the opposite sides of the transducer diagonals. For reasons of symmetry, these signals need to have the same phase and amplitudes when the sphere centre lies on the acoustic axis.

The electrical part of the set-up uses an arbitrary waveform generator, a high-frequency power amplifier, and a self-made unit with sixteen independent channels used for sending multiplexing, transmit-receive-switching and amplification of the analogue receive signals. For excitation, quasi-harmonic sinusoidal waves with a frequency of 3 MHz and four oscillations were used, and in each measurement, a complete matrix of 16×16 signals with all combinations of transmitting and receiving transducer were recorded. This allows focusing in send- and receive mode independently in signal post processing. The analogue signals were digitised with a high-frequency digitiser by Gage Applied Technologies at a sample rate of 50 MS/s and a nominal resolution of 14 Bits.

3. Results of curvature measurements

3.1. Spherical reflectors

The results of the measurements with spherical reflectors are shown in Fig. 2. In the topmost panel, the normalised echo signal energy (i. e. the sum of the squared echo signal sampling values) is plotted versus the focus setting Foc. Note, that different focusing settings in send and receive mode were used: A fixed 'focus' pointing to an infinite distance was used in *focusing in transmit mode*, whereas a variable focus (marked on the x-axis of Fig. 2) with the focus position Foc was used in *focusing in receive mode*. This was done to realise similar conditions as in the light experiment. The panel on the bottom plots the focus position that yields a maximum in the echo signal amplitude. It shows that the focus setting generating a maximum in the echo signal always points behind the sphere surface (z = 50 mm), where the focus lies the closer to the sphere surface the smaller the sphere is.

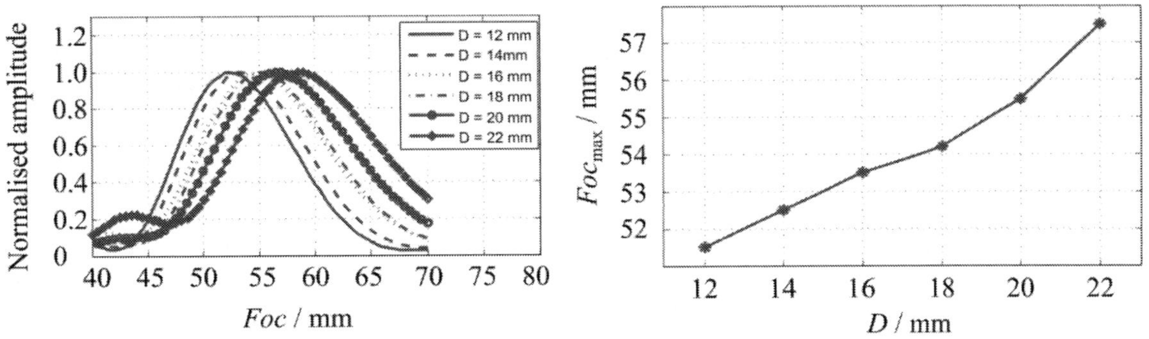

Figure 2. Top panel: Echo signal energy as a function of focus setting for different sphere diameters D (distance between transducer and sphere surface: $z = 50\,\mathrm{mm}$). Bottom panel: Focus setting generating a maximum in the echo signal as a function of sphere diameter.

3.2. Ellipsoidal reflectors

A test measurement aiming at the measurement of ellipsoidal reflectors was also done with a reflector having axes of 12 mm and 22 mm, respectively. For that purpose, the transducer diagonals were adjusted parallel to the ellipsoidal axes. As in the experiments with the spherical reflectors, the transmit signal was realised with a fixed 'focus' pointing to an infinite distance in *focusing in transmit mode.*

Unlike the measurements of the spheres, however, two different receive signals each comprising the eight signals of the receivers on the diagonals were calculated, where the same delay times for focusing as in the sphere measurements were chosen. As can be seen from comparing Fig. 3 with Fig. 2, the maxima in the plot nicely represent the diameters of the axes.

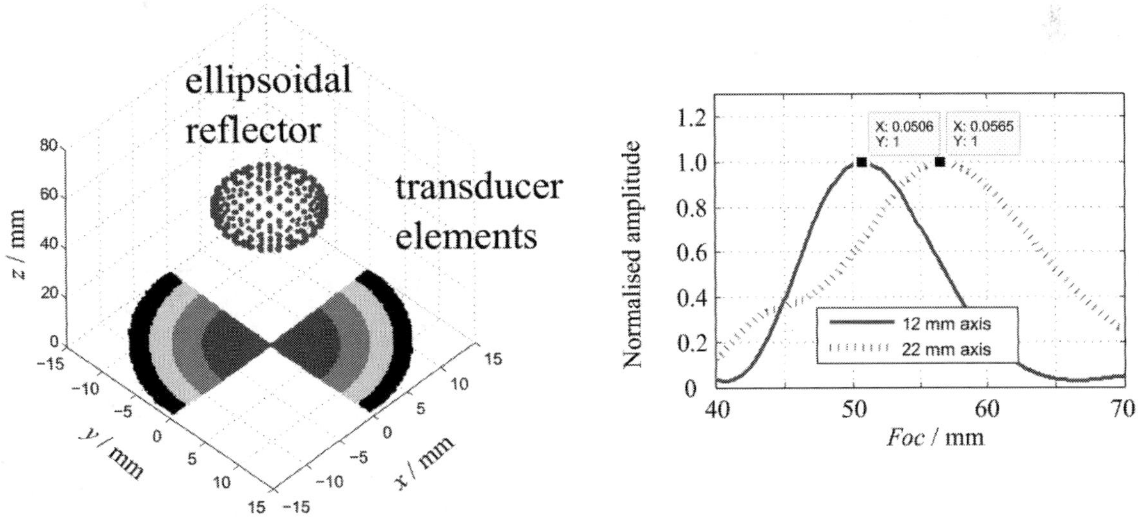

Figure 3. Left panel: For measuring the reflector curvature of an ellipsoidal reflector, solely the eignt signals from the opposite sides of the diagonal were used for sound reception. Right panel: By that way, the two lengths of the ellipsoidal axes can be distinguished by analysing the respective receive signals.

4. Summary and future prospects

The results of the curvature measurements verify the feasibility for the measurement of spherical curvatures, where reflectors with diameters between $D = 12\,\mathrm{mm}$ to $22\,\mathrm{mm}$ were used. Owing to the use of a novel quarter element transducer with 16 elements, promising results for extending the method to ellipsoidal curvatures could be shown.

Unfortunately, the measurement accuracy for the curvature radius of approximately $0.1\,\mathrm{mm}$ to $0.5\,\mathrm{mm}$ did not improve, although the transducer adjustment in the x- and y-plane could be controlled precisely by comparing the signals on the quarter elements. Distance variations of the spheres are planned next, where distance steps much smaller than one wavelength are to be used to discover possible interference effects. The general work aims at more sophisticated tasks that come closer to practical requirements: concave curvatures, buried spheres, detection and compensation of non-radial irradiation of the sphere.

Acknowledgments

The authors would like to thank Deutsche Forschungsgmeinschaft (DFG) for financial support of the ongoing research project KU1075/11-1, Falk Schellhorn for the design of electronic circuits, and Sonaxis SA for transducer manufacturing.

Appendices

5. References

[1] Kühnicke E and Trier H G 2009 Assembly and method for determining the combinations of curvature radii and distances to acoustic boundary areas using ultrasound *E.U. Patent EP2090229*

[2] Lenz M, Sorber J, Kühnicke E, Gerlach G, Rödig T and Trier H G 2009 Non-scanning measurement of local curvature with an ultrasound annular array *IEEE Ultrasonics Symposium* pp 2726–2729.

[3] Lenz M and Kühnicke E 2011 Toward non scanning measurement of eye lens curvature with ultrasound *13th World Congress of the World Federation for Ultrasound in Medicine and Biology (WFUMB)*

[4] Brown J A, Démoré C E M and Lockwood G R 2004 Design and fabrication of annular arrays for high-frequency ultrasound *IEEE Transactions on Ultrasonics, Ferroelectrics, and Frequency Control* **51(8)** pp 1010–1017

[5] Brown J A, Lockwood G R 2005 Design of Sparse Annular Arrays for High-Frequency Imaging *IEEE Ultrasonics Symposium* pp 125–128

Catalase-positive microbial detection by using different ultrasonic parameters

S K Shukla, C Durán and L Elvira

Centro de Acústica Aplicada y No Destructiva Evaluación
Serrano 144, 28006 Madrid, Spain. Tel. - +34 91 561 88 06

E-mail: shiva.kant@caend-upm.csic.es

Abstract. A method for rapid detection of catalase enzyme activity using ultrasonic parameters is presented in this work. It is based on the detection of the hydrolysis of hydrogen peroxide molecule into water and oxygen induced by the enzyme catalase. A special medium was made to amplify changes produced by catalase enzyme during the hydrolysis process. Enzymatic process can be monitored by means of ultrasonic parameters such as wave amplitude, time of flight (TOF), and backscattering measurements which are sensitive to oxygen bubble production. It is shown that catalase activity of the order of 10^{-3}unit/ml can be detected using different ultrasonic parameters. The sensitivity provided by them is discussed.

1. Introduction

Rapid and reliable detection is important for the quality control of those industries (water, food, dairy, pharmaceutical…) handling and producing complex organic substances which are susceptible being contaminated by living microorganisms. Microbial contaminations may spoil the products which results on important economical losses. Several techniques are available such as pH measurement, electric impedance measurements, ATP luminescence-based measurement, colorimetry, flow cytometry, epifluorescence, microcalorimetry, nucleic acid technologies etc [1, 2, and 3]. Ultrasound seems to be an emergent technology to characterize biochemical and biotechnological processes. Living microorganisms arise elastic changes in the medium where they are present; therefore, ultrasound based technologies can be used for the characterization of microorganism induced processes.

The proposed technique is applicable for catalase-positive microorganisms. Catalase is an antioxidant enzyme which can be found in nearly all living organisms except some anaerobic bacteria. As a first step, we emphasized our study on the catalase activity evaluation. The enzyme catalyses the following reaction:

$$2H_2O_2 \xrightarrow{Catalase} 2H_2O + O_2$$

Different ultrasonic parameters (wave amplitude, time of flight, and backscattering) were measured and compared. Catalase-activity has been characterized down to the 10^{-3} unit/ml.

2. Materials and Methods

2.1. Measurement medium

The analysis medium for the ultrasonic tests is a key point to obtain a good performance with the technique proposed. Three important features must be fulfilled by it:

- Being stable in the presence of hydrogen peroxide which is a highly reactive molecule.
- Being able to trap forming bubbles
- Being fluid enough to allow a rapid mixing of the hydrogen peroxide solution.

The selected medium was based on American Bacteriological Agar (cat.1802.00; CONDA). A 1% w/w suspension of agar in distilled water was made, autoclaved at 120°C to avoid cross contaminations and cooled at room temperature inside an ultra violet flow chamber.

2.2. Ultrasonic measurement technique

The ultrasonic measurement technique is based on a "pulse-echo" method.

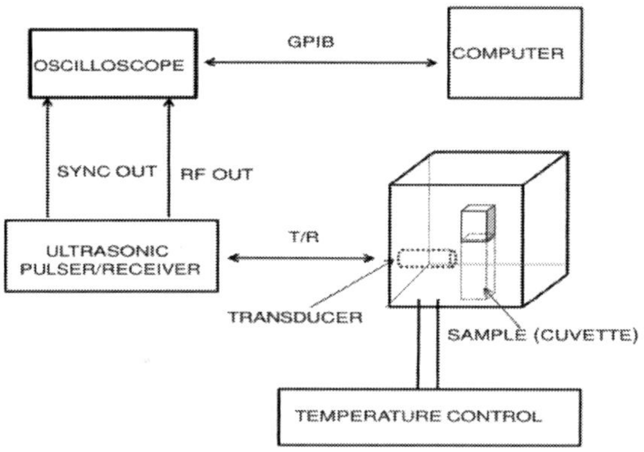

Figure 1. Experimental arrangement of the measurement system.

Ultrasonic pulses are emitted into the sample and after being reflected from the sample and container walls, the waves reach the transducer. As shown in figure 1, the transducer is mounted horizontally into an almost cubic container (51mmx51mmx53mm) which is partially filled with water at 30°C.

Inside it, 3ml plastic commercial cuvettes (which are commonly used for spectrophotometric measurements) are placed and play the role of sample containers. A typical laboratory arrangement formed by a transducer, a pulser, an oscilloscope and a computer was used for the experiments. A 7.8 MHz Doppler wideband transducer is driven by a Panametrics ultrasonic pulser/receiver, model 5052UA, emitting 300V spikes at a 5 kHz repetition rate and receiving the signals coming from the transducer. The signals are acquired by a Tektronix TDS 2014 oscilloscope wired to the ultrasonic pulser. The measurement process is controlled by the computer through a software application programmed with Matlab. This software captures the traces acquired by the oscilloscope via GPIB connection, performs the signal processing and stores the results.

For the tests, 1 ml of the agar medium is mixed with 1 ml of the sample which must be analyzed (containing catalase). Then 100 µl. of hydrogen peroxide solution 10M in distilled water is added and shaken three times vigorously for a good homogenization of the hydrogen peroxide. Immediately, the cuvette is placed inside the container for ultrasonic continuous monitoring.

Figure 2 shows the wave acquired. At the left of the window (<10µs), excitation noise and cuvette front wall echoes appear, between 10 and 15 µs, backscattering echoes of bubbles can be detected, and finally (>15µs) reflection from the back wall which was used for wave amplitude and time of flight analysis. Each experiment was repeated four times for every catalase concentration and the average was obtained.

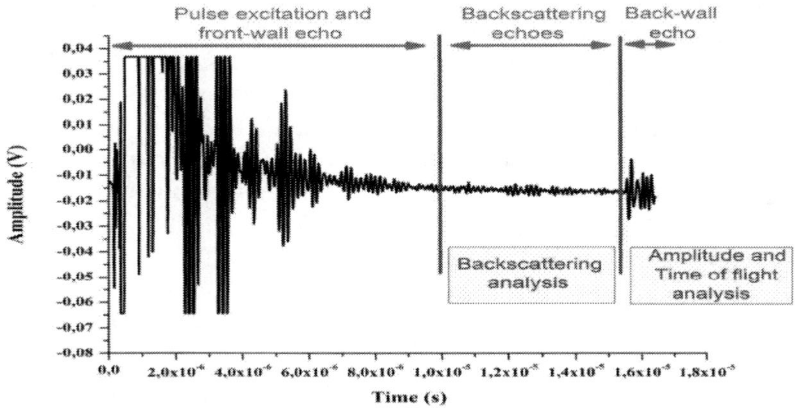

Figure 2. Ultrasonic echoes acquired during the hydrolysis of H_2O_2.

3. Results

Due to the presence of catalase, the hydrolysis of hydrogen peroxide molecules starts the production of bubbles. These bubbles are trapped in the agar gel medium and interact with the ultrasonic wave. To get reference of the enzyme activity evaluation, blank measurements were also done. Wave amplitude and time of flight of the back wall echo signals and bubble backscattering were analyzed.

3.1. Amplitude

As soon as the hydrolysis starts, attenuation increases continuously because of bubble production and consequently, the amplitude decreases. Results are shown in figure 3(a). For higher catalase concentrations, the curve slope becomes increasingly negative. The deviation from blank measurement shows that catalase activity can be detected using this method down to 10^{-3} unit/ml.

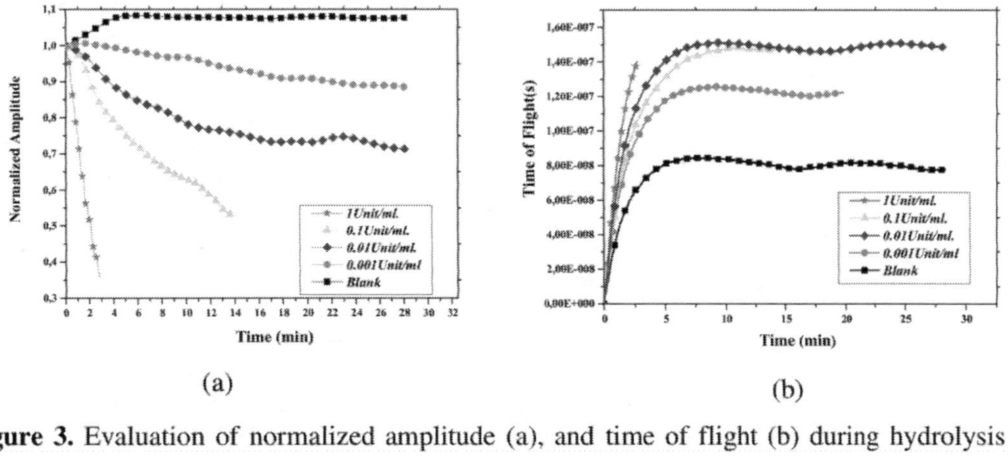

Figure 3. Evaluation of normalized amplitude (a), and time of flight (b) during hydrolysis of H_2O_2.

3.2. Time of flight (TOF)

It is well known that time of flight is too much sensitive to temperature variations. The first 7 min rise in TOF of figure 3(b) corresponds to a thermal instability stage. After reaching the working temperature, the curve behavior changes due to thermal oscillations. These oscillations interfere on the characterization of the enzyme activity.

3.3. Backscattering from bubbles

As bubbles grow the backscattered echoes increase their amplitude being the backscattered energy calculated. Higher concentrations raise plenty of bubbles which increases reflected energy. As was shown for the back-wall echo amplitude, the deviation from blank measurement shows that catalase activity can be detected using backscattering down to 10^{-3} unit/ml.

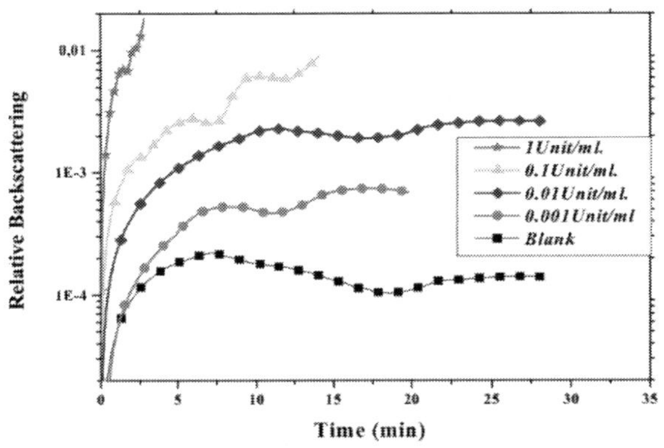

Figure 4. Backscattering evaluation during the hydrolysis of H_2O_2.

4. Conclusions

An ultrasonic based technique to detect catalase activity is shown. For that purpose, a special medium to enhance ultrasonic sensitivity was developed. Different ultrasonic parameters were calculated showing that both the back-wall echo amplitude and bubble backscattering give a good sensitivity. The method can be used to detect positive-catalase microorganisms.

Acknowledgements

This work was funded by the Spanish Ministry of Science and Innovation (CICYT DPI2010-17716) and the JAE-CSIC Predoctoral Program/European Social Fund (JAE-PreDoc2009).

References

[1] Fung D 2009 Rapid Methods and Automation in Food Microbiology: 25 Years of Development and Predictions *World Congress Book: Global Issues in Food Science and Technology* vol 11 ed G Barbosa-Cánovas *et al* (USA: Academic Press) pp 165-176

[2] Pishawikar MS, Singhal RS and Kulkarni PR 1992 Rapid non-microbiological methods for detecting microorganisms in foods *Trends Food Sci. Technol.* **3** 165-169

[3] Vanne L, Karwoski M, Karppinen S and Sjiiberg A-M 1996 HACCP-based food quality control and rapid detection methods for microorganisms *Food. Cont.* **7**(6) 263-216